The Benchmarking Book:
A How-to-Guide to Best
Practice for Managers
and Practitioners

The Benchmarking Book: A How-to-Guide to Best Practice for Managers and Practitioners

Tim Stapenhurst

AMSTERDAM • BOSTON • HEIDELBERG • LONDON
NEW YORK • OXFORD • PARIS • SAN DIEGO
SAN FRANCISCO • SYDNEY • TOKYO

Butterworth-Heinemann is an imprint of Elsevier

Butterworth-Heinemann is an imprint of Elsevier
Linacre House, Jordan Hill, Oxford OX2 8DP, UK
The Boulevard, Langford Lane, Kidlington, Oxford OX5 1GB, UK

First edition 2009

British Library Cataloguing in Publication Data
A catalogue record for this book is available from the British Library

Library of Congress Cataloging-in-Publication Data
A catalog record for this book is available from the Library of Congress

ISBN–978-0-7506-8905-2

For information on all Butterworth-Heinemann publications
visit our website at elsevierdirect.com

Printed and bound in United Kingdom
09 10 11 12 13   10 9 8 7 6 5 4 3 2 1

Contents

Part II: The Benchmarking Process

Introduction and Process Overview

Phase 1: The Benchmarking Process: Internal Preparation

3 Selecting a Project and Commissioning the Team

4 The Team Begins Work: Honing the Project Charter

5 Identifying and Selecting Benchmarking Participants

6 Metrics and Data

Appendix A 2 : Querying the Quartile

Glossary   433

Index   445

List of Figures

Chapter 21

Chapter 22

Chapter 23

Chapter 24

Appendix A1

Appendix A2

List of Case Studies and Examples

Page	Case study/example name	Summary
8	Tupolev Tu-4 Aircraft	The Russians reverse-engineer a crashed American B-29 bomber and developed the Tupolev Tu-4 bomber on their findings.
9	Why Xerox Invented Benchmarking	Summarizes the history of how and why Xerox developed benchmarking as a business improvement tool.
27	Xerox and L.L. Bean Distribution Benchmarking	Xerox visited L.L. Bean after reading about their distribution process and later developed their own system based on what they had learned.
27	Formula 1	A manufacturing organization visited a Formula1 racing team to learn how to set production lines effectively and efficiently.
32	Review Benchmarking	Review Benchmarking process example also shows the importance of understanding participants' working environment.
65	Solomon Refinery Benchmarking	Facility benchmark study.
66	Valve Trialling	Product benchmarking: an organization buys different makes of valves and installs them to see which perform best.
72	Team Charter: Order Fulfilment	Abbreviated example of a project charter.

Page	Case study/example name	Summary
117	Weighting Factor Normalization Gas Processing 1	Use of a weight factor to normalize for different types of equipment at processing facilities.
118	Weighting Factor Normalization Airline Maintenance 2	Worked example of comparing performances using weighting factors.
118	Weighting Factor Normalization Gas Processing 2	Worked example of comparing performances using weighting factors.
120	Deriving a Weighting Factor for a Hospital	Example of how to derive Weighting Factors for different treatments at hospitals.
121	Applying the Weighting Factor to Hospital Treatment	Worked example of comparing performances using weighting factors for hospital treatments.
127	Transport Arrival Delays	Calcvlatina a ranked performance.
128	Scoring Using a Business Excellence Model	Shows how to benchmark performances using the results of a Business Excellence Award model assessment.
132	Choosing Between Categories, Weighting Factors and Modelling: Cancer Treatments	Worked example using categorizing, weighting factors and modelling for benchmarking cancer survival rates.
133	Infrequent Activities: Maintenance	Illustrates the problem of benchmarking major infrequent maintenance activities when reporting over a fixed period which may or may not include the major activity.
136	Benchmarking Infrequent Maintenance Events	Describes how participants learned how to experiment with reducing major maintenance activities.
140	Revising the Scope: Information Technology	A project team revises the scope of a project at metric selection stage.

Page	Case study/example name	Summary
309	Smoke Screen	One manager was challenged to provide examples of organizations where benchmarking had worked, but even having provided examples, was challenged to provide more and more.
310	Short cut to Cutting Costs	Describes how one manager wanted to benchmark manning levels to ascertain what an appropriate level would be. His managers said they didn't have time to benchmarking. Being concerned for the deepening financial problems of the company, he decided to act by just cutting staffing.
311	We are Doing Well Enough (Xerox)	Xerox thought they were "doing well enough" before they were nearly swept away by Japanese competition. (from Prophets in the Dark xiii)
331	Citigroup IT Benchmarking	IT benchmarking study using Global Information Partners as consultant.
341	Drilling Performance Review	Project benchmarking club for drilling oil wells.
367	Dundee City Council	Describes a public sector body's approach to benchmarking in general and how they benchmarked a One-Stop-Shop facility with another council.
368	Recycling	Describes how a local authority monitor domestic recycling rates and adopt practices from better performing authorities.
370	Servicing of Aircraft and Hospitals	Explains that one authority benchmarked hospital servicing with aircraft servicing.

Preface

When I started out in benchmarking in 1993 I knew very little about the subject. A good place to learn, I thought, was to read some books. The first book I read was Business Process Benchmarking by Robert Camp, the recognized father of benchmarking. Useful though his book is for an organization wanting to benchmark by visiting other organizations (called One-to-One benchmarking in this book), there was little mention of groups of organizations wanting to work together to benchmark and improve. What I needed was help on facilitating benchmarking meetings, developing metrics for benchmarking studies, expediting, validating and analysing benchmarking data... if fact I needed practical help and advice on all aspects of running successful benchmarking studies. That was in 1993, and I believe there still is a dearth of benchmarking books giving this type of practical guidance.

Together with various colleagues I have learned many different ways to benchmark – simply by responding to clients' needs and being open to new methods. I learned, for example, how to run benchmarking clubs; how to benchmark using a database of performance levels; how to organize benchmarking visits and how to facilitate learning between organizations.

My aim in writing this book is to provide a practical "how-to" guide for benchmarking and the sharing of Best Practices. It is based on experience and practice, illustrated with many case studies (based on actual situations we encountered) and examples (hypothetical situations developed to illustrate a point).

One of the many things I have learned is that there is no single "right" way to benchmark. Each study needs to be tailored to meet the needs of the organizations involved in it. However, there are common phases and tasks that need to be considered by all benchmarking studies. We need, for example, to determine our objectives, decide with whom we want to benchmark and select which performance metrics, if any, we want to compare. For some benchmarking studies some of these steps will be very easy or may even be omitted, whilst the same tasks may require considerable time, effort and research in other studies. What is important is that each task is addressed appropriately. The aim of this book is to explain and provide practical guidance on each of these tasks, highlighting pitfalls for the unwary and giving tips to help make every study a success.

Acknowledgements

As the author of this "how to" book I am very aware that the information it contains is a conglomeration of what the many people I have worked with over the last 16 years have said or done. Even the ideas and practices that I developed were usually developed in conjunction with, or at the prompting of others. There are too many people to list them all, and in any case the source of some ideas, practices and methods are lost in the mist of time.

To start at the beginning, I owe much of my benchmarking experiences to Peter Rushmore, and am grateful to his and Helen's agreeing to my using the Drilling Performance Review as an excellent project benchmarking club case study. My thanks also to Alexander Janssen and his team at Juran Institute: we have worked together for over 10 years and many of my experiences of benchmarking derive from the work we have done together.

There are also many participants of the benchmarking studies and learning events I have been involved with whose ideas, requests and observations are included throughout this book.

The aim of this book has always been to illustrate the wide variety of benchmarking methods. One key method of doing so has been to elicit a cross section of case studies from both participants and consultants, in different business sectors, using different benchmarking methods. In addition to the Drilling Performance Review which is used to benchmark projects, my thanks go to Ray Wilkinson at the Best Practice Club for submitting a case study of how on-line communities can benchmark, learn and improve together. My thanks also to Caleb Masland of Information Management Forum for the IT benchmarking study with Citygroup, a major international financial services organization. Many people see benchmarking as appropriate only for industry, and so I am grateful to Paul Carroll for his case study explaining several different benchmarking methods used by city councils and for his explanation of the One-Stop-Shop case study.

Thanks to Michael Conway at the British Quality Foundation who encouraged me in the writing of this book and allowed me to carry out a benchmarking survey of the British Quality Foundation's membership.

A common aim in benchmarking is to learn from other organizations. Experience has shown that sometimes this simple idea works and sometimes it does not. I wanted to elicit the views of current management thinking on benchmarking and on the importance of understanding why practices may not be transferable between organizations. Gordon Hall's knowledge of management thinking is extensive and he kindly offered to contribute a chapter on the importance of understanding how and why copying practices may fail.

Finally, my gratitude for her patience, hours of dedication to meticulous proof reading and advice, to my wife Pat, who is herself an experienced benchmarking professional. Without her this book would not have been written.

Introduction

The aim of this book

If you want to know *how* to benchmark, this book is for you.

If you have picked up this book the chances are that you are, or plan to be, involved in benchmarking. It may be that you have participated in benchmarking projects and wonder if there is a better way of reaping the promised benefits. My aim and hope in writing this book is that it will provide a guide to successful benchmarking.

This is a practical "how to" book about how to run effective benchmarking projects and clubs. The aims of the book are to:
   ✓ Explore the different methods of benchmarking (summarized in Chapter 2, with examples throughout the book).
   ✓ Highlight unusual uses and applications of benchmarking that go beyond the traditional concept of comparing performance levels and practices.
   ✓ Provide a road map that will guide both participants and facilitators in completing successful benchmarking studies and projects. ⇨ Part 2.
   ✓ Provide details of how to use data analysis tools and charts for benchmarking purposes. ⇨ Chapter 14, Appendix A1.
   ✓ Provide a selection of detailed benchmarking case studies ⇨ Part 4, along with vignettes and learning points from a wide range of organizations, to demonstrate how benchmarking has been used.
   ✓ Finally, to encourage those not currently involved in benchmarking to use this tried and tested tool to help improve their own organization.

To achieve this aim I have:
   ✓ Kept theory to a minimum.
   ✓ Arranged the material in a logical order following a typical benchmarking project so it can be used as a handbook.
   ✓ Included examples and short case studies in the text, and stand alone case studies as the last section of the book ⇨ Part 4.

This book is aimed at:

- ✓ Managers who want to understand how to use benchmarking to improve their part of the organization.
- ✓ Representatives from participant organizations or independent facilitators responsible for managing and steering benchmarking studies.
- ✓ Anyone involved in the benchmarking process.

Currently available books on benchmarking tend to:

- Focus on only one type of benchmarking e.g. what we call in this book "one-to-one" benchmarking where the organization initiating the study seeks to visit one other organization.
- Focus on specific aspects of benchmarking such as Strategic Benchmarking.
- Provide little detail on exactly *how* to run and manage a benchmarking study.

This book aims to fill that gap and provide the reader with a practical guide for use at all stages of running a benchmarking study.

The structure of the book

Part 1 provides an introduction to benchmarking, including the role of benchmarking in improvement activities, and an explanation of different types of benchmarking.

Part 2 describes in detail how to successfully carry out a benchmarking project, following the benchmarking process outlined in the introduction to Part 2.

A typical benchmarking project consists of three phases:

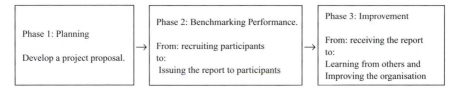

Phase 1, Planning, consists of all the internal preparation from project inception up to the point at which we begin inviting organizations to join the study.

Phase 2, Benchmarking Performance, begins by recruiting participants onto the study and continues by honing the project plan with participants, gathering and analysing data and ends, frequently, with issuing a report.

Phase 3 Improvement, it is the responsibility of each participant to use the data and information presented in the report to drive through improvements. These improvement activities are different to performance level comparison activities and are covered in phase 3.

Part 3 Explains the two managerial aspects of benchmarking. Firstly we explain how to manage and support benchmarking activities within an organization including highlighting some of the legal considerations, benchmarking protocols and the use of consultants. Secondly, we explain the project management activities and responsibilities.

Part 4 consists of four detailed benchmarking case studies. The cases have been carefully selected to demonstrate benchmarking from both the facilitators/consultants and the participants' view. They are drawn from local government, public service and industry and illustrate several different methods of benchmarking.

How to use this book

This book has been designed for the general reader interested in benchmarking to read from the beginning through to the end. It begins in Part 1 with an explanation of what benchmarking is, how organizations use benchmarking and the benefits it brings. Part 2 describes in detail, step by step, how to run a successful benchmarking study, from identifying projects through developing metrics, collecting, validating and analysing data, working with and learning from participants. Part 3 discusses both how to manage individual benchmarking projects and how to promote and manage benchmarking activities within an organization.

The book is written with the benchmarking practitioner in mind. Once the reader has found his way around, it can be used as a reference book for helping with issues that may arise during benchmarking studies. These include dealing with confidentiality issues, organizing Best Practice meetings and overcoming resistance to benchmarking.

Some readers may have specific needs or reasons for picking up this book and the table below suggests ways in which you may wish to use it.

If this is you….	Try reading this….
I know very little about benchmarking. I want an overview of what benchmarking is and how it can help me improve performance in my organization.	This book was written in a specific sequence for you. • It begins by explaining what benchmarking is, the benefits of benchmarking and its role in process improvement in Part 1. • At the beginning of Part 2 is an overview of the benchmarking process. This may be all you need, or if you want more detail you can dip into the appropriate pages in the rest of Part 2. • For information about managing benchmarking activities within your organization and for how to manage specific projects, see Part 3 • To see how benchmarking works in practice, Part 4 provides several detailed case studies.
Many people in my organization believe that benchmarking is just another fad. How can I persuade them otherwise?	See the benefits of benchmarking Chapter 1 and the case studies in Part 4. Start with small simple projects, perhaps implementing ideas that you have read or heard about outside the organization. Once these can be shown to be effective, progress to a more formalized method of identifying improvements from external sources – i.e. benchmarking. Some organizations benchmark, but because of the connotations of the word within their organization call it something else. Good luck with changing their minds, remember that doing so may take a long time!

If this is you….	Try reading this….
You can't benchmark what we do because every time we provide a service or product it is different	• Many organizations have the illusion that because their business is "different" they cannot benchmark, and so miss out on the benefits that benchmarking can bring. The Drilling Performance Review case study is an example of benchmarking projects where every project is different. The Citygroup benchmarking case study shows how IT services can be benchmarked even though each participant's IT service is tailored to its own needs. See also the Formula1 case study where a manufacturing organization benchmarked with a Formula1 team. • For methods of normalizing to take account of differences in participants operations see Chapter 7.
We tried benchmarking and found that copying other organizations' practices didn't work.	There is a real temptation to copy what works in another organization without thorough investigation. Sometimes such copying will work, sometimes not. See Chapter 19 on Copying Without Understanding.
I am tasked with benchmarking my section, but I know nothing about benchmarking.	• Read Part 1 which explains what benchmarking is, the benefits of benchmarking and its role in process improvement. • Read the summary at the beginning of Part 2 for an overview of the benchmarking process. • Read some of the case studies throughout the book and in Part 4 to see how benchmarking has been implemented. • Get a feel for the detailed benchmarking process in Part 2 by skimming through it, ignoring any sections that may not be relevant to your situation.

If this is you….	Try reading this….
	• Review "Methods of benchmarking" in Part 1 to help focus on what type of study you want to develop. • Work through Part 2 to develop a detailed project plan.
How do I select what to benchmark?	There are many tools for helping you decide what to benchmark – see Chapter 3 on selecting projects.
I am involved in a benchmarking study run by a consultancy, but I don't really know if they are doing a good job or how I can improve the study.	• Part 2 especially explains in detail what the facilitators should be doing. Check to see if they are. • Read the Drilling Performance Review and Citigroup IT case studies. Your facilitators are probably delivering a different service, but are they as committed to improving the service they provide?
We are concerned about the quality of data used for analysis. What should we do?	Data integrity is a very real concern when benchmarking. Ensuring that data is consistent, correct and complete is covered in detail in Part 2, especially Chapter 13.
It's not possible for us to benchmark because we won't be allowed to show other organizations, especially our competitors, what we do nor give them data on how we perform.	This is a common concern. There are several potential solutions depending on your situation. Database benchmarking, overviewed in Chapter 2, would only require sharing data with an independent benchmarking consultant. Other organizations need not even be aware that you are benchmarking. If you want to work with a known group, methods of anonymizing or not sharing data are discussed in Chapter 15. Lawyers may tell you, incorrectly, that you cannot share data. Be persistent in ensuring that what they are telling you is correct. See Chapter 21

PART I

A Background to BM

INTRODUCTION

In Part 1 of this book we aim to provide a backdrop for the rest of the book which deals with the practice of benchmarking.

The first question to answer is what exactly is benchmarking? In Chapter 1 we explore different ideas of benchmarking as proposed by those who have developed and used it. To help clarify what benchmarking is, we also look at some practices that are not benchmarking.

Benchmarking developed as a solution to a problem. Though practices akin to it had been around for many years, it is Xerox that we have to thank for formalizing it and developing it into a crucial performance improvement tool. It is appropriate, therefore, to include a description of how benchmarking enabled Xerox not only to survive but to become a market leader in a competitive market during the 1970s.

Benchmarking has moved on over the last 40 years and there are now many reasons why organizations benchmark, and many benefits that they gain. Performance improvement, budgeting, testing ideas, technical problem solving and resolving disputes are just a few of the reasons why organizations benchmark today.

There has been a surge in interest in Six Sigma and other improvement methodologies in the last few years. The connection between these and benchmarking is highlighted at the end of Chapter 1 and shows that they are closely related. Indeed, it has been said that benchmarking is a short cut improvement process as it identifies best practices without us having to try to invent them ourselves.

Many books on benchmarking propose one specific process. That is a shame, for in reality the benchmarking process that an organization uses needs to suit the specific objectives of the project. When writing this book I realized that the only way to keep the size of the book within reasonable limits was to follow the same path of focusing on one specific, albeit generic process. However, a key message throughout the book is that there is no one right way to benchmark. To highlight this fact Chapter 2 overviews several different methods of benchmarking, and acknowledges that there are almost endless variations. Despite the very different types of benchmarking the process of each can be seen to be a special case of a generic process which is explained in detail in Part 2 of the book.

Chapter 1

What is Benchmarking?

The chances are that if someone is able to do what you are doing better, faster and/or cheaper, they have different practices than you have. Discovering what those practices are, adapting them to your situation and adopting them is very likely to improve your performance.

INTRODUCTION

How do we define 'benchmarking'? What are its origins? Why do organizations benchmark? What benefits does it bring? What can I benchmark? How does benchmarking relate to Six Sigma and process improvement?

These are typical questions that people first ask when they become aware of benchmarking and all of them are addressed in this chapter. There is also a brief summary of how and why Xerox developed and used benchmarking as a key survival tool in the face of fierce Japanese competition.

This chapter also explains the two aspects common to many benchmarking studies:

1. Comparison of performance levels to ascertain which organization(s) is achieving superior performance levels.
2. Identification, adaptation/improvement, and adoption of the practices that lead to these superior levels of performance.

1.1 WHAT IS BENCHMARKING?

Developing a single all-embracing definition of benchmarking is not easy. It is commonly applied to a wide variety of activities that organizations undertake to compare their performance levels with others and/or identify, adapt, and adopt practices that they believe will improve their performance. Before presenting the definition of benchmarking that forms the backdrop of the book, let us look at some of the things that those involved in benchmarking have said of it.

For some, benchmarking needs only to involve the comparison of performance metrics, and needs not include an element of process improvement. This would certainly be the case for the organization with the best performance levels.

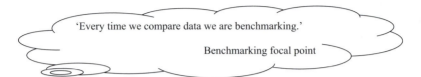

'Every time we compare data we are benchmarking.'

Benchmarking focal point

For some people, benchmarking is a continuous process rather than a one-off process.

'…the continuous process of measuring products, services and practices against the toughest competitors or those companies recognized as industry leaders.'

David T Kearns
CEO (former) of the Xerox Corporation

This definition specifically highlights that benchmarking can apply to the products, services, and practices of an organization. More broadly, benchmarking can be applied to any area where we want to compare performance and/or learn from others. The definition also specifies that we want to benchmark against either our toughest competitors, so that we know where our strengths and weaknesses are in relation to them, or industry leaders so that we are aware of the highest performance levels currently being achieved.

Some consider benchmarking as the comparison of practices, while for others, and perhaps most commonly, it includes both the comparison of performance and practices.

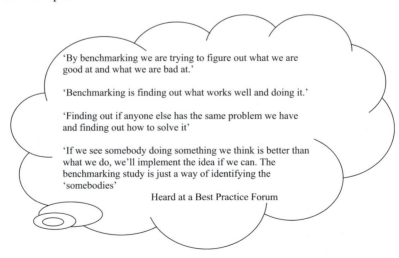

'By benchmarking we are trying to figure out what we are good at and what we are bad at.'

'Benchmarking is finding out what works well and doing it.'

'Finding out if anyone else has the same problem we have and finding out how to solve it'

'If we see somebody doing something we think is better than what we do, we'll implement the idea if we can. The benchmarking study is just a way of identifying the 'somebodies''

Heard at a Best Practice Forum

Notice that these four comments on benchmarking do not focus on the method, only on the required end result.

One interesting definition from Roger Milliken is:

'Stealing shamelessly.'

Roger Milliken
CEO Milliken

Benchmarking definitely is not stealing, at least not without permission! But it may entail adopting and adapting ideas, practices or methods, with permission, from other benchmarking participants.

As these definitions and quotes illustrate, there are different ways to view benchmarking. In this book we take the view that benchmarking consists of two basic phases (Figure 1.1): (phase 1 is a preparation phase)

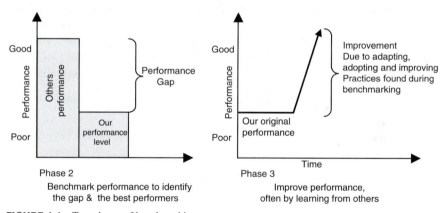

Phase 2
Benchmark performance to identify
the gap & the best performers

Phase 3
Improve performance,
often by learning from others

FIGURE 1.1   Two phases of benchmarking.

1. **Benchmarking performance** (i.e. data) to:
   - quantify performance levels of different participants,
   - identify the gap between participants, often between the best performer(s) and other participants,
        in order to:
   - quantify the potential gain for each participant to operate at the level of the best performer(s).

2. **Changing our practices** to improve our performance, possibly, but not necessarily by learning from other participants (⇨ Chapter 17).

From the above discussion we can produce a useful and comprehensive definition of benchmarking that applies to many benchmarking projects:

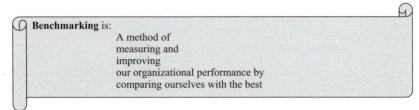

> **Benchmarking** is:
>
> A method of
> measuring and
> improving
> our organizational performance by
> comparing ourselves with the best

In Part 2 of this book we focus on the comparison of metrics and a variety of improvement activities, including methods of sharing practices. We acknowledge the many other reasons for, as well as methods and ideas of benchmarking by including case studies, discussions, and examples that do not adhere to this definition.

1.2 ... AND WHAT BENCHMARKING IS NOT

If statistics is the art of manipulating data to tell the boss what he wants to hear, then perhaps benchmarking is the art of comparing your organization's performance with other carefully selected organizations to ensure that the conclusion is what you wanted it to be before you started benchmarking. That would, of course, be a misuse of benchmarking and statistics, but sometimes it seems that these are the aims.

In order to clarify what benchmarking is, it helps to consider what benchmarking is not. Benchmarking:

✗   Is **NOT** industrial tourism whereby companies visit one another, enjoy a day out, or even a trip half way around the world but no objective comparison or analysis takes place. Such activities lead to benchmarking being seen as irrelevant and a perk for the favoured few. However:

✓   it **IS** planned research with a high return on investment.

✗   Is **NOT** a staff appraisal tool. This will lead to resistance to benchmarking by falsifying data, delaying the project, or discrediting the study (⇨ The result of using benchmarking as an appraisal tool Chapter 20). However:

✓   it **IS** helpful to identify where and how to improve the processes.

✗   It is **NOT** a copy and paste activity. Copying what someone else has done in their organization and expecting it to work assumes:

  - that your organization has a similar culture, a similar operational environment, and similar issues as the one from which you are copying (⇨ Chapter 19);

  - that the organization you are copying from has the optimum solution. However:

✓   It **IS** a potential source of ideas, information, methods, practices, etc. that it may be appropriate to adapt, adopt, and implement.

✗   Is **NOT** a one-off event. At best this will lead to achieving a competitive edge today, but that is likely to be eroded as other organizations continue to improve.

✓   It **IS** part of a culture of striving to be the best, or amongst the best, at what the organization does. (There are exceptions however, for example benchmarking can be used to solve specific problems or justify decisions ⇨ Part 4 'Dundee City Council Case Study'.)

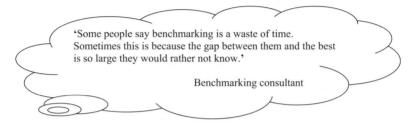

'Some people say benchmarking is a waste of time. Sometimes this is because the gap between them and the best is so large they would rather not know.'

Benchmarking consultant

1.3 A BRIEF HISTORY OF BENCHMARKING

Benchmarking may seem like a management fad from the 1970s, 1980s, and 1990s. It is not. It is not a fad and it is not from that era.

How did the idea of benchmarking evolve? What are its origins?

Perhaps the idea started when a man first looked at a neighbour's hut and thought 'That design lets in less rain than mine, perhaps I should build one like that'; or when he looked at his neighbour's crops and asked 'Why are his plants producing more fruit than mine? What does he do that is different to me'?

One can imagine a king, in the years before records began, sending out spies to observe what kinds of weapons his enemy was developing; or seeking to discover what features enabled another nation's ships to travel faster. We can picture a wise ruler sending envoys to foreign countries to learn what remedies they used for certain ailments. Our imagination tells us that somewhere near the start of the road to modern civilization men realized that they could learn from each other's discoveries and thereby improve their own situation.

As industry became more developed and organized, it became clear that large profits could be made by developing better products, and producing goods in ways that were faster and more efficient. Now there were companies vying for customers. In order to stay on top, it was necessary to be aware of the competition – what features made the competitors' products more desirable; what manufacturing processes were they using to produce their goods more cheaply; where were they getting their raw materials, or indeed, what raw materials were they using?

Nowadays there are legal mechanisms to deter companies from stealing designs and discoveries from one another, and organizations go to considerable lengths to guard those secrets which give them competitive advantage in the marketplace. In the unforgiving climate of the global market, not keeping up with competitors often means going out of business. This seems too contrary to the idea of sharing information for the greater good of the industry. In this environment how can organizations learn from each other?

1.3.1 Reverse Engineering

In the industrial world reverse engineering appeared as method of covert benchmarking. Not only did organizations look at the competition and try to improve their products and services, they acquired competitors' products, dismantled them and learned how to equal or if they could, improve on what they learned.

Reverse engineering is nowadays, of course, often illegal and the use of information gained by reverse engineering is protected, typically by patent (⇨ Chapter 21).

Case Study: Tupolev Tu-4 Aircraft

One well-known example of reverse engineering was the development of the USSR's Tupolev Tu-4 bomber aircraft. In 1944, three American B-29 bombers on missions over Japan were forced to land in the USSR. The Soviets, who did not have a strategic bomber as advanced as the B-29, decided to dismantle and study both the design and components of the B-29. The Tupolev Tu-4, a close copy of the B-29, flew in 1947.

1.3.2 Japanese Industrial Visits

After WWII Japanese industry was all but non-existent. Many of their factories had been bombed, and most had been focusing on the war effort. As part of their effort to establish a vibrant manufacturing industry Japanese industrialists visited American factories. This gave them both an insight into American manufacturing practices – i.e. what their future competition was doing and how they were doing it – and ideas that they could use in their own factories.

The Japanese had been warned by Dr W.E. Deming (ironically an American) not to simply copy what they saw, but to understand why it worked and to adapt and improve on the practices and ideas they discovered before adopting them (⇨ Chapter 19). America and the West in general, did not perceive Japan as a threat and were quite happy to show off their industries.

1.3.3 The Story of Xerox

Up until the 1970s benchmarking practices were somewhat haphazard and certainly not widely seen as a management improvement tool. Xerox developed and established benchmarking as a tool to drive out waste, drive down costs, and drive up quality. Current benchmarking thinking and practices are firmly based on what Xerox did over 30 years ago.

Case Study: Why Xerox Invented Benchmarking

The phenomenal growth of Japanese cars, radios, computers, cameras, motorbikes ... the list is almost endless forced many Western companies out of business. But not one company: Xerox. They fought against the seemingly unstoppable Japanese expansion and survived. The story of Xerox is documented in the book Prophets in the Dark, well worth reading as an account of how one organization avoided extinction, as well as an instructive book on benchmarking.

In the 1960s, Xerox was experiencing phenomenal growth. For example, profits in 1961 were $2.5 million. By 1968 they had jumped to $128 million. At one time some managers were hiring 50–100 people per month.

However, by the late 1970s Xerox had a string of problems. The Japanese had just entered the photocopier market and the Xerox 3300 copier was designed to push them out of it. The X3300 was so bad it had to be recalled and redeveloped. It was released again a year later, but could hardly be considered a success. Earnings and margins were dropping rapidly. Xerox became aware that the Japanese were using worker participation and began to study other American companies such as Lockheed and AT&T to discover what they did. As a result they tried to introduce worker participation and then Quality Circles – both unsuccessfully. Eventually they visited Japan to compare key data. They discovered that, for example:

● Xerox's overheads were double those of the Japanese.
● The Japanese were carrying six to eight times less inventory than Xerox.
● The quality of incoming goods at Xerox was 95%, in Japan it was 99.5%.

After the visit they reported that 'In category after category the difference was not 50% or anything like that; it was almost always over 100%' (p. 122, Prophets in the Dark).

Of course, knowing the performances of their own and other organizations did not solve the problem, but it did two things:

1. Made Xerox realize why they were about to go out of business and hence
2. Gave them the shock they needed to act.

Xerox finally conceived the idea of Business Effectiveness, a strategy to improve the competitiveness of the company. It embraced two underpinning thoughts:

1. Employee involvement and
2. Benchmarking.

Their goal was superiority in all areas – quality, product reliability, and cost. Starting in 1979 they identified which company was the best at distribution and used it as the standard to shoot for. The same would go for manufacturing, engineering, marketing, and so forth (p. 123, Prophets in the Dark). Their plan was to find out:

1. Who was best at doing something (benchmark the data).
2. Find out how the 'best' achieved their superior performance. (Learn)

These two concepts still form the cornerstones of benchmarking today. Xerox did it by a series of site visits. The most famous was probably that of L.L. Bean where they benchmarked distribution, but at around that time they also visited John Deere, IBM, Texas Instruments, Motorola (who later popularized Six Sigma) and Burroughs.

Identifying specific landmarks after events at Xerox is difficult. Robert C. Camp, who was heavily involved in benchmarking at Xerox wrote the first definitive book on benchmarking: Benchmarking: The Search for Industry Best Practices that Lead to Superior Performance in 1989. His latest book was published in 2007. Also in 1989, BP and nine other oil and gas exploration companies initiated a benchmarking study of wells drilled in the North Sea. This study, which is an example of project benchmarking, is still in existence in 2009 with over 200 participants from around the world and has spawned other spin-off studies (⇨ Part 4 'Drilling Performance Review').

In the 1990s benchmarking activity increased markedly. In the UK, for example, the report by Coopers & Lybrand, the CBI, and the National Manufacturing Council, "Survey of benchmarking in the UK 1994" identified that 78% of The Times top 1000 companies claimed to be benchmarking. In 1992, the American Productivity and Quality Centre (APQC) established the Benchmarking Clearinghouse (IBC). The business models that were developed throughout the 1990s and 2000s such as the Baldridge Award and the European Foundation for Quality Management (EFQM) Excellence Award imply the need for or specifically require that an organization be involved in benchmarking activities.

Benchmarking has now spread to all five continents. One small bench-marking study, for example, includes participants from Australia, Asia, the Middle East, Europe, and the USA. Some organizations still refuse to benchmark: sometimes because national regulations prevent them from divulging data, sometimes because they think they have nothing to learn or for some other reason, but perhaps most often because they do no know how to benchmark.

In the last few years benchmarking has no longer made the headlines, but that is quite possibly because it has now become an established part of business life. There has been a growth in the types, methods of and reasons for benchmarking, many of which are discussed in this book.

Applications of benchmarking have also grown as organizations become familiar with the technique and find a wider variety of uses. For example:

- Consumer magazines are well known for testing and comparing (benchmarking) competing **products and services** and often include users' experiences and comments.
- Investors are well known for analysing and comparing the **financial** performance of companies before investing or recommending others to invest.
- Organizations often compare **functions** such as maintenance, information technology, or human resource management.
- Organizations operating on discrete **facilities** such as factories, ports, airports can benchmark the whole facility.

- Organizations also benchmark **processes** such as purchasing, recruitment, or research and development.
- Perhaps the most recent development is strategic benchmarking where organizations benchmark long-term **strategies** and approaches in an attempt to find those that seem to result in the greatest success. Strategic benchmarking typically focuses on areas such as product development and delivery, customer services, and core competencies.
- Organizations also benchmark to resolve specific **problems** such as how to reduce the number of faults of a certain operation.
- **Project** benchmarking, the benchmarking of stand alone events or projects such as construction projects is also possible.

The emerging message is that any aspect of an organization where performance can be compared or where it may be possible to learn from other organizations has the potential to be benchmarked ($\Rightarrow$ Chapter 3 for a fuller discussion).

1.4 WHY DO ORGANIZATIONS BENCHMARK?

There are many reasons why organizations benchmark. For Xerox it was a matter of survival. Fortunately, few benchmarking studies are initiated in such dramatic circumstances. Some of the more common reasons for benchmarking include:

1.4.1 As Part of an Improvement Culture

It is interesting that despite Xerox's benchmarking successes, the 1980s are known mainly for the growth and development of improvement ideas, methodologies, and philosophies such as Total Quality Management, the Quality Management Standard ISO9000, Supply Chain Management, and Quality Circles amongst many others. Perhaps the spread of these ideas was itself a result of awareness that Japan was steadily encroaching on traditionally Western manufacturing havens and an effort to respond to the higher quality and lower prices that Japanese goods offered. Part of this response was an attempt to find out what the Japanese were doing differently to the West and copy at least some of these practices. A typical example of this was the adoption of Quality Circles to foster employee involvement. $\Rightarrow$ Chapter 19

A shortcoming with most of these tools is that while they are excellent in themselves and have a role to play in improving the way organizations are run, they:

✗ do not suggest where to focus improvement activities,
✗ do not identify appropriate performance levels, and
✗ do not suggest what practices are likely to lead to optimum performance levels.

Benchmarking, on the other hand, addresses the issues:

- How competitive are we?

     For example, it may cost us $20 to raise an invoice, and we can use business process re-engineering to reduce that cost to $12. What we do not know is how competitive that is with other organizations. Perhaps our competitors only spend $10 to raise an invoice, in which case we are still uncompetitive.

- Where should we focus improvement activities?

     As organizations implemented and became familiar with improvement methodologies, another issue arose. Managers realized they did not have the resources to work on improving every aspect of the business at the same time. They needed to focus improvement activities in those areas where the return on investment (i.e. the ratio of cost of improvement activities to benefit achieved) would be greatest. In order to do this they needed to estimate the post-improvement performance level. Sometimes it was possible to do this or alternatively to calculate a theoretical performance level, but often organizations did not know what improved performance level to expect until they had carried out some analysis. A solution to this quandary is to benchmark potential areas for improvement. This aims to:

1. Discover the potential performance level in each area, based on what other organizations are achieving.
2. Calculate the benefit for each area of performing at the potential performance level.
3. Calculate the expected return on investment for improving each area.
4. Prioritize areas for improvement, based, at least partially, on the return on investment.

1.4.2 To Short-Cut the Improvement Process

Some organizations view benchmarking as a possible short-cut to improvement. Why, they argue, expend effort on analysis, re-designing processes, re-training and other costs and even then perhaps not achieve the same levels of performance as others. It would be better to identify current best practices, adapt and improve them and then expend effort on implementing changes with a high degree of certainty that the performance will be at least amongst the best that we have found.

1.4.3 Target/Budget Setting

Many organizations set targets. Unfortunately, many set arbitrary targets, frequently of the form of a 5% or 10% improvement on current or previous performance levels. One use of benchmarking is to help identify appropriate target performance levels based on what others are achieving.

1.4.4 As a Driver for Improvement

If we benchmark against competitors, the results can be a very strong driver for improvement because if an organization cannot perform at similar or superior levels to competitors in enough key areas then market forces will eventually force them out of business (⇨ 'Benchmarking as a driver for improvement 1 and 2' case studies Chapter 16).

Even when benchmarking against non-competitors, poor comparative performance can be the shock required to persuade management that they need to examine their organization's performance and look for ways to improve.

1.4.5 To Solve Problems

One use of benchmarking is to solve a specific problem. Participants will look for others carrying out similar work to see what types of equipment they use and what problems they encountered (⇨ Part 4 'Dundee City Council Case Study, Formula1 Case Study').

1.4.6 Requirement of Business Excellence Models

Business Excellence models such as the Baldridge Award in the USA, European Foundation for Quality Management Excellence Award in Europe require or at least imply the need for benchmarking (⇨ Chapter 2).

1.4.7 To Build up a Network of Like-Minded People

One of the benefits of benchmarking that many people cite is that they become part of a network of people interested in improving the way they run their organization. They use this network to discuss any topic of mutual interest, to form focus groups or simply to talk with like-minded people. Though this is seldom the main reason for benchmarking, it is often a useful outcome that is fostered through benchmarking-related activities such as participant forums, websites, and meetings.

1.4.8 To Justify a Proposal

One organization I interviewed when researching this book explained that one of the main reasons for benchmarking within their organization was to test draft proposals. They would search for a benchmarking partner who had implemented, or tried to implement something akin to their proposal. The findings of the studies resulted either in the proposal being dropped or the findings being included in the proposal which was then submitted to management for approval.

Benchmarking has also been used to help justify manpower and/or expenditure increases as well as decreases (⇨ Benchmarking Manning Levels Chapter 16), and as a method for joint venturers to satisfy themselves that the organization responsible for operating facilities on their behalf is managing them efficiently and effectively (⇨ Review Benchmarking Chapter 2).

1.4.9 To Target a Competitor's Weak Points

While it is not the usual stated reason for benchmarking, there have been instances where it was used to find a competitor's relative weak points as perceived by customers. The benchmarking organization would then ensure that their performance was better in these areas and market on those relative strengths.

1.5 HOW EFFECTIVE IS BENCHMARKING?

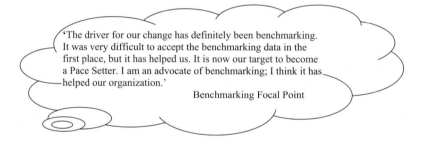

'The driver for our change has definitely been benchmarking. It was very difficult to accept the benchmarking data in the first place, but it has helped us. It is now our target to become a Pace Setter. I am an advocate of benchmarking; I think it has helped our organization.'

Benchmarking Focal Point

The extent to which benchmarking 'works' is dependent upon whether the organization is prepared to invest in making it work. Results do not come without effort and investment, and benchmarking is not a panacea for all issues. However:

**The evidence is irrefutable that
applied thoughtfully benchmarking does yield benefits**

This book and others written on benchmarking describe many examples of successful benchmarking. What seldom appears in the literature are examples of organizations that try benchmarking and gain no benefit, but they do exist. Although this book does not make a specific study of these companies, a common reason for benchmarking failing them is that they had no driving desire or reason to benchmark. A participant of one benchmarking club decided not to participate in future studies because they never used the results. A manager in a different organization stopped benchmarking because senior management had blamed him for the organization's relatively poor showing in a previous study. Such an organization will never benefit from benchmarking as long as senior management uses it as a tool to appraise and criticize.

The contrary view, expressed by forward looking organizations, is that there is always room for improvement and they can always learn something from others. For them benchmarking is simply a vehicle through which they improve.

'The benefit of benchmarking is that people talk. We have been involved in studies for over 10 years and we have built up relationships and contacts.'

Heard at a Best Practice Forum

1.6 HOW DOES BENCHMARKING RELATE TO SIX SIGMA AND PROCESS IMPROVEMENT?

All organizations develop processes by which their inputs are converted into outputs. Put simply, the inputs are raw materials and the outputs are the products, services, and/or information that they provide to their customers.

Benchmarking projects are typically initiated to help organizations improve their processes, resulting in higher quality products or services, reduced costs, and/or other benefits.

However, benchmarking is not the only improvement methodology available to organizations. Many methods have been developed over the last few decades in the never ending drive for improvement of which Six Sigma is probably the most popular.

In this section we examine the relationship between benchmarking and Six Sigma as an illustration of how benchmarking is intimately linked with organizational improvement.

We begin by summarizing the use of benchmarking as an improvement tool, and then identify how and where benchmarking fits in with improvement methodologies in general, using Six Sigma as a specific example.

In Chapter 16 we discuss how benchmarking may in turn lead to the use of improvement methodologies in order to improve performance and close the gap on the better performers in a benchmarking study.

1.6.1 Benchmarking as an Improvement Tool

Benchmarking has been used as an improvement tool for many years, and the fundamental idea behind its use is simple:

1. Define the project, i.e. the area of the business to be improved.
2. Find an organization that does what you want to do better than you can.
3. Find out what practices the organization uses that makes them better.
4. Adapt and adopt their practices to your situation.

The method of achieving improvement through benchmarking is a major theme running through this book.

1.6.2 The Role of Benchmarking in a Six Sigma Improvement Project

Many organizations use process improvement methodologies to improve the performance of their organization. These methodologies follow a similar pattern from project identification through identifying root causes, selecting and implementing solutions to ongoing monitoring of process performance. We have chosen the Six Sigma DMAIC (Define, Measure, Analyse, Improve, Control) improvement process as a popular and typical example to show how and where benchmarking can be used to help in the improvement process. Figure 1.2 gives the steps and brief explanation of the DMAIC process along with comments summarizing typical roles of benchmarking.

In addition, what is commonly called Step 0, project identification, has been added to the table.

It is also interesting to note that the benchmarking process explained in this book can be mapped onto the DMAIC process. The correspondence is not exact on a step-by-step basis, but all the elements of the DMAIC process are contained within the benchmarking process:

Step 0, Project selection, is described in ⇨ Chapter 3.
Step 1, Define, corresponds to finalizing the team charter (⇨ Chapter 4).
Step 2, Measure, corresponds to the benchmarking performance levels which is described in the rest of Part 2 phase 1 of the book.
Step 3, Analyse, is described in Chapters 14 and 15.
Step 4, Improve is described in Chapters 16–18.
Step 5, Control is not specifically discussed in the book, but is considered as part of the improvement phase discussed in Chapter 18.

Step	Brief description of Six Sigma activity	Role of benchmarking
0	Project Selection Management is responsible for setting the criteria for selecting projects, reviewing candidate projects and selecting those that will be carried forward. The team is commissioned.	Benchmarking studies have a useful role to play by: 1. Identifying areas of weakness compared to other participants (i.e. identifying potential projects) and 2. Quantifying the potential benefit of achieving the performance levels of the better performers. Since the potential benefit of improving the process is usually a key criteria for selecting improvement projects, benchmarking can play a key role in project selection.
1	Define. The team review and finalize the project requirement (i.e. the team charter). Much of this phase is honing, validating and adding details to the information passed to them when they were commissioned. This includes: • Defining the scope of the project. • Documenting the reasons for selecting the project. This usually includes evidence that the process needs to be improved. • Estimating (or validating if it has been estimated beforehand) the project benefits/impact on the organization. • Finalizing the Team Charter and gaining management approval. Any quick improvements that have been identified and can be made at this stage are implemented.	Benchmarking has a limited role to play. However, benchmarking may have identified quick and simple improvements that can be made. These may have already been made outside the improvement project, or may be incorporated during the appropriate phase.
2	Measure. The Measure phase of the project gathers information about the area being studied. Typical activities include: • Mapping the process, identifying key inputs and outputs. • Develop cause/effect relationships to identify potential causes of poor performance. • Developing and implementing a measurement system for gathering the data to investigate the relationships between the causes and effects. • Establishing the process capability.	Talking with benchmarking partners who have experienced similar problems in the past may provide theories of causes that the team only needs to confirm with their own data. Process maps may be available from other participants allowing comparisons to be made with the aim of identifying potential improvements. Benchmarking data can provide comparison process capability information from other participants.
3	Analyse. The team analyses the data gleaned in the Measure phase to establish the relationship between the causes and the effects. The output of this phase is proven root causes of current poor performance levels.	Some benchmarking studies include detailed analyses of data and these may be of help, or data may be available from these studies for the team to carry out their own analyses.
4	Improve. The team generates and evaluates potential solutions, testing those that it believes will resolve the root causes. Those solutions to be implemented are selected and the implementation planned.	Benchmarking can play a key role in finding solutions that other organizations have implemented. These solutions can be reviewed, adapted and adopted as appropriate. Where other participants have implemented similar solutions to those being planned, they may be able to provide advice on implementation.
5	Control The team designs and implements controls to monitor performance to confirm expected performance improvements and ensure that the improved performance is maintained.	Benchmarking has a minimal role to play, but participants who have implemented similar solutions may offer advice on effective controlling and monitoring as well as experience of problems that occurred after implementation.

FIGURE 1.2   Relationship between Six Sigma and benchmarking.

SUMMARY

There are many different definitions of benchmarking. The one we present here reflects many, but not all, benchmarking activities.

Benchmarking is:

A *method* of
Measuring and
Improving
Our organization
By *comparing* ourselves with the best.

Benchmarking usually consists of two aspects:

1. Comparison of performance levels to ascertain the gap between 'us' and the 'best' and to ascertain from which organizations we are likely to be able to learn the most.
2. Studying how the best or better performers achieve their superior performances and then adapting and adopting their practices as appropriate.

Modern benchmarking activities have become established since the 1940s when governments reverse engineered military equipment. In the 1950s and 1960s Japanese industrialists visited American factories to gain insight into mass production methods. In the 1970s Xerox visited leading organizations in order to discover what levels of performance were possible and to discover how to achieve these levels of performance. Since then the use and applications of benchmarking have grown rapidly and encompass product/service, financial, process, functional, facility, strategic, and project benchmarking.

Organizations benchmark for a variety of reasons, including:

✓ To enhance improvement culture.
✓ To short cut the improvement process.
✓ As a driver for improvement.
✓ As an aid to planning/budgeting/target setting.
✓ To solve specific problems.
✓ As part of a submission for Business Excellence Awards.
✓ To build up a network of like-minded people.
✓ To justify proposals.

Benchmarking is a useful tool in process improvement and problem solving. Its key uses are to:

✓ Help select and prioritize projects.
✓ Search for appropriate solutions.
✓ Identify appropriate target performance levels.

Methods of Benchmarking

INTRODUCTION

It is true to say that there is no one single right method to benchmark, though it is also true that there is a benchmarking process that most benchmarking studies follow to a greater or lesser extent.

That there are so many benchmarking methods is both important and useful: it is important because it helps us to avoid the trap of thinking that there is only one way to benchmark and therefore force-fitting our needs into a method that may not be appropriate for us, and it is useful because it encourages us to focus on using the most appropriate method to achieve the objectives of our own particular project.

In this chapter we illustrate the wide variety of methods available by outlining seven of them. In practice, a benchmarking study may adhere closely to one of the methods outlined here, or may follow a mix of methods. We exclude club benchmarking as this will be explained in detail in Part 2 of the book. The names given to different types of benchmarking, along with their definitions, are indicative of the nature of the type of study and the distinctions drawn between them are somewhat arbitrary.

In this section we discuss the following benchmarking methods:

1. **Public Domain** benchmarking as, for example, published in consumer magazines or newspapers.
2. **One-to-one** benchmarking where one participant visits one other participant. This is by far the most common benchmarking method discussed in books on benchmarking.
3. **Review** benchmarking which is typically carried out by a team visiting each participant, identifying relative strengths and weaknesses, best practices and perhaps making recommendations and even facilitating improvement activities.

4. **Database** benchmarking in which a participant's data are compared to a database of performance levels.
5. **Trial** benchmarking is carried out by trialling and/or testing products and services from other organizations and comparing them against your own products and services.
6. **Survey** benchmarking, usually carried out by an independent organization surveying customers to ascertain customers' perception of relative strengths and weaknesses compared to competitors.
7. **Business Excellence Models** benchmarking in which an independent assessor scores aspects of the organization according to a Business Excellence Model such as the Baldridge Award or the European Foundation for Quality Management (EFQM).

For each method we discuss, as appropriate:

1. The **concept** behind the method.
2. **Examples** of a typical benchmarking study.
3. The **participants.**
4. Who **controls** the study.
5. The **risks** from the viewpoints of the participants, customers, and others who may be involved.
6. The **learning potential**, i.e. the opportunities for participants to learn from the study.
7. Its **duration**.
8. The **benchmarking team**, i.e. who may be involved in carrying out the study.
9. Typical **uses** of this type of study.
10. A typical **process** for running the study. Many of the steps outlined here are discussed in detail throughout the book.

2.1 PUBLIC DOMAIN BENCHMARKING

2.1.1 Concept

In public domain benchmarking the benchmarker collects data from public sources, analyses it, and provides a report.

The metrics used in these studies are usually 'output' metrics, i.e. metrics that measure the product in terms of the experience of the customer. For example, when comparing mobile phone services, metrics are likely to include phone tariffs and network coverage as they are critical to the customer. Metrics important to the running of the business, such as employee satisfaction, purchasing, billing, manufacturing, distribution, and marketing, will not be compared.

2.1.2 Examples

Probably the most common and visible example of this type of benchmarking can be seen in hobby and consumer magazines. They provide independent reviews of products such as cars and cameras, as well as services such as banking and airlines. Consumer magazines also survey their subscribers to compare the customer experiences of products and services.

For an airline study the metrics reported may include punctuality, pricing, and seat pitch for standard flights. For 'soft' metrics, such as comfort, the benchmarker will often survey recent users of the airline (called survey benchmarking ⇨ 2.6 below). The results may be presented as tables or charts such as the ranked bar chart in Figure 2.1. Charts are generally preferred to tables as they are easier for the reader to interpret.

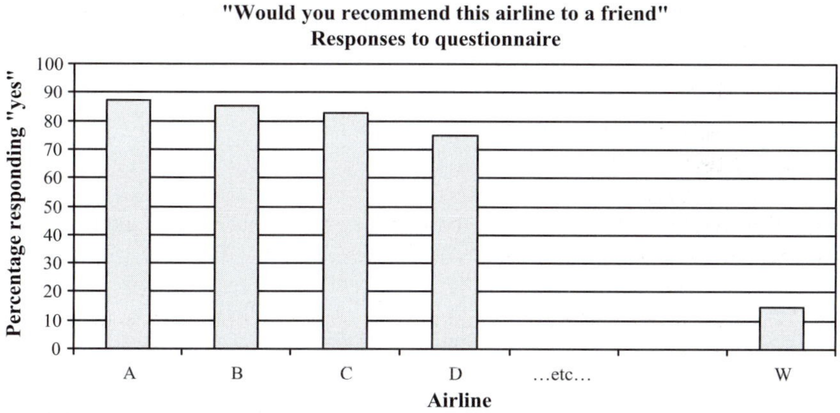

"Would you recommend this airline to a friend"
Responses to questionnaire

FIGURE 2.1   Public domain benchmarking sample bar chart from the airline industry.

With the development of the internet in recent years, comparative performance data for a wide variety of products and services are freely available.

Similar benchmarking exercises are often carried out by newspapers. Usually the reports relate to public services and are aimed at highlighting actual or perceived differences in performance between participating groups. Typical examples include: schools being compared by reporting their ranked exam pass rates; hospitals being ranked with respect to waiting times for operations; and the police or fire brigade services compared by response time to emergency calls. In these cases it may be a legal requirement for the services to report their data which are then held in the public domain so that everyone has access to them and can analyse them.

Organizations may also carry out such studies, usually to compare products and/or services offered by other organizations, typically competitors.

2.1.3 Participants

The selection of participants is controlled by the benchmarker and the participants usually have no influence over membership. When comparing data in the public domain, or when products or services are purchased for the study, the participants may not even be aware that they are part of the study until the results are published. Where the benchmarker invites participants to supply products, the potential participants may refuse, and so be excluded from the study.

2.1.4 Control of the Study

All aspects of the study are usually controlled by the benchmarker. Where the participants are requested to supply products, refusal implies that the potential participant believe their products or services will fare badly in the study. Where data are required to be reported by law or are otherwise in the public domain, most benchmarkers will use that data and probably not seek any further information or data.

2.1.5 Risks

Risks can be considered in relation to the three groups that have an interest in the study:

1. The organization carrying out the study (the benchmarker).
2. The customers of the study, usually the public.
3. The participants of the study and any others affected by its results.

1. **From the viewpoint of the benchmarker**, the risks are very low and generally limited to that of a lawsuit that may result if they are unable to substantiate published findings.
2. **The risk to the customers** of these studies is the potential for incomplete or inaccurate analysis leading to dubious conclusions, or the omission of potential participants from the study.

   Publications of 'league tables', usually found in newspapers, are often criticized as being irrelevant. For example, a league table may list schools in descending order of exam pass rates. This implies that the schools with higher pass rates are the more successful while those at the bottom of the table are the poorest performing schools. While pass rates may be correctly reported, the data need to be interpreted carefully as there is usually no measure of, for example, social background, which is known to be a key influence on academic achievement. Therefore, the tables can hardly be used, for example, to determine which school is 'better' than other.

   Another risk is that participants manipulate their processes in order to improve their position in the tables. For example, a hospital may reduce the maximum waiting time for a particular treatment by treating first those who have been waiting the longest, regardless of urgency. In schools, the proportion of pupils passing public exams can be increased by entering only those pupils who are thought very likely to pass.

Public domain benchmarking has the great disadvantage that when comparing products, often only one sample from each participant is benchmarked. For some aspects of a product, such as physical attributes or recommended retail price, one sample will suffice. However, where the attribute being benchmarked is subject to variation, such as reliability, the participant may be lucky in that the sample chosen for the study happens to be one of the better ones, or unlucky if it has a fault.[1] For example, when testing the durability of a car, the mean time to failure of a certain component may in reality be 1000 hours but any individual item may be likely to fail between 900 and 1100 hours. If the sample tested fails after 900 hours that is the figure that will be reported. Selecting an appropriate sample size and carrying out the appropriate analysis in these situations will be beyond most people except a trained statistician.

Where product review is based partly on the personal experience of one or a small number of reviewers, the results may be biased due to personal preferences.

Where a large group of users is surveyed for their experience of a product or service, the results are far more likely to reflect the experience of future customers, albeit the respondents may not be an unbiased cross-section of users. It is with this type of situation that database benchmarking merges with survey benchmarking as described below.

3. **From the viewpoint of the participants**, their risk is that if the study shows them as a poor performer their image may be affected.

In summary, while Public Domain benchmarks may provide interesting and useful data, caution must be exercised when interpreting the data.

2.1.6 Learning Potential

The learning potential for participants of public domain benchmarking is usually low. However, there may be some valuable information regarding how the participant's products and services compare to other organizations. This can provide a driver, and a good starting point, for improvement.

---

1. The same problem can also occur in other forms of benchmarking. However, where participants choose their own metrics they would normally be aware of the problem and compare data from many samples.

2.1.7 Duration

The duration of a public domain benchmarking study can vary from perhaps a few days, where data are already in the public domain and only need to be extracted, formatted, and reported, to several months, where extended trials or tests are required.

2.1.8 Benchmarking Team

The benchmarking study may be carried out by one person, for example where data are in the public domain and all that is required is to carry out some analysis and write a report. More complex situations, requiring expert knowledge, laboratory testing or trialling products or services may require a team of people.

Where customer surveys are required, the task of developing, administering, and analysing them may be subcontracted to an organization specializing in surveying.

2.1.9 Uses

The main use for most public domain benchmarking is to provide information to the public to help them make informed decisions about products and services that they use.

The participants, willing or unwilling, will have access to valuable information regarding how their products and services relate to others in the marketplace. Those that come out well in the report will be able to use the results to market their services and products. Those that come out poorly should use the information to help drive improvement within the organization.

2.1.10 Benchmarking Process

Each benchmarker will have their own detailed process for acquiring, analysing and reporting data for a benchmarking study and the steps may vary between studies carried out by the same benchmarker. However, the overall process steps will be similar and are outlined below and shown in Figure 2.2.

1. **Determine the objectives and scope of the study.** For a typical consumer magazine benchmarking study, this is usually to determine the best product or service from the customer's viewpoint giving reasons why the 'winner' was selected. A second objective is to comment on favourable and unfavourable aspects of at least the better if not all the products/ services reviewed.

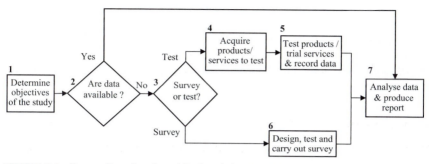

FIGURE 2.2   Process flow chart for public domain benchmarking.

Investigative journalists may also carry out comparative studies or surveys. However, their objectives are usually to highlight disparity in public services such as medical services, schools, or policing. These surveys are likely to focus on differences between the best and worst performers.

2. **Determine whether data already exist.** There may already be data available on the subject to be studied. This might be from government sources, the internet, or other published material. In these cases the benchmarker acquires the data and moves to Step 7.

3. **Decide how to obtain data.** If data are not available, then there are typically two choices open to the benchmarker.

   i. **Test**: The benchmarker may experience the product or service first hand and report on the findings (Step 4) or

   ii. **Survey**: The benchmarker may survey current or past users of the product or service (Step 6).

   Combining both approaches allows the benchmarker to corroborate his test results with the survey findings.

4. **Acquire products.** The benchmarker may acquire the product or service as a member of the public in order to experience the products or services without revealing his intent. Alternatively, the benchmarker can ask the participant to supply samples for testing. This has the advantage or reducing the cost, but the product/service cannot be said to have been randomly selected.

5. **Testing.** However the products and/or services are obtained, testing/trials need to be designed and carried out. Clearly the nature and extent of the tests vary widely, from a simple reporting of manufacturers' claims and perhaps the taking of some measurements, to field trials over time. The benchmarker may carry out tests himself and/or may invite members of the public to trial the product and provide feedback.

6. **Survey.** If a survey is to be carried out it is advisable to appoint a specialist survey organization because carrying out a reliable survey in any but the simplest of situations is a complex process fraught with difficulties for the unwary.
7. **Analysis.** Once the data have been gathered they are analysed, conclusions drawn, and a report issued. The complexity of the analyses will depend on the objectives of the study and the quantity and type of data collected. In most cases the report will include the objectives of the study, how it was carried out, and where the data came from. Finally, it will identify performance levels. In some cases the report may also include the result of investigations as to differences in performance.

2.2 ONE-TO-ONE BENCHMARKING

2.2.1 Concept

One-to-one benchmarking was the original benchmarking method developed by Xerox in the 1970s and 1980s, and is the probably the method most written about in books since then. The concept behind one-to-one benchmarking is straight-forward:

1. Find out which organization(s) is the best, or amongst the best at performing the aspect of your business which you want to improve.
2. Visit them in order to ascertain their level of performance and to learn how they achieve that performance.
3. Study their practices, adapting them where necessary, and improving them where possible.
4. Finally, adopt the new practice into your organization.

It is not necessary, or usually possible, to be sure that the target participant is the best. With this type of benchmarking it is frequently enough to know that the target participant is:

- Performing significantly better than your are.
- Amongst the best in the area being benchmarked.

To ensure that the target participant is amongst the best performers it would be necessary to gather and compare data from a range of organizations. While data may be available in some areas (for example, because of public domain benchmarking or benchmarking clubs), it would be difficult to persuade orga-nizations to take part in a data gathering study in order for you to decide which one(s) to visit.

However, it is usually possible for the initiating organization to visit a number of target organizations that are thought to be amongst the best and compare their practices and performances, selecting the practices that are most appropriate to adapt, improve, and adopt.

2.2.2 Examples

Case Study: Xerox and L.L. Bean Distribution Benchmarking

The book Prophets in the Dark by Kearns and Nadler has many examples of one-to-one benchmarking. For example, a published article highlighted the efficient distribution process used by the large mail order distributor L.L. Bean. As it appeared to be much better than Xerox's distribution process, managers at Xerox decided to visit L.L. Bean to find out more. They discovered, for example, that L.L. Bean could pick and pack goods three and a half times faster than Xerox. Xerox later developed a distribution system based on what they had learned from L.L. Bean.

It was because Xerox *believed* that L.L. Bean was amongst the best at distribution that they visited them to learn about distribution processes. L.L. Bean may not have been amongst the best at billing, purchasing, order tracking, or any other process. The key is to determine what the target organization excels at and learn about that aspect of their business.

Case Study: Formula 1

A potato chip manufacturer was concerned about the length of time required to set up production lines when changing from one product to another.

One creative individual within the organization who was keen on Formula 1 racing realized that the experts at quick setups were the people working on the cars during pit stops. They could refill a car, change the wheels, and complete various other tasks in a matter of seconds. At his suggestion the chip manufacturer decided to visit one of the Formula 1 teams and learned from them how they were able to achieve such fast 'set-ups'.

They learned, amongst other things, that their old approach of minimizing the cost of machine set-up was not appropriate as it resulted in longer set-up times and loss of production due to idle time. While having idle production lines was not a problem when utilization was low, it became very inefficient when utilization was high. They realized that their most cost effective approach would be to ensure that equipment, parts, and people were available to complete the setup in minimum time, thus increasing the production of the lines. They also learned that to do this they needed to plan carefully. They needed to ensure that appropriate resources would be available where and when required. They also learned that they needed to train all those involved in the set-up process to work together as a team.

In this example, it was not important to select the 'best' Formula 1 team. It is likely that all Formula 1 teams are working to similar principles.

See also ⇨ Part 4 'Dundee City Council'.

2.2.3 Participants

As the name implies one-to-one benchmarking is usually carried out between two organizations: the initiator (the organization initiating the benchmarking study) and the target (an organization with which the initiator wants to benchmark, usually because it is perceived as a high performing organization in terms of the scope being benchmarked).

With one-to-one benchmarking it is possible that the initiator will embark on a series of paired benchmarking studies both within and across industries and geographical regions. These will provide a number of practices which the initiator can review and compare before implementing. Each study can be tailored to build on the information learned from previous studies.

2.2.4 Control of the Study

The study is controlled by the two participants. They will agree timescales, information to be exchanged, and all other legal and logistical aspects of the study.

It is usual, but not necessary, that the initiator visits the target organization to gather information and experience how that organization operates. Unlike most other forms of benchmarking, the target organization may be content to share practices and data without asking for or expecting a reciprocal visit or information.

2.2.5 Risks

There are some unique risks with this type of benchmarking study. Because there may only be one or two visiting representatives, the information fed back to the initiator's management may be incomplete, subjective, or anecdotal. Typical reasons for this include lack of a structured method of collecting all relevant information or a fear of reporting to management how poor their own performance is.

Another risk, which has given benchmarking a bad reputation in some areas, is that the initiator does not make any changes following the visit. This may happen for a variety of reasons including lack of commitment to change, internal resistance to change, or more urgent issues arising. When benchmarking visits are not followed up with actions, others within the organization will see them as expensive trips for the privileged few.

Conversely, a mistake that some organizations make is to identify a practice that has helped the target participant and simply implement it without considering whether it is appropriate. For example, in a typical car manufacturing plant, the lack of inventory is noticeable. A visiting organization may therefore cut its own inventory only to discover that the production lines are continually stopped because they have run out of components. It is necessary to understand that the minimal inventory in most car plants is only effective if supported by excellent planning, scheduling, and relationships with suppliers that promote the delivery of on-specification parts, on time, every time. Only once this is achieved can inventory levels be reduced (⇨ Chapter 19).

On the positive side, the financial risk of the benchmarking study is relatively low because the manpower effort required to complete the study is low.

2.2.6 Learning Potential

The learning potential of these visits is potentially very high if an appropriate target organization has been chosen and the visit well planned. In particular, in cross-industry benchmarking it may be possible to identify quantum differences in performance levels as illustrated by previous examples.

2.2.7 Duration

Benchmarking studies of this nature are usually very quick to complete. Once a target organization has been identified it is only a matter of setting up and agreeing the terms of the study. Where there are site visits, one day is often sufficient, but a visit could last several days.

2.2.8 Benchmarking Team

The benchmarking team typically consists of a small group of people from the initiator, possibly with an external consultant who can be used to:

- Provide an independent report.
- Assist the initiator to prepare appropriately for the visit and so ensure that it is as productive as possible.
- Facilitate post-visit activities, typically improvement activities.

The other members of the team are likely to include those with responsibility in the area being benchmarked; front line staff involved in the process and, especially if there is no external consultant, an internal benchmarking consultant/ advisor. It may also be advisable to include staff with specialist skills such as health and safety, legal, and commercial depending on the subject being benchmarked.

2.2.9 Uses

One-to-one studies are best carried out when the initiating participant knows what they want to learn and from whom they want to learn it. In some situations, as explained in the examples above, potential benchmarking targets may be easy to identify. In addition, paired benchmarking can often be the natural next step after some other form of benchmarking. For example, the results of a public domain or benchmarking club study may suggest to a poorer performer a suitable target organization to visit.

2.2.10 One-to-One Benchmarking Process

Typical steps in a paired benchmarking study are shown in Figure 2.3.

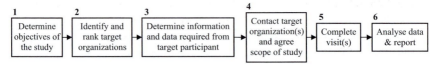

FIGURE 2.3   Process flow chart for one-to-one benchmarking.

1. **Determine the objectives.** These will typically include the identification of best practices that lead to superior performance in the area being bench-marked.

    If the target organization is known to be a significantly better performer in the area being benchmarked, it may not be necessary to gather data to demonstrate its superior performance. Alternatively, it may only be neces-sary to determine a rough approximation of the performance level. However, if several target participants are being visited, the collection and analysis of data to compare their performance levels becomes more important as an aid to understanding the effects of different practices.

    It is likely that an objective of the benchmarking study will be to quantify the costs and benefits of operating at similar levels as the target organization, or at least of adopting the ideas, philosophies and practices of the target organization. In the example of the chip manufacturer visiting the Formula1 racing team, they will probably never be able to set up their equipment in the few seconds that it takes to service the car in the pit lane, but they could estimate the costs and benefits of completing the tasks on the critical path efficiently with no delays.

    Where there are several target organizations involved, the study becomes very similar to a review benchmarking study (discussed below), with the exception that in a review benchmarking study it is usual that all participants will want to learn from each other.

2. **Identify and rank potential target organizations.** In some situations po-tential target organizations or industries may be obvious. For example, for security of on-line accounts the finance industry would be an obvious target. Sources for identifying potential target participants include news items, trade magazines, books, and personal experience. There will be a number of criteria for ranking the potential targets in order of preference including

perceived performance level, perceived similarity of benchmarking subject, commercial considerations, and location.

3. **Determine information and data requirements**, and document exactly what is to be learned and hence what information is needed from the target organization. This will usually require that the initiating organization understand their own processes and their weaknesses.

4. **Contact target.** Having ascertained clear data and information needs, the next step is to contact the preferred target organizations to ask for their cooperation. The target participants may well want to satisfy themselves so that you have a clear idea of the information that you require to ensure that they can identify the appropriate people to answer the questions and ensure they have any data, demonstrations, documents, etc. prepared for the visit.

5. **Complete visit.** If the visit has been well prepared, it should provide the required information. It is important to adhere to appropriate codes of conduct, e.g. to be punctual, polite, respect the decision of the target participant to withhold information or data, and to be prepared to share information and data from your own organization.

6. **Analysis and reporting.** After the visit, the initiating participant should write a report outlining what information was shared, conclusions, and recommendations for further action. The target organization may request a copy of the report, or at least the parts relating to itself.

2.3 REVIEW BENCHMARKING

2.3.1 Concept

Review benchmarking is the name given to a benchmarking study when a person or group visits a number of participants with the remit of reviewing certain activities at each facility and comparing the findings. The review may consist entirely of data comparison or of working practices but is more likely to include both. The objectives for a typical study are likely to include:

1. To identify the performance levels of each participant.
2. To quantify the performance gaps between each participant and the best performer and/or other performance standards of the group such as the average.
3. To identify differences in working practices that lead to differences in performance levels.
4. To recommend courses of action for each participant with the aim of improving their performance.

In this section we consider a review benchmarking situation where a group of participants initiate their own benchmarking study for their own benefit. However, some organizations, such as Best Practices in the USA, carry out research to discover best practices with the aim of publishing and selling their findings. This approach can be considered as a type of review benchmarking because a person or group gathers information from different organizations, and the aim is to publish a report.

ⓘ The Best Practices website is at www.best-in-class.com.

2.3.2 Example

Case Study: Review Benchmarking

A processing plant was owned by several companies in a joint venture (JV). It had been designed by one company and was being operated by another. Several of the owners had similar processing plants elsewhere in the world and were concerned that this plant was not being operated as efficiently as their other plants.

The owners contracted a group of consultants to visit the jointly owned plant along with other plants and comment on manning and cost levels, as well as practices. The report focused on the JV plant but also commented on the relative efficiency of each of the other plants. The report included performance gaps, estimated benefits of operating at the performance level of the best performer as well as recommended actions.

The need to understand the working environment is illustrated by the differences in work shift patterns. The manning levels of one plant, which appeared to be high, were traced back to legal restrictions on the length of shift and shift scheduling. This resulted in more lost time due to shift changeover than at other plants. Similarly, inspection costs were significantly higher at one plant. On investigation this was found to be due to increased inspection activity, which itself was due to government regulations.

2.3.3 Participants

In the above example the participants were connected through business. However, this does not need to be the case. Any group of organizations, within or across industries may agree to take part in a review study.

If there are only two participants, the study becomes similar to a one-to-one study. As the number of participants increase, the study will require more careful planning and control and will take longer to complete.

2.3.4 Control of the Study

Usually this type of study is controlled equally by the participants. However, as in the case study above, it is possible that the group of participants agree to the objectives of the study and leave the details to one participant.

2.3.5 Risks

The risks of this type of study are generally low because the participants have total control and can specify all the aspects of the study including participants, use of a consultant, metrics, processes, etc. to be benchmarked. Unless representatives from each participant are included in the review team, the participants will not be involved in site visits thereby missing out on gaining first hand insight into how others work.

The costs can be high, depending on how the study is run, and careful planning and project management is important in any but the smallest of studies to ensure the objectives are met.

2.3.6 Learning Potential

The learning potential depends on how the benchmarking study is run. If only data are to be shared, the learning potential is relatively low, but the study is cheaper. If processes, equipment, and other aspects are to be compared the potential benefit can be very high as participants can learn in detail about how other participants achieve superior levels of performance. If appropriate, this type of study can be followed by one-to-one studies or other information sharing activities such as Best Practice Forums.

2.3.7 Duration

The duration of this type of study is dependent on the number of participants and the depth of the study but they are usually completed within weeks or a few months. Generally these studies are carried out in more depth than the one-to-one studies outlined earlier, and will include both the collection of data and the documenting of working practices along with a comparison.

2.3.8 Benchmarking Team

Probably the most appropriate benchmarking team for this type of study is one that includes members from each participant. This helps to ensure that each participant's needs are met and, if appropriate, that a member of each organization visits all the other participants.

A useful approach, particularly for medium size studies with up to about six participants, is to employ consultants for the secretarial and analytical work of the study such as report writing and statistical analysis. Participants will manage the consultants and take part in visits, contributing to the report according to their knowledge and skills. The advantage of using consultants is that they can help ensure that the study is carried out efficiently and effectively in such a way as to maximize the benefit to the participants. They may also bring specialist skills to the study such as data gathering, analysis, and reporting. In addition consultants can help ensure independence and objectivity in the study.

For larger studies it may not be possible to have a member from each organization visiting all the facilities and the whole study may be left in the hands of a consultant with a participant steering committee directing their activities.

2.3.9 Uses

Review benchmarking is particularly useful when a small group of participants want to benchmark areas where differences in performance level are due to subtle operational differences. This will normally require an in-depth understanding of, for example, commercial, legal, or engineering issues that will not be readily identifiable from data alone. Where such information is commercially sensitive it may be necessary for an independent consultant to carry out the study.

Because review benchmarking is usually relatively expensive, it is often used only once as it is believed that there are significant differences in performance levels and that the poorer performers will be able to benefit from the study.

Internal review benchmarking studies can also be used on a routine basis to help spread best practices throughout the organization and strengthen relationships between participants. Such studies may be seen as being similar to audits, but the focus is firmly on working and learning together to improve the organization.

2.3.10 Review Process

The review benchmarking process, shown in Figure 2.4, is similar to the one-to-one benchmarking process which is explained above.

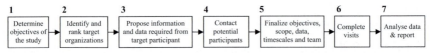

FIGURE 2.4   Process flow chart for review benchmarking.

1. **Determine scope and objectives.** The initiating participant(s) will determine the objectives of the study. These are likely to include:

   - Ascertaining relative performance levels,
   - Quantifying the potential benefit of performing at the level of the best performer (or other standard such as top quartile),
   - Identifying best practices.

2. **Identify potential participants** and rank them in order of those that are preferred to be in the study. If this study is being carried out as a result of

previous benchmarking, or is an internal study, or involves organizations that are in some way linked, it is likely that the participants will already be identified.

3. **Draw up a list of proposed information and data requirements.** It is necessary to do this before contacting potential participants as they will want to know what is being proposed.

4. **Contact potential participants** and invite them to take part in the study.

5. **Finalize plan.** It is likely that the details of the study will need to be reviewed in a series of meetings, to ensure that participants understand the nature of the study and its benefits. Participants may not be able to take part for a number of reasons, the most usual being inability to supply the proposed data and confidentiality concerns.

6. **Complete visits.** The benchmarking team visits the participants to collect the data and ensure it, and other information, is properly understood.

7. **Analyse and report.** The findings of the visits are written up and issued as per the agreement. As a precaution, it is advisable to issue a draft report so that participants have the opportunity to correct any inaccuracies and/or add further appropriate information.

2.4 DATABASE BENCHMARKING

2.4.1 Concept

Most types of benchmarking compare performance levels of a group of two or more participants. However, the term 'database benchmarking' is applied to a situation where the benchmarker, usually an independent consultant, has built up a database of performance levels from many organizations over time. As each participant joins the study their data are added to the database and their performance compared to that of other participants in the database.

2.4.2 Examples

On-line database benchmarking is offered by some websites, for example the Benchmarking Exchange, offers several benchmarking studies where data can be entered on-line. At the time of writing, a survey on Supplier Management is open for public use, while others are only open to members.

Solomon Associates offer to benchmark organizations in the oil and gas industry against their databases of manning and cost data of refineries and hydrocarbons processing facilities. Information Management Forum offer tailored IT benchmarking against their database (⇨ Part 4 'Citigroup Case Study').

ⓘ The benchmarking exchange can be found at www.benchnet.com.

2.4.3 Participants

The number and types of participant can vary widely. Participants may be given a complete or partial list of other participants, but individual performance levels of the other participants are unlikely to be divulged.

2.4.4 Control of the Study

The consulting organization, i.e. the database owner, controls most aspects of the study. Participants have little or no influence over the data collected, data definitions, or metrics calculated as they have to fit with the data held by the consultant.

2.4.5 Risks

The risks of this type of study lie with the participant. They need to take on trust that their data are compared with data supplied to the same definitions with organizations with whom it is meaningful to compare.

The quality of the results is largely dependent on the integrity of the consultant and accuracy of the supplied data. At one extreme, the consultant will provide detailed definitions of required data and may help with the data collection; they will also validate the data before entering it into the database. At the other extreme, there may be no clear definitions, and no validation.

Benchmarking studies open to the public where the participant enters data on-line and receive instant feedback are particularly suspect.

A reputable consultant should be able to provide useful information with appropriate pointers as to where improvement efforts should be focused and may even facilitate improvement activities.

The cost for the participant should be relatively low both in terms of financial outlay (e.g. of joining a study) and in time collecting data, but will increase if consultants are involved in data collection and improvement activities.

2.4.6 Learning Potential

The learning potential varies widely and depends on how the study is administered. Generally, the greater the involvement of the consultant, the higher the integrity of the results and the learning potential. This is especially true where the consultant is involved in improvement activities. At the other extreme, free on-line benchmarks will nearly always be unverifiable, provide ball-park figures only, give no advice and they may require that your name be added to a mailing list.

2.4.7 Duration

This type of study is very quick to complete, perhaps a few weeks, and will depend mainly on the speed with which the data can be supplied to the consultant. Since all the consultant has to do is validate one set of data and compare it with a database, the elapsed time between data submission and report should be days or a few weeks at most. On-line studies usually give results as soon as the data have been entered.

2.4.8 Benchmarking Team

The team, if there is one, is likely to consist of those involved in supplying the data to the consultant, the initiator of the project, and the consultant. On-line studies can often be completed by anyone with access to appropriate information in minutes or a few hours.

2.4.9 Uses

The key use of this type of benchmarking is where a reputable consultant is known to have a database of high quality data and also has a successful track record in helping organizations to improve.

2.4.10 Benchmarking Process

The first step is to determine the scope and objectives of the study. The next step is to identify a suitable reliable consultant with a database containing the data and information needed to meet the objectives of the study. Thereafter the process will depend on the consultant's norms.

2.5 TRIAL BENCHMARKING

Trial benchmarking can be considered as similar to public domain benchmarking, the main difference being that the initiating organization carries out the benchmarking itself.

2.5.1 Concept

Trial benchmarking is carried out entirely by the initiating organization and usually other participants will not be aware that they are being benchmarked. The organization carries out the study simply by using the services or products that it wants to benchmark. Many organizations will do this on a regular basis to monitor competition and identify new ideas.

2.5.2 Examples

To benchmark how customers experience airlines, members of the initiating airline simply fly with other airlines. They evaluate the aspects of performance in which they are interested, for example, quality of food, friendliness of staff, seat comfort.

An organization selling goods over the internet may compare its own website against those of competitors to evaluate such things as ease of navigation, the quality of information provided, ease of ordering, and product cost.

Organizations sometimes publicize their findings in the form of advertisements in which they claim to be in some way better than their competitors.

Trial benchmarking can also be carried out by customers to evaluate different products and/or services, though the suppliers of the benchmarked products and/or services may not be aware that they are being benchmarked (⇨ Chapter 3 'Valve Trialling Case Study').

2.5.3 Participants

The organization can choose the most appropriate organizations to benchmark against, usually its competitors. These participants will be unlikely to know that they are being studied.

2.5.4 Control of the Study

The initiating organization controls all aspects of the study.

2.5.5 Risks

There are very few risks associated with this type of benchmarking.

2.5.6 Learning Potential

The learning potential is usually limited to measuring the output of the processes rather than analysing the processes themselves. The organization may identify areas where they perform less well than competitors but will gain limited information on ways to improve. An exception, however, is in benchmarking websites where ideas for organization, functionality, colour schemes, etc. may be used to improve the organization's own website.

2.5.7 Duration

The duration of these studies varies widely depending on how the product/service is being studied. Trialling paint may take months or years to test its durability, while trialling internet purchasing may take hours.

2.5.8 Benchmarking Team

Benchmarking, for example, product features or static services such as internet purchasing, may only require one person. Trialling products or services that may vary, such as a help desk or the ease of use of a product, requires subjective judgment and may require a number of people.

2.5.9 Uses

An organization can use this type of benchmarking:

✓ To monitor competitor performance and identify where improvements need to be made.
✓ To identify potential product/service improvements.
✓ To identify potential participants for a detailed benchmarking study.
✓ To identify its own areas of high performance and market to these strengths.

2.5.10 Benchmarking Process

The benchmarking process is illustrated in Figure 2.5.

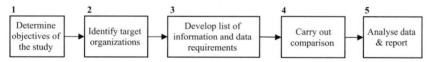

FIGURE 2.5   Process flow chart for trial benchmarking.

1. **Determine scope and objectives.** The organization will determine the objectives of the study. These are usually limited to:

   - Ascertaining features and performance of products and/or services, but may include:
   - Quantifying the costs and benefits of improving the product and/or service.

2. **Identify potential target organizations** and rank them in order of those that are preferred to be in the study. For competitor benchmarking potential participants will usually be well known. For non-competitor benchmarking studies, e.g. for internet ordering processes, some research may be required.
3. **Develop a list of information and data requirements.**
4. **Carry out comparison.** Acquire/experience products and/or services as planned.
5. **Analyse and report.** The findings of the comparisons are analysed and documented as required.

2.6 SURVEY BENCHMARKING

Survey benchmarking is similar to trial benchmarking and can also be considered as a subset of public domain benchmarking. It is therefore treated only briefly here.

2.6.1 Concept

In the world of consumer goods and services, the customer's perception of a product or service is an important, if not the most important, aspect for the supplier to focus on. Regardless of the actual product or service quality, if it is perceived by potential customers to be good they are more likely to buy it than if they perceive it to be poor.

One way to ascertain perception is to carry out a customer survey. This may be carried out by interview, post, telephone, or even email, and the aim is to compare customer perceptions of a range of products and/or services.

2.6.2 Examples

Most people have been questioned in the street, or have received a questionnaire asking about their experiences. Hotels, for example, frequently email guests with a survey questionnaire asking for their opinion of the hotel compared to those of competitors.

2.6.3 Participants

The organization can choose who to benchmark against, and the participants will be unlikely to know that they are being benchmarked.

2.6.4 Control of the Study

The organization has overall control of the study.

2.6.5 Risks

There are very few risks associated with this type of benchmarking. However, designing and administering effective surveys of customers, or potential customers, and interpreting the results require statistical expertise. For this reason specialists in customer surveys should always be consulted for all but simple surveys.

2.6.6 Learning Potential

The learning potential is usually limited to measuring the output of the processes rather than the processes themselves, and so the potential to learn how to improve are limited.

2.6.7 Duration

Eliciting feedback from every customer, or a percentage of customers can be a continuous process. A one-off study with the involvement of a survey consultancy may take up to several months.

2.6.8 Benchmarking Team

The team is likely to consist either of an in-house group whose job is to monitor customer perceptions and/or an outside surveying organization.

2.6.9 Uses

An organization can use this type of benchmarking:

✓ To monitor customer perception of competitor performance.
✓ To identify potential product/service improvements.
✓ To identify potential participants for a detailed benchmarking study.
✓ To identify its own areas of high performance and market to these strengths.

2.6.10 Benchmarking Process

The benchmarking process is illustrated in Figure 2.6.

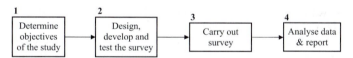

FIGURE 2.6   Process flow chart for survey benchmarking.

1. **Determine scope and objectives.** The organization will determine the objectives of the study. These are usually limited to ascertaining strengths and weaknesses of competitor products and services. However, the survey could be used to gather additional information such as complaints, suggested improvements, new products, and services.
2. **Design, develop, and pilot test survey.** Develop a questionnaire and select its method of delivery (e.g. email, post, telephone). Determine the sample size, sampling method, and other survey details. Pilot the survey to ensure that the results produce the required information.
3. **Carry out survey.**
4. **Analyse and report.**

2.7 BUSINESS EXCELLENCE MODELS

2.7.1 Concept

A Business Excellence Model is a set of interrelated criteria that aims to capture all key aspects of any successful organization. The model is designed so that the extent to which an organization adheres to these criteria reflects its success.

Excellence Models provide a mechanism for comparing the performance of any group of organizations by scoring each one against a standard and comparing the scores. However, direct comparison between organizations is seldom if ever carried out.

2.7.2 A Brief History of Excellence Models

In the aftermath of World War II, Dr Deming was invited by the Japanese Union of Scientists and Engineers to help Japan rebuild its destroyed industry. In 1951 the Japanese honoured Dr Deming's influence on their industry by developing and naming their national quality award after him. The award is given to organizations or divisions of organizations that have achieved distinctive performance improvement through the application of Total Quality Management (TQM) in a designated year.

In the early 1980s, many industry and government leaders in the USA were concerned at the apparent inability of American industry to compete with Japanese imports. In 1982 and 1983, several studies were sponsored by the American government to find ways of increasing American competitiveness. They concluded that their industry needed to adopt what was then called TQM and as a way of encouraging organizations to do so the Baldridge Award was established in 1985. The award was named after the chairman of the committee that established the award, Malcolm Baldridge (ⓘ quality.nist.gov for more details).

In 1991, the European Foundation for Quality Management launched the European Quality Model out of a concern that Europe was losing its competitive edge within the global economy (ⓘ BQF 1998 A Guide to the Business Excellence Model – Defining World Class, British Quality Foundation, London).

In each case applicant organizations are assessed against criteria and those most closely meeting the criteria win the award.

These two awards are the most widely known, but there are many other awards. For example, in the UK there are various regional awards and other countries such as Dubai, have developed their own awards.

2.7.3 How Excellence Models Work

All Excellence Models work on similar principles. Criteria are developed which, it is believed, if effectively implemented will lead an organization to operate at world class standards. As an example we consider the Baldridge Award. In developing the model its creator, Malcolm Baldridge, concluded that there were seven criteria critical to the success of any business. (From Criteria for Performance Excellence, 2007.)

1. **Leadership:** how the organization's senior leaders guide and sustain the organization. It includes the organization's governance and how the organization addresses its ethical, legal, and community responsibilities.
2. **Strategic planning:** how the organization develops strategic objectives and action plans; how these are deployed and, if necessary, changed; and how progress is measured.
3. **Customer and market focus:** how the organization determines the requirements, needs, expectations, and preferences of customers and markets. It includes how the organization builds relationships with customers and determines the key factors that lead to customer acquisition, satisfaction, loyalty and retention and to business expansion and sustainability.
4. **Measurement, analysis, and knowledge management:** how the organization selects, gathers, analyses, manages and improves its data, information and knowledge assets and how it manages its information technology. It includes how the organization reviews these aspects of its business and uses these reviews to improve its performance.
5. **Workforce focus:** how the organization responds, engages, manages, and develops the workforce to utilize its full potential in alignment with the organization's overall mission and action plans. It includes the organization's ability to assess workforce capability and capacity in order to build a workforce environment conducive to high performance.
6. **Process management:** how the organization determines its core competencies and work systems, how it designs, manages and improves its key processes for implementing these work systems to deliver customer value and achieve organizational success and sustainability. It includes the organization's readiness for emergencies.
7. **Results:** how the organization's performance and improvement in all key areas – product and service outcomes, customer-focused outcomes and performance levels – are examined relative to those of competitors and other organizations providing similar products and services.

Each criterion is further broken down into a number of sub-criteria.

The EFQM criteria are similar as illustrated by the EFQM model in Figure 2.7 (ⓘ Further details are available from www.efqm.org).

The categories are similar to the Baldridge Award (which also has a diagrammatic model). The EFQM model uses the concept that good enablers (e.g. how an organization plans, manages itself and its staff, and how it designs and operates work processes) are required to produce good results. As with the Baldridge model, each of these criteria is further subdivided.

The Dubai Quality Award has similar criteria and is based on eight criteria (ⓘ www.dqa.ae).

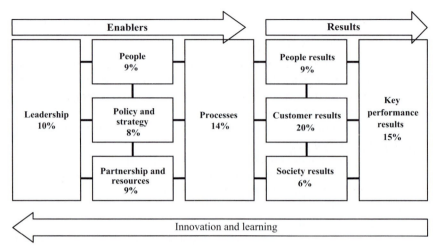

FIGURE 2.7   The EFQM model.

2.7.4 Excellence Models and Benchmarking

From a benchmarking perspective, some or all of these criteria and sub-criteria can be used to develop metrics for benchmarking purposes.

However, this was not the purpose for developing awards. The most common use is for organizations to assess the degree to which they meet the criteria. Areas in which their performance is weakest (i.e. their scores are lowest) can then be targeted for improvement, possibly, but not necessarily, with the help of benchmarking.

Assessments are carried out against the criteria and points awarded depending on the extent to which the criteria are fulfilled. The points can then be summed to give a total out of 100 for the EFQM model or out of 1000 points for the Baldridge Award. If the organization wishes it can enter for the corresponding award and if it receives the highest score it wins the award for that year. Assessments usually include not only a score against each criterion and sub-criteria but also a report discussing performance against the criteria.

As the assessment criteria are published and public training courses available anyone can assess a company against the model. Some organizations train their own staff with the aim of self-assessment, while others use independent consultants to carry out the assessment.

2.7.5 Examples

Examples are generally difficult to identify as they are seldom published. However, ⓘ further information about self-assessment for the Baldridge Award can be found at baldrige.nist.gov. ⇨ Chapter 5 case study Benchmarking with a Business Excellence Award Winner.

2.7.6 Participants

The purpose of the models is to compare an organization's own performance against a standard, in order to identify the weakest areas and improve in those areas. When viewed as benchmarking there is only one participant – the organization being assessed, unless a study is initiated in which participants agree to share assessment information and learning.

2.7.7 Control of Study

The organization is in control of the study, but may request that a consultant or the award body carry out the assessment.

2.7.8 Risks

There are no real risks in entering for excellence awards. There is usually a fee and the time required to gather the information and make a submission can be large.

2.7.9 Learning Potential

The learning potential is generally limited to identifying where performance is poor, unless shared within a benchmarking study.

2.7.10 Duration

To gather data and submit it usually takes weeks or months, but clearly this depends on the size of the organization and the availability of data and information.

2.7.11 Uses

Excellence Models are a good method of reviewing all of an organization's activities and identifying weak areas.

Organizations that enter for and do well in the awards benefit from good publicity.

2.8 A COMMENT ON CHOOSING AN APPROPRIATE BENCHMARKING METHOD

The choice of benchmarking method will often become clear as an organization prepares for its benchmarking projects.

The most appropriate method in any given situation will depend on a number of factors including:

- Confidentiality requirements (principally the extent to which data can be shared openly, shared but anonymized or not shared).
- Whether the participant knows which organization(s) are the best performers or whether comparative levels of performance need to be ascertained.
- Scope of the study.
- Whether a benchmarking club already exists with a similar scope to the proposed study.
- Experience of benchmarking with the organization.

All these factors and the wide variety of ways of addressing them leads not so much to selecting one of a number of benchmarking methods but rather to tailoring each benchmarking study to meet the needs and constraints of the participants.

Many of these issues are discussed throughout the book.

SUMMARY

We have illustrated the wide variety of benchmarking methods by selecting and discussing seven of them in some detail. Six methods are compared in the table (Figure 2.8).

The benchmarking method used for any particular study should be designed to meet the objectives of the participants, and will not necessarily exactly follow any of the example methods discussed here.

The details of setting up and running a benchmarking club is discussed throughout the book and is not summarized here.

Each of the benchmarking methods discussed follows a similar ordered set of steps that need to be considered for every benchmarking study. Not all steps are appropriate in all studies, and in some situations a step may be carried out by a consultant or other party. The steps are:

✓ Identify the scope and objectives of the study.
✓ Select performance metrics to be compared.
✓ Identify, rank, and select potential participants.
✓ Invite participants and finalize project plan.
✓ Collect and validate data/information.
✓ Analyse data and/or report.
✓ Take further action as indicated by the report.

All these steps are discussed in detail in the remaining part of this book.

Methods	Public Domain	One-to-One	Review	Database	Survey	Business Excellence Models
Participants	Organizations within the industry.	Cross industry offers best learning opportunities and is usually possible, at least for generic processes such as purchasing, IT, finance, warehousing. Within industry is common and may be the only possibility for industry specific subjects. Internal studies possible, but yield least benefits.		Any organization that chooses to supply data.	Any organization within the industry.	Only the organization wanting to benchmark against the Model.
Control of the study	Benchmarker.	Participants agree all aspects of the study such as timescales, data/information to be exchanged and use of consultant.		Consultant.	Own organization.	Own organization.
Risks	"League" tables may mis-represent the true performance of the participant. When benchmarking products, if only one product is sampled it may not be representative of the average performance.	May not select the organization with the best performances or practices to benchmark against. Organization's practices may not be appropriate to adopt.		Rely entirely on consultant's integrity. Can be difficult to check other participants' identity, data quality or applicability.	Few. But survey usually needs to be administered by specialists.	None.
Learning opportunity	Low. Usually only benchmark data, but may identify what aspects of a product or service need to be improved.	Can be high especially if, as is usual, working practices are shared and appropriate target participants are visited. Greater learning opportunities occur if several paired studies are undertaken.	Can be high if, as is usual, working practices are shared, and because there are more participants than paired benchmarking, the learning opportunity is even higher. However, if participants are only drawn from the same industry, learning opportunity may be less.	Usually low, depending on the specific service offered by the consultant. However, some consultants will tailor the study and help with improvement activities.	Usually limited to customers' perception of competitors.	Usually low.
Duration	Usually quick, but participant usually has no control over timing.	Usually quick, within weeks, once agreement has been reached with the target.	Usually several months depending on the number of participants in the study.	Usually quick once data have been submitted.	Usually a few weeks or months, but can be continuous.	Usually quick, though often a continuous process.
Benchmarking team	Independent benchmarker, usually magazine or newspaper.	Usually participants perhaps with support from consultants.	Usually participants perhaps with support from consultants. Could be consultant managed by participants.	Consultant.	Either in-house or external specialist.	Own organization.
Uses	Mainly to provide information to the public. However, participants can glean useful information about their competitors and possibly use the results for marketing.	When the initiator knows who the "best" targets are and wants to learn from them.	When a group of participants, usually in the same industry, want to learn from each other.	Participant wants a quick and easy guide as to where to focus improvement activities.	Monitoring customer perception against competitors.	Identify where improvements are needed.
Advantages	Cheap/free. If results are favourable, can be used for marketing. If unfavourable can highlight areas that need improving.	Cross industry or internal: competitors not aware that you are benchmarking. Best opportunities for learning.		Quick guide to comparative performance levels.	Best way to ascertain customer sentiment.	Quick, anonymous, reviews the whole organization.

FIGURE 2.8 Summary of some typical features of different benchmarking methods.

The Benchmarking Process

Introduction and Process Overview

INTRODUCTION

Part 2 of the book considers in detail the process of running a benchmarking study.

This introduction provides an overview of the process. It provides a useful preview of the following chapters and it is recommended to read it before moving on to the rest of Part 2. The overview can also be used as a review and as a contextual guide whilst reading Part 2. The process map referred to throughout the book is provided in Figure 3.

No One Benchmarking Process Fits Every Study

Whenever one benchmarking process is presented as 'The' process to follow, it always raises the question as to whether one process can fit all situations. Usually the answer is no. It is unlikely that one process can meet all the requirements of all organizations in all situations. However, sometimes, as with this benchmarking model the answer is ... 'yes ... but'. With the process described here the 'but' is that this is not a process that should be followed blindly, rather it is a series of steps and tasks that need to be considered. The extent and manner to which the steps need to be followed will vary depending on the nature of the study. For that reason the case studies and examples included in the book do not always follow every detail of the presented process.

As an obvious example, it may have been decided that the study will be an internal study. In this situation the steps to identify, invite and work with participants are greatly simplified.

Development of the Process

The benchmarking process presented has evolved with experience of being involved in a variety of both internal and external studies over the past 15 years in roles varying from general facilitation to managing help desk services, analysis and report writing as well as leading site visits, data gathering activities and best practice sharing. It is also based on information gleaned by talking and working with participants, consultants, engineers, clients and other people who have been involved in benchmarking studies. The process developed has been used effectively in different benchmarking situations such as running benchmarking clubs and one-off studies to address specific issues. The benchmarking process is presented for the most complex situation, that is the development and running of a benchmarking club.

A benchmarking club is:
>   a group of organizations
>   that benchmark together
>   on a regular basis.

While the membership of the club is likely to change over time, most participants remain in the club for many years, reflecting the stance that benchmarking should be a regular, not a one-off, activity.

PROCESS OVERVIEW

Basis of the Model

The basis for using the model is that an individual within an organization has identified a need for benchmarking. Where a benchmarking study already exists that meets most or all of the needs, the best course of action will probably be to join the study rather than attempt to set up an alternative and perhaps competing one. In this situation the internal preparation will be minimal. If there are no appropriate studies in existence, the first major phase of the project will be internal preparation (Figure 1). The deliverable of this part of the process is a well thought out benchmarking proposal to initiate a benchmarking study. The proposal is likely to include the purpose and objectives of the study, performance indicators (PIs), timescales, potential participants, contacts, that is: everything required to present to management a full and persuasive project proposal.

The second major phase, data comparison, can be considered to consist of two parts. The first is to recruit and work with participants in order to finalize the details of the benchmarking study. Most of the work involved will be reviewing, honing and agreeing what was developed in phase 1. The second part of this phase consists of data collection, validation and report writing. This results in information, usually but not necessarily in the form of a report issued to participants.

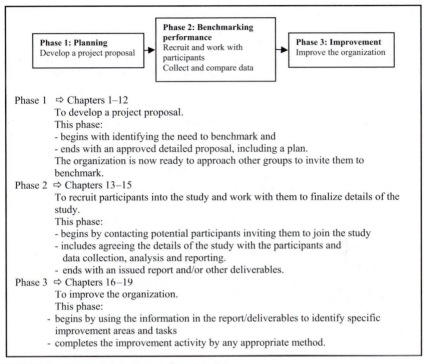

FIGURE 1.   The three main phases of benchmarking.

The final phase is for the participants to use the information contained in the report to improve their own organization.

All of these activities may be carried out by one person or a team commissioned to work on the project. The people working on the project may change as the project progresses. We assume, as will often be the case, that there is one team with unchanging membership for phases 1 and 2. ⇨ Chapter 3 for a more detailed discussion of team membership

We now provide an overview of each step of the three phases in the benchmarking process.

Phase 1: Internal Preparation

Selecting a project and demonstrating the need for benchmarking ⇨ *Chapter 3*

Most organizations will not be prepared to allocate significant resources to any project, including a benchmarking project, unless they believe that it will benefit the organization. A challenge with benchmarking is that one of its uses is to identify areas of weakness in an organization. If there is no apparent weakness, management may be reluctant to invest in a study. In these types of situations the first step is to identify the need and persuade the decision makers that benchmarking is a key tool in their part of the business. Often this can be done by carrying out an investigation to identify a particular weakness in the organization, estimating the cost of that weakness and explaining how benchmarking could improve the situation. Methods of identifying and selecting projects along with stating the business need for the study are discussed in Chapter 3. Chapter 4 discusses ideas for gaining management support and dealing with objections.

Developing a Persuasive Benchmarking Proposal

Having obtained the resources, usually predominantly time, the next step is to commission a team or person to develop a complete detailed compelling benchmarking proposal that management will want to support.

The team's first task will be to develop or review its Charter. A typical Charter will include as a minimum the scope of the study, business reason for the study (the justification), project objectives and team members.

The scope can be based on, for example, a product/service, process, facility, function, specific issue. ⇨ Chapter 3

The justification for the study will be based on the weakness already identified and should include potential benefits of the study. The potential benefits need to be attractive enough to gain support from management and types of expected benefit can also be used to help recruit participants.

It is also important to identify the objectives of the study ⇨ Chapter 4. The objectives identify what we plan to achieve with the study and ultimately act as a yardstick against which the success of the study is measured. Typical objectives include:

- To identify the performance level of comparable organizations.
- To determine a reasonable performance level target based on what others are achieving.
- To quantify the gap between our current performance and the target performance levels.
- To identify practices that will help us achieve the target performance level.
- To implement appropriate changes to improve performance (this objective may be the subject of the next phase of the study and not included at this point).

As illustrated in Figure 2, three parts of the Charter, scope, justification and objectives, have key roles to play throughout the study.

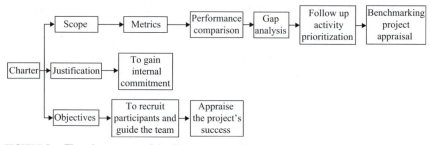

FIGURE 2.   Three key aspects of the Charter are used throughout the project.

The development and honing of the Charter is discussed in ⇨ Chapters 3, 4.

Once the Charter is finalized, the team's next task is likely to be to develop a project proposal to meet the objectives detailed in the Charter. The proposal evolves from the Charter and is a key aspect of the study as it is used not only to gain management support, but also as a communication tool to others in the organization, as a key tool in recruiting participants, as a guide to the team throughout the project and ultimately the success of the project will be judged against it.

Selecting Potential Participants ⇨ Chapter 5

A key issue in any benchmarking study is whether to benchmark internally, that is with other groups in the same organization, or externally. When benchmarking with external participants we need to identify with whom we want to benchmark and why. For external studies will need to consider issues such as: whether to benchmark within our industry across other industries; with or not with our competitors; and what geographical coverage is appropriate. These issues may affect the metrics that we are either legally allowed or commercially willing to benchmark.

Develop Metrics ⇨ Chapters 6–10

Metrics enable us to measure and compare performances between participants and quantify future improvements. There are a number of sources of potential metrics, most notably the analysis of the mission/vision/aim or similar statements relating to the area being benchmarked and currently reported metrics.

For many benchmarking studies the metrics are the embodiment of the scope of the study and run through its course (Figure 2). The metrics will be used to achieve the objectives of the study, typically to compare performance levels, quantify performance gaps and potential gains. The potential gains will be a key input in deciding where to focus improvement activities and, after improvements have been implemented, the metrics will be used to quantify the gains actually achieved. Eventually these gains will be compared to the cost of the benchmarking study which will form the basis of deciding whether or not it was a success.

Management and potential participants will usually see the metrics as the key element of the study. If only for this reason it is important to ensure we develop a well thought out, collectable, representative and complete set of metrics covering all aspects of the scope part of the Charter.

Metrics are not always necessary to complete a successful benchmarking study. In some situations, usually where we are identifying a completely new approach to running an aspect of the organization, or where a target participant is known to have a superior performance level, it is not necessary to measure. ⇨ Cases Studies Dundee City Council and Formula1 are two such examples.

Resource Planning ⇨ Chapter 11

The final aspect of the planning stage is to determine the resource requirements and timescale for the study. These are necessary for any sizeable project not only for management approval but also to set expectations and guide the project. The additional use of a project plan unique to benchmarking is that the plan will be a useful tool when inviting participants to join the study. The required resources will usually include manpower and budget, but there may also be a requirement for external facilitators and/or experts.

Joining an established study

The sections above summarize how to develop a study internally where no appropriate external study exists. However, it is likely and recommended that having defined the scope and objectives of the study we would search for studies already in existence that would fulfil our need.

The key advantages of joining a well-run study are that they will already have honed effective metrics and definitions and will have a variety of participants who are committed to benchmarking. If the other participants have been benchmarking for any length of time, they will have used study findings for improvement activities and may already have identified and implemented some best practices. In addition there will be a positive atmosphere of wanting to learn and improve together. Where a vibrant successful benchmarking club already exists it is difficult for management to reject joining it.

Gain Management Approval ⇨ Chapter 11

The final step of the planning phase of a benchmarking study is to gain management approval for the project plan. Like any other proposal, it must demonstrate to management not only that its benefits outweigh the monetary and other costs but also that it provides a better investment than other projects vying for funding and resources.

Regardless of the need for management approval, at this stage of the project it is very useful to have a complete review of the proposal before inviting other participants to join the study. This will help ensure a well thought out complete and enticing invitation is offered to potential participants, thus increasing the likelihood of their acceptance.

Inviting Participants to Join the Study ⇨ Chapter 12

Unless joining an established study, once the project has been approved, the next step is to invite participants to join the study. The basic approach to inviting participants will be similar for both internal and external participants, however, with internal participants we are able to use managerial influence to encourage participation.

At this stage in the process potential participant organizations have already been identified. However, new potential participants can be invited at any stage in the process. Most participants will be recruited at this step as we will need to work with them to finalize the details of the study.

Phase 2: Benchmarking Performance

The second major phase of the benchmarking study begins when potential participants work together to finalize and implement the study. Frequently this will be achieved through a series of participant meetings The first of these meetings, often called the kick-off meeting ⇨ Chapter 12, is usually different in format to follow-on meetings as the focus is on building relationships and discussing the overall aims, objectives and running of the project. In later meetings the aim shifts to being working meetings with the aim finalizing the details of the study including metrics, definitions, timescales, legal agreements and all other related aspects.

Data Collection and Validation ⇨ Chapter 13

By this stage the participants will have agreed what data (including information) they want to collect and will have developed definitions. The next step is to determine how best to capture the data. The most likely method will be to design a data collection pack likely to include a spreadsheet for data collection and/or a questionnaire for each participant to complete and return for validation and analysis. However, other collection methods include hard copy forms, data gathering visits by one or more facilitators and web interfaces. Whatever method is used, it is important to design data and information collection documents carefully to help ensure that the data and information submitted are as required.

It is likely that however carefully the data collection documents have been prepared, participants will have questions. An excellent method of answering these queries efficiently is to set up a Help Desk facility which can be contacted by phone or email. The Help Desk plays a variety of roles at this stage of the benchmarking process. Not only does it respond to queries, but also because of the queries it receives, it will know how data collection is progressing, be aware of and be able to respond to shortcomings in the data collection files and expedite timely submission of data. By the nature of the questions the Help Desk may also gather information about participants' operations which can help during analysis and reporting.

Once the data have been submitted, the next step is to validate the submission resolving any omitted items, apparent inconsistencies or seemingly spurious values. During the validation phase it is also likely that valuable information regarding the participants' operation will be discovered which should be retained for the analysis.

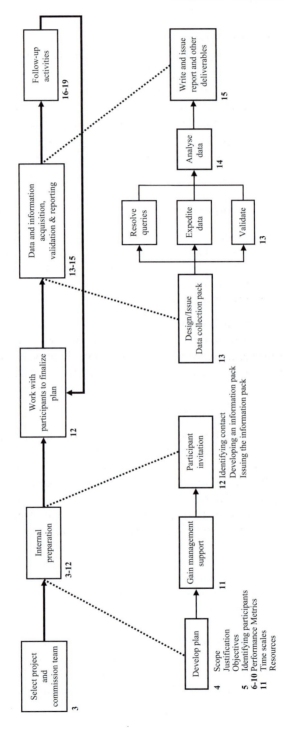

FIGURE 3.   Generic Benchmarking Process Map (Chapter shown below the boxes).

Analysis and reporting ⇨ *Chapters* 14, 15

Once all the data have been collected and validated the analysis can begin. The amount and depth of analysis and subsequent reporting will vary greatly depending on the terms of the study. In some studies there may be a minimum of analysis as, for example, is often found in magazines that review products and services. In other studies there may be a great deal of analysis including for example, performance level comparison, performance gap analysis, testing of theories, evaluation of practices and recommendations for improvement.

Phase 3: Improving the organization

Up to this stage participants are unlikely to have benefited from the benchmarking study. It is only once they have received the report or other study deliverables that help them identify where and perhaps how to improve that they can start the important task of improvement. There are many ways in which participants may choose to drive improvement including:

- Initiating Best Practice Forums.
- Taking part in Information Trades with one or more other participants.
- Carrying our further in-depth studies before taking improvement action.
- Implementing internal activities such as process improvement, process reengineering.

Deciding Which Parts of the Process Map to Use

The process map provides a comprehensive route map for a initiating a new benchmarking club or study. However, each study and each club may choose to modify the map to meet its own specific objectives and needs. In most cases studies will simplify some of the steps in the map and emphasize others.

Benchmarking studies carried out by consultants will also follow a similar pattern. A project proposal with scope, reasons for the study, objectives etc. is still likely to be required, but many of the other tasks will be carried out by or led by the consultant.

The Benchmarking Process:
Internal Preparation

Selecting a Project and Commissioning the Team

INTRODUCTION

The first steps in any project are to define and scope it and commission people to work on it.

In this chapter we discuss:

- ✓ Types of benchmarking project.
- ✓ Methods of identifying and selecting projects.
- ✓ Stating the business reasons (justification) for the study.
- ✓ Selecting and commissioning the benchmarking team.
- ✓ The importance, contents, and development of a Project Charter for carrying out a benchmarking study.

Selecting a project is dependent on a number of factors which are discussed here but project selection also needs to take cognizance of factors such as likely participants ⇨ Chapter 5.

DELIVERABLE

The deliverable for this step is a person or team commissioned to progress the benchmarking project. Their remit may be to carry the project through to completion; alternatively, it may be to take the project part of the way to completion, typically through to gaining management approval.

3.1 TYPES OF BENCHMARKING PROJECT

The first step is to select a benchmarking project. Traditionally, people have found it useful to identify different types of benchmarking project, though there is no one defining list of types nor globally accepted group of definitions. Also, as we will see, there is a great overlap between different types of benchmarking. What is important is selecting an appropriate project; identifying the type of benchmarking is only useful in as far as it helps communication between those involved in the study.

We can choose a project based on, for example:

- A specific **process** such as purchasing, manufacturing, maintenance, design, training, complaints, IT support.
- A specific **group**, **facility**, or physical **area** such as a manufacturing or processing plant, an airport, hospital, office, school, legal department. Within the unit or units we may benchmark the whole unit or specific aspects, for example, the catering provision at each hospital.
- Benchmarking of the **final products or services** though common, is not usually carried out as a benchmarking study between the suppliers. Usually this type of benchmarking is carried out by magazines or other independent groups as discussed under database benchmarking, or competitor analysis carried out internally and is discussed in ⇨ Chapter 3.
- **Activity** benchmarking is the benchmarking of a specific activity or small group of activities, such as how to maintain a pump, prepare a metal surface for soldering, or design a workspace conducive to effective working. This type of benchmarking can be considered as a subset of other types of benchmarking.
- **Functional** benchmarking is the benchmarking of a function such as the purchasing department or warehouse. The function may extend over more than one facility and more than one process.
- **Generic** benchmarking focuses on an end result and is divorced from the industry or subject being benchmarked. For example, the need for a 'smooth surface' applies to both lipstick and bullets.
- A specific **project** such as implementing a software system, construction project, implementing Balanced Score Cards, disaster management or outsourcing.
- . . .or any combination of these or other criteria.

It is important that the project is clearly defined so that other people, both within the organization and the organizations of potential participants, will easily be able to identify what it is proposed to benchmark.

Definitions

- **Process benchmarking** is the benchmarking of a process such as purchasing, warehousing, maintenance.
- **Facility benchmarking** is the benchmarking of complete site or facility such as a factory, airport or refinery.
  (note that a warehouse, for example, could be benchmarked as a process OR as a facility)
- **Product/service benchmarking** is the benchmarking of a product or service or combination (i.e. the output of a process), usually as experienced by the customer. Examples include domestic goods, cars, travel (e.g. by boat, plane or train)
- **Activity** benchmarking is the benchmarking of a specific activity or small group of activities.
- **Functional benchmarking** is the benchmarking of a function such as the purchasing department or warehouse. The function may extend over more than one facility and more than one process.
- **Generic benchmarking** is benchmarking between different functions or processes and focuses on achieving specific results, e.g. fast turnaround.
- **Project benchmarking** is the benchmarking of activities carried out for a limited time to create a specific product or service such as construction project or software development.

Each type of benchmarking is discussed in more detail below.

3.1.1 Process Benchmarking

Process benchmarking is the term given to studies where partial or complete processes are benchmarked. Typical examples include benchmarking the purchasing, design, order fulfilment or recruitment processes. Either the whole or part of the process may be benchmarked. For example, in the purchasing process, the process from identification of the need for a product or service through to the fulfilment of the need, including payment, may be benchmarked or just, for example, from placement of a purchase order to arrival of goods on site.

Such a process may involve different locations, facilities, and groups, or may be confined to one particular location.

Example

⇨ Chapter 2 'Xerox and L.L. Bean Distribution benchmarking'

The benchmarking of a process generally allows a detailed study to be made of one process and can often be carried out cross-industry. For example, many business support processes such as purchasing, warehousing, billing, and recruitment are not industry dependent. While it is usually more difficult to benchmark cross-industry, it is also where the greatest opportunities for learning exist.

Performance improvement usually requires changes to processes. By benchmarking the whole process it will be easier to understand how performances of different parts of the process interact and investigate the likely effects of changes in one part of the process on other parts.

On the other hand, processes can be highly complex involving several groups, departments, or even locations. A purchasing process may be considered to begin (and end) with the originator and include warehousing, supplier auditing, engineering, and other departments or groups. In addition, processes often contain sub-processes that are rarely carried out but are nevertheless part of the process. In the purchasing example, tendering, disposition of goods that fail goods inward inspection and qualifying new suppliers may all be little used parts of the process.

In order to simplify the benchmarking of complex processes, one possibility is to benchmark only part of the process. For example, rather than benchmark all aspects of a purchasing process only the awarding of contracts may be benchmarked (Figure 3.1).

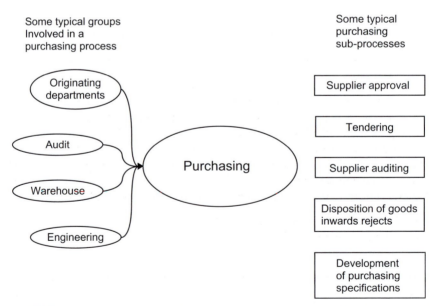

FIGURE 3.1   Processes can be complex covering several departments and with seldom used tentacles.

3.1.2 Facility Benchmarking

Many studies seek to benchmark self-contained facilities such as factories, power plants, or airports. While it is possible and common to benchmark part of a facility such as one of its processes, departments, or products there are advantages to benchmarking the whole facility. These include:

✓ The total cost and/or hours required to produce a unit of output can be compared.
✓ Where tasks are not clearly split between groups they can be more easily benchmarked if the whole plant is included. For example, operators may carry out planned maintenance routines or other duties in some plants, but not others. Similarly, the maintenance team in one facility may carry out planning and/or purchasing and/or warehousing tasks, whereas in another these tasks may be carried out by other groups.

In these situations benchmarking the whole facility helps management decide on which areas to focus improvement activities.

Example: Solomon Refinery Benchmarking

Many refineries take part in the Solomon Associates Refinery benchmarking study. The study compares performance between peers and measures each participant against the best in the industry. Taking part in the study allows participants to optimize such areas as operating and maintenance costs, identify performance gaps and so concentrate improvement efforts in the areas that will yield most benefit to the participant's plant (ⓘ www.solomononline.com).

Benchmarking a specific facility or group often results in looking at many different activities and gives a very good overview of where its strengths and weaknesses lie. However, to study a complete facility or group in depth may result in a huge project; as a result each activity is usually benchmarked only at a high level. For example, within an airport there are many areas including passport control, luggage handling, check in, catering, shopping, plane management, recruitment, and purchasing. Clearly to cover all these in detail would be a highly complex project. Three common solutions to this problem are to:

• Benchmark all activities at a high level or
• Initiate separate studies for selected areas or
• Limit the scope of the study.

One common difficulty with benchmarking relatively small facilities is that many tasks require small amounts of time and gathering accurate data is difficult. For example, in a production facility of 50 people activities such as HRD, purchasing, or warehousing may not require a full time worker. In addition, some people may be involved in several tasks. For example, operators in a factory may also have other duties such as HSE responsibilities, internal auditing, or planning. Measuring activities requiring such small amounts of time is difficult and likely to be error prone, and the error can result in seeming significant differences between participants. For example, an operations supervisor working 1800 hours a year may estimate that he spends 130 hours a year on purchasing activities when he actually spends 100 hours. This is only a 30 hour under estimate which is insignificant compared to1800 hours, but represents a 30% over estimate.

3.1.3 Product and Service Benchmarking

An alternative to benchmarking a whole process or facility is to benchmark specific products and/or services. Typical service examples include IT help desk, after-sales service, or customer queries. Typical product examples include domestic goods, compressors, and houses. Studies focused on products and services usually focus on the output from the organization rather than the process of producing it (which would be process benchmarking).

Product and service benchmarking is frequently carried out by consumer magazines. Information from these publications can be very useful for identifying potential areas for improvement (⇨ Chapter 2).

When organizations want to benchmark customers' or the public's experiences of products and services they will often commission independent customer surveys. These can be targeted at investigating specific issues (⇨ Chapter 2).

The key advantage of product and service benchmarking is that it provides information on the critical part of a process – how the customer experiences the product or service. The main disadvantage is that there is limited opportunity for learning which practices may lead to producing better products or services.

Case Study: Valve Trialling

One organization wanted to benchmark valves from different manufacturers and so decided to purchase and install several valves from different suppliers as a trial. The key metric was the failure rates of the different valves.

A statistician was used to design the test plan deciding on the number of valves and the length of time required for the field trials and to carry out the required analysis.

3.1.4 Activity Benchmarking

Activity benchmarking is the benchmarking of individual specific activities or tasks. It is generally limited in scope and can be considered to be related to other forms of benchmarking. For example, it is related to process benchmarking in that processes can be considered as consisting of individual steps or activities and activity benchmarking is the benchmarking of these individual tasks. It can be considered to be related to functional and facility benchmarking because if all the activities in the function or facility were benchmarked in the same study, then the whole function or facility would be benchmarked. Being of limited scope, activity benchmarking can usually be carried out in great detail. The term activity benchmarking also refers to the concept of benchmarking activities in contrast to, for example, benchmarking of products, customer perception, or financial performance.

3.1.5 Functional Benchmarking

Functional benchmarking is similar to facility benchmarking where the subject of the study is not a whole facility, nor necessarily a whole process but a specific function, group or subject which may extend over several sites and/or several departments. Typical examples include finance and warehousing ($\Rightarrow$ Part 4 'Citigroup').

3.1.6 Generic Benchmarking

Generic benchmarking focuses on a general required result and the process to achieve it.

For example, if an organization needs to make good decisions quickly in real time, processes it may consider looking for best practices with stock traders, emergency response teams, or armed forces. A challenge for many organizations is how to make repetitive jobs interesting. Typical repetitive tasks are found, for example, on production lines and so maintaining interest and quality of work can often be benchmarked cross-industry from car manufacturing, to food processing, to mail sorting and so on.

With this type of benchmarking we are seldom comparing our performance with others by using metrics; rather we are investigating how other industries successfully manage what for them may be key aspects of the business.

Example: Formula 1

The Formula1 Case Study described in Chapter 2 described how a manufacturer learned how to set up production lines from a Formula1 racing team. The aim was 'fast setup'

Generic benchmarking is perceived as being difficult because it requires a leap from the obvious route of comparing like with like to learning from others who may carry out very different tasks. However, if successful the gains can be spectacular.

3.1.7 Project Benchmarking

Benchmarking projects is not as common as other types of benchmarking, and does have some unique difficulties. The most cited problem of project benchmarking is that no two projects are the same and therefore cannot be benchmarked. For example, no two construction or software development projects are likely to be the same. While this its often true, there usually are aspects of projects that can be benchmarked, see, for example, the ⇨ Part 4 'Drilling Performance Review'. Methods of normalizing for the differences between projects are discussed in Chapter 7.

3.2 METHODS OF IDENTIFYING AND SELECTING PROJECTS

It may be that the subject for the benchmarking study has already been chosen, but if not, the selection process is made up of two aspects:

1. Identifying potential benchmarking projects.
2. Selecting which project(s) to progress.

When selecting a project, management are likely to consider questions such as:

? Who are our key customers? What are their needs? How well do we meet those needs? What do our customers say about us? What do independent reviews say about us?

? What is our mission? Our vision? What are our values? Our policies? How well do we achieve them?

? What current problems or issues do we have that other organizations may be facing or have faced in the past? Is it possible that we could learn from them?

? What are our highest cost areas/processes/plants, etc.?

? What are our Critical Success Factors?

? What information do our reporting/monitoring systems give us about our performance? Specifically, where does our performance appear to be poor?

? What are our performance levels as judged by business models such as Baldridge or the EFQM Excellence Model?

? What are our Costs of Poor Quality? Analysing the Costs of Poor Quality is a useful starting point for selecting benchmarking projects.

From questions such as these, management can draw up a list of potential benchmarking projects that are likely to deliver the highest return on investment (Figure 3.2).

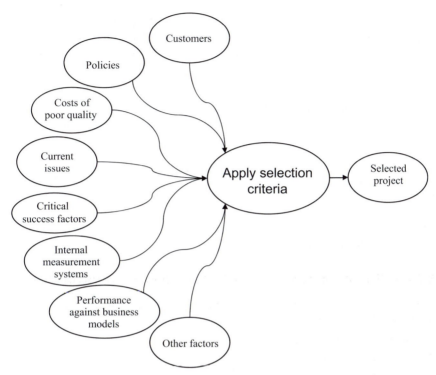

FIGURE 3.2   There are many aspects we can consider to generate a list of potential projects. These can be analysed against project selection criteria to select benchmarking projects.

Organizations are likely to have project selection procedures in place and potential benchmarking projects will be evaluated against them. Where such selection criteria do not exist or do not address benchmarking projects the following considerations may be useful:

- The benchmarking experience of those who will be involved in the project. Particularly in early studies, one of the overriding considerations is that the benchmarking project should be easy to complete with a high probability of success. Benchmarking soft issues such as research or design is seldom easy, if only because measuring these activities is difficult. It is much easier to measure, for example, production rates, costs, hours, and errors per unit. Once the organization in general, and those who will be involved in benchmarking in particular, have gained experience of benchmarking, more complex issues can be benchmarked.
- Preferred areas for benchmarking are those which are key to the success of the business and where performance is believed to be poor.
- The opportunity for gain needs to attract support from management and other potential participants.

- Key people in the area being studied need to be supportive of the project. Where people are hostile to benchmarking they may resort to non-cooperative behaviours such as delaying or falsifying data.

Selecting a project is in practice tantamount to scoping it. For example, if it has been decided to benchmark the admissions procedure in a hospital, the scope will be the admissions procedure. If it has been decided to benchmark the whole hospital, the scope will be all hospital activities. Later we will discuss defining the scope more carefully (⇨ Chapter 4); at this step a clearly defined scope is not necessary.

3.3 STATING THE BUSINESS REASON FOR THE STUDY: THE JUSTIFICATION

For all projects and initiatives it is important not only to know what is to be done but also to know why. Benchmarking projects are no different. If a project selection process has been followed, then it is likely that the reasons for selecting the project(s) are already known. If there has been no selection process, perhaps because the benchmarking study is seen as a means of solving a particular problem or because there is only one obvious subject, then it is useful to develop a reasoned explanation as to why the project is important. This can:

✓ Help gain support from those funding and/or sanctioning the study.
✓ Help those involved in the study understand why the study is important.
✓ Be used, albeit in perhaps a modified form, to help persuade other organizations to join the study.
✓ Be a guide for those involved in the study.
✓ Be a yardstick against which to measure the success of the study.

A standardized project proposal procedure will normally include reasons why a particular project is proposed. In an organization where there is no such procedure a good starting point is to consider the people whose support is necessary: budget holders, managers, and those who will be impacted by the project. Couching the reasons for the study in terms of their interests will help ensure their support. As managers tend to think in financial terms it is useful to highlight expected financial benefits such as cost savings or increased revenues. Even non-monetary benefits such as environmental factors, safety, or improved customer relations can often be translated into financial impacts.

With benchmarking projects it is often difficult to predict the benefits. This is because until we have benchmarked performance levels we do not know how much improvement, if any, is achievable. However, there may be some evidence from published data or other sources, as to the performance level of others and this information can be used to estimate possible benefits.

The justification is likely to include references to:

✓ The current performance level of the organization.
✓ Why the current performance level is not acceptable.

✓ Why it is important to know comparative performance levels of ourselves versus other organizations.

The statement should be factual and should not imply a solution or blame.

3.4 SELECTING AND COMMISSIONING THE BENCHMARKING TEAM

Once the outline of the project is complete the next step is to select a team, or an individual, to carry it forward. As with most projects, before embarking on a benchmarking project it is important to develop a plan and this will be the first task of the team. The number of people and the length of time required to finalize a detailed proposal or plan will depend on such factors as:

- The experience of those involved.
- The scope of the project. When it covers several groups or areas it is likely to involve representatives from each.
- The number and complexity of metrics. Where there are highly developed industry standards, the time required for developing and defining metrics will be less.
- The type of participant. Developing a list of preferred participants will be easier within one area of an industry, e.g. multinational airlines are easily identifiable. Identifying suitable participants across industries, e.g. for a purchasing study, will take longer.

When selecting the team some points to consider are:

✓ The team should include members from the key area(s) being benchmarked in order to help ensure buy-in to the project, its results and any resulting follow up activities.

✓ The team should include people with complementary and necessary skills for completing the project. For example, it is likely that the project will require team facilitation, statistical, inter-personal, administration, benchmarking skills as well as knowledge of the topic being benchmarked, and possibly other skills depending on the nature of the project.

Once the team has been selected, the next task is to co-opt them onto the project. One very successful method of doing this is by use of a Project Charter.

3.5 PROJECT CHARTERS

One very useful document for commissioning and guiding benchmarking (as well as other) projects is the Project Charter. This encapsulates the scope, objectives, resource requirements, and authorization in a short, preferably one page, document.

The key aspect to the Charter is the authorization. Many benchmarking projects will require a team of people drawn from different disciplines. Many, if not all of them, will be working on the benchmarking project on a part-time basis and will continue their normal activities at other times. The project may take many months to complete and experience shows that often managers will pull members off the team when other issues arise. As people are prevented from taking part in the project, so the team loses resources, becomes less able to complete the project and may end up disillusioned. Sometimes the team disintegrates and the project progresses no further.

To help overcome this problem, the Charter includes the signature of the line managers of the team members as well as the project sponsor. Once a manager has committed in writing to releasing people to work on a specific project he/she is far less likely to pull them off the project. If this does happen the project sponsor, usually a senior manager, will be on hand to resolve the issue.

Charters can be used for any project of any duration and complexity. The example below is for a team commissioned to carry out a benchmarking study of the Order Fulfilment Process.

Once the teams are briefed and the team Charter signed they are ready to start work on the project (Figure 3.3).

Team Charter: Order Fulfilment

The problem
The recently completed Business Review showed that our Order Fulfilment Process requires 13 hand-offs, involved 27 databases and takes on average 15 elapsed days to complete. The average processing cost is $x. The actual average processing time is estimated to be 11 hours. In addition many errors occur during the process, though these are not routinely measured.
Experience of staff who have worked for other organizations and informal contacts indicate that we could significantly reduce elapsed time, processing cost and errors by improving our order fulfilment process.
The process and databases that we use have been developed over the 34 years of the company's existence and there has never been a systematic review of the whole process.

The mission of the team will be complete when they have:
• Quantified the gap between our and other similar organizations' Order Fulfilment Process performances in terms of elapsed time, cost, processing time and errors.
• Recommended appropriate performance levels and estimate the benefits of working at these levels
• Recommended a way forward for improving our processes
The first deliverable is a complete project proposal to be submitted for approval by xx/xx/xx

Resources requirement
The main resource requirement for producing the project plan will be one full time team leader and three team members, one from each group involved in the process, 1 day per week.

Name	Signature/date		Manager	Signature/date	
Team Leader:					
H. Sullivan	*Harry Sullivan*	13 March	Jo Last	J. Last	15 March
Team Members:					
Alex Avengo	*A Avengo*	13 March	Phil Hurst	Philip Hurst	15 March
Kit Kitson	*Kit Kitson*	14 March	Clare Davis	C. Davis	14 March
Fay Fenton	*F. Fenton*	14 March	H. Hewell	*Harry Hewell*	16 March
Project Sponsor					
Fred Bosman	Freddy Bosman	17 March			

FIGURE 3.3 Sample Project Charter.

SUMMARY

Benchmarking projects are typically based on:

✓ **Processes.** A detailed study to be carried out of a complete process. With business support services, benchmarking can be carried out across different industries. However, for complex processes it may be preferable to exclude some aspects of the process to keep the study within manageable proportions.

✓ **Facilities.** A high level study of a facility gives a good indication of overall strengths and weaknesses. However, for larger facilities and in-depth study for each area of the facility may result in a very complex study.

✓ **Product/service.** Benchmarking projects of services and products are usually carried out independently by consumer magazines or by organizations commissioning customer or public surveys. Organizations sometimes benchmark products or services by trialling and comparing them.

✓ **Activities.** Benchmarking of individual tasks/steps or activities is called activity benchmarking. It can usually be carried out in great detail because the scope is limited.

✓ **Functions.** Similar to facility benchmarking, but a function may be spread over one or more locations and include more than one process.

✓ **Generic activities.** Generic studies focus on a general required result and are based on practices from different industries and different processes. They are the most difficult to complete, but can result in huge gains.

✓ **Projects.** Benchmarking projects is difficult because no two projects are the same. However, projects can be successfully benchmarked and there are some helpful normalizing techniques to overcome inherent project differences.

Project identification and selection there are many methods of identifying potential projects including identifying areas of perceived weakness, areas of high cost, and areas critical to the success of the organization.

Where there is a choice of projects organizations usually draw up a list of selection criteria as a basis for project selection.

Stating the business reasons for the study is important in order to elicit support from those who will be impacted by the project. They can also be used to encourage participants to join the study.

The benchmarking project team is selected on their experience, skills, and knowledge of the subject being benchmarked. Selecting members from areas impacted by the project will help elicit cooperation from those areas.

Project Charters (also known as Terms of Reference), may include:

• The scope of the project.
• Reasons for carrying out the project (Justification).

- Resource requirements.
- Mission/objectives of the project.
- Agreement from appropriate managers to release personnel to work on the project.

Charters are a key communication and authorization document explaining to others the scope and reasons for the study and guiding the project team.

✝ ACTIVITIES

1. Develop a list of methods for identifying potential benchmarking projects within your organization.
2. Use the list developed in 1 to draw up a list of potential projects.
3. Develop a list of criteria and a methodology for selecting benchmarking projects from the list developed in 2.
4. Use the criteria to select a project or projects that will be taken forward.
5. Draft a Project Charter to include team members, subject of the study, reasons for the study along with any other pertinent information that is available according to your organization's project commissioning process.

Chapter 4

The Team Begins Work: Honing the Project Charter

INTRODUCTION

Once the project team has been commissioned, their first task will be to develop a complete detailed project proposal. As a first step in developing the proposal, the team review and clarify the Project Charter. This will be signed off by management when they are satisfied that the terms of the Charter are aligned with the business needs. The Charter will be an important component of the final project proposal.

The project proposal should be developed in conjunction with the standards and norms of the organization and may include:

1. Scope of the subject being benchmarked.
2. The business reason (justification) for the study.
3. The objectives of the study.
4. The proposed performance metrics and definitions (⇨ Chapters 7–10).
5. Proposed participants (⇨ Chapter 5).
6. Timescales (⇨ Chapter 11).
7. Resource requirements (⇨ Chapter 11).
8. Report contents (⇨ Chapter 15).

The project proposal is a key document in the benchmarking study because it may be used for:

✓ Persuading management to support the project (⇨ Chapter 11).
✓ Persuading participants to join the study (⇨ Chapter 12).
✓ Guiding the project.
✓ Measuring the success, or otherwise, of the project.

DELIVERABLE

The deliverable of this step is a finalized Project Charter, the terms of which are fully understood by the project team and signed off by management.

4.1 SCOPING THE PROJECT

The Team Charter will have outlined the scope in general terms. For example, it may name the process or facility or type of project to be benchmarked. However, this will normally need to be more carefully specified by defining boundaries and what is to be excluded as well as included.

For process benchmarking, a useful method of specifying the boundaries so that there is no confusion as to what is included or excluded is to complete the sentences:

- The process begins when ...
- The process ends when ...

In addition, the team may want to include a high level flow chart, context diagram or other representation showing the bounds of the project. They should also consider whether or not all of the process is to be included.

Example: Maintenance Process Benchmarking Study Scope

'The maintenance process begins when a maintenance task is first flagged on the maintenance management system, and ends when the maintenance system shows the maintenance task as being closed out'.

From this statement we conclude that there is a computerized maintenance system; the process of determining maintenance frequency is excluded from the project, and the maintenance is complete only once the system shows it as being complete – i.e. completing the physical maintenance is not tantamount to completing the maintenance process.

For a facility the same two questions can be answered, but in this situation the definition will be a physical entity.

Example: Processing Facility Benchmarking Scope

For a hydrocarbons processing facility:
'The facility begins with and includes the inlet valve(s) and ends at and includes the delivery valve(s)'.

A useful chart for showing links between, for example, processes, functions, groups, facilities, or departments is the context diagram (Figure 4.1). Its role is to show those areas that are impacted by the subject being benchmarked.

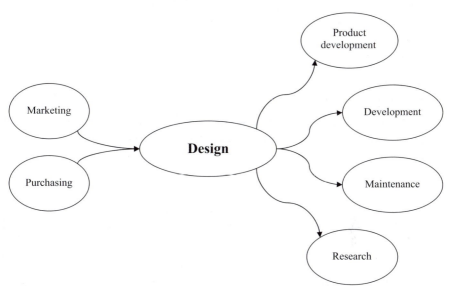

FIGURE 4.1   A simplified design function context diagram showing groups requiring information (right) from the design function and those providing information to the design function (left).

For process benchmarking the context may also be shown by highlighting where the steps to be benchmarked fit within the larger process and/or how they relate to other processes, see for example Figure 4.2.

A floor or ground plan can be a means of identifying which areas are included and excluded from the scope. For example, Figure 4.3 shows a simplified plan of a warehouse in which the white areas are to be benchmarked and the greyed areas are excluded from the study.

The purpose of these and similar charts and diagrams is to show simply and quickly how the area being benchmarked relates to other areas either physically, conceptually, or procedurally.

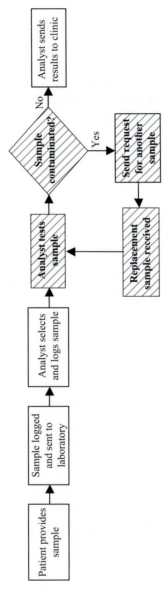

FIGURE 4.2   Context flow diagram for testing medical samples. The hatched steps are those to be benchmarked.

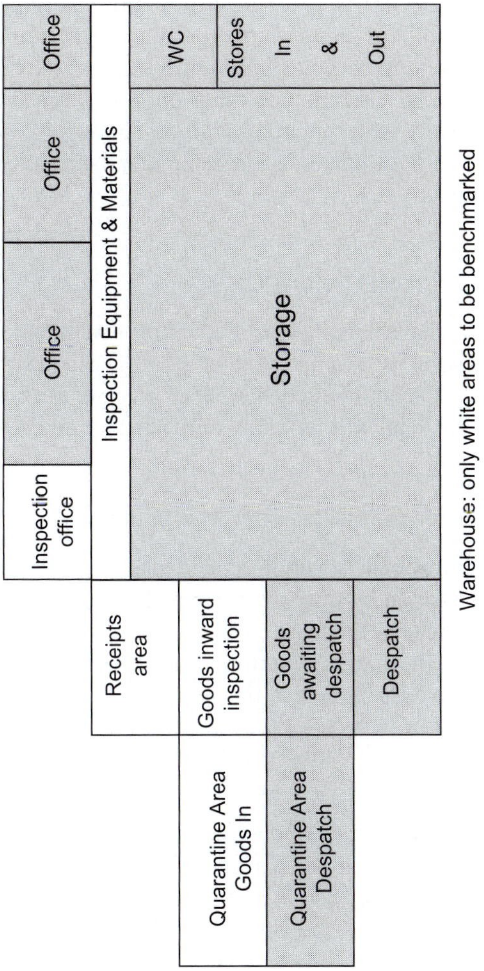

Warehouse: only white areas to be benchmarked

FIGURE 4.3   Simplified ground plan showing which areas are to be benchmarked and which are excluded.

4.2 STATING THE BUSINESS REASON FOR THE STUDY

The scope defines what is to be benchmarked: the reason explains why it should be benchmarked.

The team will probably already have been given the reasons for the study in their Team Charter or similar brief. However, they need to understand and be convinced of the need for the study, not only to support their own commitment but also to help secure support from others in the organization. As a result they will want to review and possibly expand the business reasons for carrying out the study. There is one other important reason why the team should consider carefully why the study is to be conducted: when recruiting other organizations they will have a clearer understanding of the potential benefits for participants.

4.3 OBJECTIVES OF THE PROJECT

In the first two steps we have defined what we want to benchmark (the scope) and explained why (the reason) we want to benchmark it. The next step is to agree the objectives of the study. Objectives may have been set prior to commissioning the team, in which case the team will probably only need to review and ensure they understand them.

Setting the objectives for a project is important because:

✓ They specify what the team is commissioned to achieve and so success of the project will be measured by the extent to which it achieves its objectives.
✓ They provide guidance to the team.
✓ They are an important aspect of recruiting participants and gaining support from others who have an interest in the study.

Where the benchmarking study will compare performance levels typical objectives include:

1. To identify performance levels of comparable organizations.
2. To quantify the gap between our performance and the best in class (or top quartile or some other standard derived from the benchmarking data).
3. To estimate the benefits to the organization of operating at the chosen standard.

The next objectives typically follow on from comparing performance levels and are usually the responsibility of each participant:

4. To select improvement projects.
5. To set performance targets.

Improvement may take place without further reference to the other participants, entirely through the organization making its own investigations and analysis. However, organizations will often want to learn from other participants how they achieve their superior performance levels. In these cases a typical objective would be:

6. To identify how other participants achieve superior levels of performance.

The final objectives for many studies are likely to be:

7. To improve our own performance to an appropriate level.
8. To quantify the improvements achieved.

Although the above are a selection of typical objectives, benchmarking may be used to achieve others such as:

- To use experience from other organizations to support proposals for implementing new procedures, plant, etc.
- To gather information and/or advice, for example when planning a new project or venture.

With reference to an example from logistics we show how these eight objectives link and relate to each other.

Example: Logistics Objectives

As an illustration we use one cost and one non-cost metric from a benchmarking study with four participants. The metrics are:

- Cost per kilometre to transport a ton of goods and
- percentage of late deliveries.

1. Benchmarking implies some form of comparison. In many benchmarking studies the first objective is to identify performance levels of all participants through the gathering and analysis of data.

   The table shows the performances of the four participants.

	$/ton km	Percentage of late deliveries
Participant A	$1.00	3.0%
Participant B	$1.10	2.5%
Participant C	$1.20	2.6%
Participant D	$1.25	2.3%

2. The next common objective follows directly from the first and is to quantify the differences between the best in class, top quartile or other standard and each participant.

In this example we select the best performer as the standard to measure against. The table shows the difference in performance between participant B and the best performer.

	$/ton km	Percentage of late deliveries
Best performer	$1.00	2.3%
Difference in performance between B and best performer	$0.10	0.2%

3. The next objective is likely to be to estimate the potential benefits of operating at the chosen performance standard.

The table includes the potential savings for B if performing at the level of the best performer. (Participant B transports 2 million kilometres per year, and the estimated benefit of reducing the percentage of late deliveries by 0.2% was $25 000.)

	$/ton km	Percentage of late deliveries
Best performer	$1.00	2.3
Difference in performance between B and best performer	$0.10	0.2
Potential benefit for B	$200 000	$25 000

Notes:

- This example used 'best in class' as the standard for comparison. There are several reasons why this might not be appropriate (⇨ Chapter 14).
- It may be necessary to look behind the figures to get a fair evaluation of performance levels. For example, mines in remote locations may have to pay travel and accommodation expenses to their workforces that are not incurred at mines located near to towns and cities.

4. Many benchmarking studies will include these types of analysis in the report. This then leaves each individual participant with the task of using the report and/or other benchmarking study deliverables to decide what aspects of the business to schedule for improvement. Those areas where the performance gap is greatest are likely to be the key areas for improvement, and by ranking these gaps a prioritized list of potential improvement projects can be

developed. However, it is not always as simple as just ranking the potential benefit. In the logistics example, participant B may factor in other considerations, such as the benefits that they expect to accrue from reducing late deliveries e.g. attracting more and losing fewer customers. Hence, it may be preferable to focus on the delivery aspect of the business rather than reducing transportation cost. For more information on selecting improvement projects ⇨ Chapter 16. The relationship between the objectives and the steps in the improvement process are shown in Figure 4.4.

5. Another common objective of benchmarking is to select, or be one of the factors in selecting, suitable performance targets. Once we know what other organizations are achieving we can use that information, perhaps modified to take account of differences in physical, commercial, legal, or other factors, to help set appropriate targets.

6. Once performance levels have been compared and improvement project(s) selected, the next likely objective for benchmarking studies is to learn how to reach the performance levels of the best, or better performer(s). This is usually achieved by identifying and studying the practices that lead to better performance levels. While it is likely that the participants in the study will aim to learn from each other, they may alternatively choose to benchmark areas of weakness in more detail with different organizations. For example, having decided to improve on-time delivery, participant B might search for organizations that are world class with respect to on-time delivery and carry out one-to-one benchmarking studies with them.

7. Having learned what practices are likely to lead to improved performance, participants need to take action within their own organization to bring about improvements and finally quantify the improvement achieved. It is only with these objectives that the real benefits of benchmarking for improvement begin to be realized: if no change is made there will be no performance improvement and the benchmarking study will have been of little benefit to the participant.

8. Finally, there needs to be a quantifiable means of showing that improvement has occurred as a result of the changes – a 'before and after' measure of the appropriate metrics.

The set of eight objectives presented above push right through to the final aim of many benchmarking studies, i.e. to improve performance. However, as discussed at the beginning of this section not all projects will include all these objectives. Other possible objectives are, for example:

- To identify best practices within the company and install them globally.
- To develop a set of metrics to be used for future performance evaluation.
- To gain benchmarking experience so that more complex studies can be initiated in the future.
- To help plan a new development/project etc.
- To solve a technical problem.

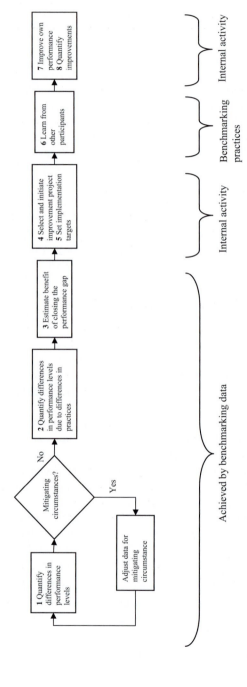

FIGURE 4.4 Relationship between the objectives and steps in the improvement aspect of the benchmarking process. (The numbers in the boxes refer to the numbered objectives.)

Example: Warehouse Benchmarking Objectives

A recent analysis of warehousing budgets across the group identified a difference of a factor of four between the lowest and highest. These differences are not readily explainable. It is proposed to benchmark internally all warehouse activities within the group in order to:

- Quantify the difference in performance levels between warehouses.
- Identify and account for legitimate operational differences.
- Quantify reducible differences, e.g. due to improving efficiency.
- Establish a set of metrics to be used throughout the group for monitoring and comparing warehouse performance and driving improvement.
- Identify best practices and set up appropriate activities for implementing them across the group.
- Set up parameters for budgeting operational costs for new warehouses.

It is intended that after completing this study we will begin external benchmarking.

4.4 MANAGEMENT REVIEW OF CHARTER

The previous three sections give an idea of the work entailed in reviewing the Project Charter. The end result will be a Charter in which the scope, project justification, and objectives are expressed with more detail and clarity. To confirm that the Charter is still aligned with the intentions of its originator(s), it should now be presented to management for review and ratification.

It may be that a steering group (⇨ Chapter 20) will be responsible for seeing a project through to the end, but different teams may be commissioned for each phase. For example, one team may produce the benchmarking proposal, another may carry out the benchmarking, and further teams may be set up to carry out improvement activities. At each phase of the study it is important to review the plan, objectives, team members, and all other aspects of the project to determine the best way forward. At the other extreme, it may be that not a team but a single person carries out all aspects of the benchmarking study.

The remaining work of developing a project proposal, e.g. selecting participants, developing metrics and definitions, planning the report structure are covered in the following chapters.

SUMMARY

The project team's first task will be to review and hone the Charter.

Tools for helping to clarify and define the scope of the project include:

✓ Definitions of the beginning and ending of a process or physical location.
✓ The use of context and similar diagrams to show relationships between the area being benchmarked and other areas.

Clear objectives for the study are important because they:

✓ Help persuade management to support the study.
✓ Help persuade organizations to join the study.
✓ Provide guidance to the project team during the project.
✓ Help measure the success of the project.

Typical objectives are likely to include some or all of the following:

✓ To identify performance levels of comparable organizations.
✓ To quantify the differences between our performance and the best in class (or top quartile or some other standard).
✓ To estimate the benefits to the organization of operating at the best in class (or some other standard) level of performance.
✓ To select improvement projects.
✓ To set performance targets.
✓ To identify how other participants achieve superior levels of performance.

The final objectives for many studies are likely to be:

✓ To improve our own performance to an appropriate level.
✓ To quantify the improvements achieved.

However, there are many other possible objectives for benchmarking studies.

If any changes are made to the Charter these need to be reviewed and approved by management before proceeding with the project.

At the end of this step the team has a well-honed, verified clear Charter and mandate to continue.

ϒ ACTIVITIES

Review a benchmarking Project Charter:

? Is the scope clearly defined? Will others understand clearly what is included/ excluded?
? Are the reasons for carrying out the study clear? Are economic factors included?

? Who will be impacted by the study? Will they be persuaded to support the study based on the information in the Charter? Why? If you think some of those impacted may not be supportive, why not? What are their concerns? Can you address their concerns in the justification for the study?

? What are the objectives of the study? Do they meet the needs of the organization? How will you determine the success of the project?

? Is there anything not currently addressed by the Charter that your organization's norms require you to address?

Chapter 5

Identifying and Selecting Benchmarking Participants

'If we are going to learn from someone,
we may as well learn from the best
rather than
someone who is merely a little better than us'

INTRODUCTION

Deciding who to invite to join a benchmarking study can be both a difficult and important decision. Typical attributes of preferred benchmarking partners include:

- ✓ World class performers in the aspects that we want to benchmark: so that we can maximize our learning.
- ✓ In our industry, i.e. with similar commercial environment, structures, performance metrics, and other aspects: so that identifying, adapting, and adopting best practices is as easy as possible.
- ✓ Located nearby: for ease of communication.

Unfortunately, it is unlikely that we will know who are the ideal benchmarking partners, and will need to carry out some research to find them.

In this chapter we:

- ✓ Investigate the different classes of participant (internal, competitor, non-competitor in the same industry and cross-industry) and the advantages and disadvantages of benchmarking with each class.
- ✓ Investigate how to develop criteria for ranking and selecting participants.
- ✓ Investigate methods of identifying specific organizations with which to benchmark.

DELIVERABLE

The deliverable of this step is a list of preferred participants, perhaps ranked, with whom we wish to benchmark and the reasons why we wish to benchmark with them.

5.1 CLASSES OF PARTICIPANT

There are several classes of participant with which we can benchmark as illustrated in Figure 5.1.

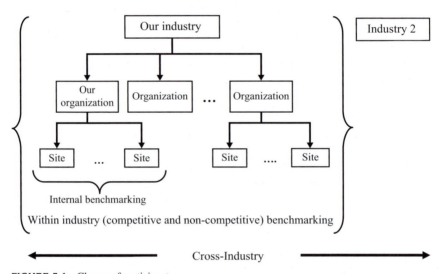

FIGURE 5.1   Classes of participant.

Within our organization we may have several sites (i.e. factories, processing plants, sites, locations, etc.) and benchmarking between them is called internal benchmarking. Benchmarking with competitors is called competitive benchmarking while benchmarking with non-competitors within our industry non-competitive benchmarking. The term industry benchmarking does not indicate whether the benchmarking study is between competitors or non-competitors or both. The term non-competitive benchmarking can lead to confusion because it can be used to describe both within and cross-industry benchmarking.

In this section we discuss each class in turn explaining some of the features, advantages, and disadvantages of each.

> **Definitions**
> * **Internal benchmarking:** carried out between groups within an organization.
> * **Competitive benchmarking:** carried out between competitors who by definition will be in the same industry.
> * **Non-competitive benchmarking:** carried out within the industry between non-competitors.
> * **Cross-industry benchmarking:** carried out between organizations in different industries and therefore between non-competitors.

The potential gains of each class of benchmarking can be summarized as:

* Benchmarking internally can lead to being the best in the organization.
* Benchmarking with competitors can lead to being better than them.
* Benchmarking within the industry can lead to being the best in the industry.
* Benchmarking cross-industry can lead to being the best in the world.

5.1.1 Internal Versus External Benchmarking

Probably the first issue that most organizations consider when deciding with whom they would like to benchmark is whether to benchmark internally or externally. Since benchmarking is conducted between at least two groups that carry out similar activities, internal benchmarking may not be possible for organizations with no duplicate or similar processes or activities. Where it is possible to benchmark internally there are a number of factors that should be considered including:

✓ Gaining agreement to benchmarking may be easier internally, especially if senior management encourage, support, and reward benchmarking activities.

✗ On the other hand, groups who are forced into joining a study may be disruptive, for example, by delaying progress or supplying incorrect data or information. While disruptive participants are encountered during external studies, it can be especially harmful with internal studies as the disruptive participant has a better opportunity to influence other participants.

✓ Where there are organization-wide standards for data reporting, the data collection, validation, and analysis should be much easier and faster than cross-organization studies where data may be reported to different standards.

✓ Transferring practices is likely to be easier in an internal benchmarking study than an external study because of the similarity of other aspects of the organization such as the culture and the process/activity/subject being benchmarked. Transferring practices may be difficult or even impossible between organizations in different industries, cultures, environments, political backgrounds, etc.

✗ The learning potential from internal benchmarking is generally far less than external benchmarking (unless best or good practices happen to lie within the organization), and so world class or industry best practices are unlikely to be identified.

In most situations organizations will prefer to benchmark externally because of the greater learning opportunity. However, from the above considerations some general guidelines can be deduced for when benchmarking internally may be preferable:

✓ As a way of gaining experience of the process of benchmarking and helping to define the objectives of an external study.
✓ Where there are different metrics, reporting standards, or practices within the organization, internal benchmarking will help to identify, document, standardize, and replicate best practices and performance metrics across the organization.

In the above two situations internal benchmarking is often viewed as a pre-cursor to external benchmarking. However, occasionally there may be no intention to progress to external benchmarking, e.g.:

● If the benchmarking subject is unique and it is not possible to benchmark externally.
● If the subject is highly confidential and the organization is not willing to divulge information openly or anonymously, even to those in other industries.
● If the organization wants to complete the study in the shortest possible time and/or at minimum cost and is willing to forego the greater learning potential of external benchmarking.

5.1.2 Within Industry Benchmarking

Many benchmarking studies are carried out between organizations in the same industry. These organizations may be competitors (competitive benchmarking), or non-competitors (non-competitive). A key advantage of benchmarking within the industry is that because processes have similar aims the transfer of practices should be relatively straightforward.

5.1.2.1 Some subjects can be readily benchmarked between competitors . . .

Non-commercially sensitive subjects, such as health, safety, and environment, are probably most easily benchmarked between competitors because many of the issues are similar. In addition, competitors may work together on specific topics such as training to establish industry norms and standards.

5.1.2.2 . . . and others not – at least not openly

Many organizations will not be willing to benchmark commercially sensitive subjects such as production, design, or development, with competitors. A solution to this problem in some circumstances is to hold the data anonymously. There are a variety of methods of doing this, such as comparing each participant against an average or 'best-in-class' value or identifying each participant's performance by code rather than by name (⇨ Chapter 15).

5.1.2.3 Most subjects can be benchmarked between non-competitive organizations

The key advantages of benchmarking against non-competitors in the same industry are that:

✓ There are few commercially sensitive concerns.
✓ There is a larger potential group of participants, and therefore a greater potential for learning.

Typical examples of non-competitive benchmarking within industries include power plants, railways, and other transportation services in non-competing areas. In many sectors, such as local government, policing, and emergency fire services, competitors do not exist and the only external benchmarking possible is with non-competitors.

5.1.3 Within Versus Cross-Industry Benchmarking

The decision whether to benchmark within or across industries and with competitors or non-competitors are closely related and depend on the subject of the study. Benchmarking studies can include competitors and non-competitors both within and between the industries.

Example: Aircraft Maintenance Benchmarking Participants

Benchmarking the maintenance of aircraft between competing domestic airlines within the USA would be a competitive benchmarking study within the industry.

If regional airlines in Europe (i.e. non-competitors with the US domestic airlines) were to join the study, the resulting study would include both competitors and non-competitors within the industry.

If maintenance of military equipment (e.g. aircraft, helicopters, ships, vehicles) were also included, the study it would cover both competitors and non-competitors across industries.

5.1.3.1 Cross-Industry Benchmarking

Some subjects can be readily benchmarked across industry ...

The key advantage of cross-industry benchmarking is that there is a greater opportunity for learning and little concern about commercial conflict. For example, an airline, a refinery, and a hospital can readily benchmark and transfer practices for support services such as IT, HRD, purchasing.

. . . And others not . . .

In contrast, industry specific subjects, such as aircraft pilot training, oil and gas production, are difficult to benchmark, at least in their entirety, between industries. In these situations it is usually preferable to benchmark within the industry.

. . . and sometimes it is not clear

Occasionally it can be difficult to determine whether a subject can be benchmarked across industries. Normally, safety is readily comparable across industry: there are no commercial sensitivities and working environments may be common, such as offices or production lines. However, some environments are unique to an industry and safety procedures are necessarily unique too, such as on an off-shore drilling rig. Even here comparison of, for example, kitchen safety may be valid across industries.

Defining what is meant by 'within' or 'cross' industry may not always be clear. For example, supermarket chains such as Walmart, Tesco, and Carrefour may be considered to be in the food retailing industry. However, many also sell domestic goods and financial services. The appropriate identification of 'their industry' for benchmarking purposes is better defined by the subject of the study (e.g. as supermarket, financial service provider, or even 'financial services sold through supermarkets').

From the point of view of identifying and implementing practices, a useful general guideline is that:

✓ Support services are best benchmarked with whatever company/companies we think are likely to have best practices that we can transfer, regardless of the industry.
✓ Industry specific subjects are best benchmarked with non-competitors (less commercial sensitivity) if possible, or with competitors if not.
✓ Highly commercially sensitive subjects may only be benchmarkable internally.

5.1.4 Comparing Different Classes of Participant – a View

We have discussed different classes of benchmarking participant and outlined some of the key advantages and disadvantages of each.

One logical view on developing and using benchmarking within an organization is:

1. To benchmark internally in order to standardize all processes and working practices, develop meaningful and comparable benchmarking metrics, and gain experience of benchmarking.
2. To regularly benchmark comparable activities, probably within, but possibly across industries in order to ensure continuing competitive performance

levels, perhaps as an aid to setting budgets and targets, and to maintain a mutually supportive network of those wishing to improve performance and/ or for other reasons.

3. Finally, we may identify specific issues that we wish to benchmark with the best performers we can find. These organizations may be in any industry and the aim is usually to identify completely new approaches or ways of thinking.

5.2 NON-ORGANIZATIONAL PARTICIPANT SELECTION CRITERIA

The above discussion and considerations may lead us to the class of organization(s) (competitor, cross-industry, etc.) with which we wish to benchmark, but there are a number of other issues that we may need to consider. In this section we will look at three typical considerations for selecting benchmarking partners: the commercial environment, the physical environment, and the size of participant organizations.

5.2.1 Commercial Environment

Commercial environment is a generic term used to encompass pay rates, employment law, legal, cultural, and other issues that affect the commercial operating environment. It is easier to benchmark with participants that have similar commercial environments.

Difficulties of benchmarking with participants from different commercial environments include:

- Although salaries vary between countries with similar commercial environments such as North America, Europe, and Australia, the differences are relatively small when compared to Asia or Latin America. This results in an immediate difficulty when benchmarking costs between one participant paying salaries that may be 20 times higher than another. In addition, the fact that pay rates are high or low may have a significant effect on operational strategies. For example, where pay rates are high, facilities may be highly automated and minimally staffed, while where pay rates are low, perhaps because of a policy of employing local staff, there may be little automation and high staffing levels.
- Legal requirements and standards differences may also have a large impact on an organization. Some countries have more stringent Health, Safety and Environment laws along with the associated inspections, reporting requirements, fines and in some cases taxes (e.g. on carbon dioxide emissions). These often result in an increase in the manning levels and costs.

> **Example: Difficulties of Cross-Economic Zone Benchmarking: Work Hours**
>
> The average number of hours worked in a particular Nordic organization has been reported as a little over 1300, with most European countries working less than 2000 hours per year. In Asia the figure is often over 2300 hours per year.

- Even though organizational culture is likely to vary between organizations within the same country, the differences between countries and regional cultures are likely to be greater.

The importance of the commercial environment to a benchmarking study will depend on the scope of the study. Benchmarking social services across different commercial zones is likely to be far more difficult than benchmarking, for example, manufacturing.

x A key difficulty in benchmarking between different commercial environments is identifying and accounting for the differences in commercial environments.

✓ If these difficulties can be overcome, the advantage is that there is a much greater potential for learning.

5.2.2 Physical Environment

> **Example: Diavik Mine, Canada**
>
> For most of the year there is no road to mines such as Diavik, 300 km north of Yellowknife in Canada. The only access is by air. Every winter, once the ice is thick enough, a road is built over the ice and all materials required for the next year's operations are transported to the mine. When the ice melts, in April, so does the road.

Physical location refers to issues such as road/rail/air infrastructure, climate, and remoteness. Operations in remote areas such as deserts or jungles may require the construction of roads, airports, labour camps, and large warehouses which would not be required for operations in hospitable populated areas.

It is not only access that may affect operations. Sand, extreme heat or cold can also bring increased costs in the form of equipment protection and the physical difficulties of working in such extreme conditions.

For temporary or smaller operations, such as oil exploration, operating in remote locations can bring the added difficulty that support staff and senior management may be at a different location, and perhaps even in a different time zone.

5.2.3 Size of Organization

The size of the organization chosen as a potential benchmarking partner may or may not be important. At one extreme, benchmarking the solution to a technical problem, organization size is unlikely to be important. Similarly, when benchmarking legal practices at branch level it may not matter whether the participant is an independent firm or part of a larger group. However, benchmarking many of the activities of an airline, such as booking or customer relations, could be difficult when comparing a multinational airline against a local carrier with perhaps only half a dozen aircraft.

Each individual situation will have to be reviewed to ascertain whether organization size is important. However, other aspects being equal, it will be easier to benchmark with and learn from organizations of a similar size.

5.3 IDENTIFYING PARTICIPANTS

Developing a list of organizations that might be suitable benchmarking participants is sometimes straightforward. When the possible candidates are not obvious the methods outlined in this section may be considered.

While using the methods below consider:

? Who carries out the same (or similar) function to the area to be benchmarked within our own organization. (We can benchmark against them and/or involve them in identifying and selecting participants.)
? Who are our competitors? (They are likely to have similar processes.)
? Who are in our industry but are non-competitors (e.g. desalination plants not competing for your business, railways in different geographical areas, hospitals)?
? Who has a similar process/product/service to us in other industries? (This is especially appropriate for support services such as IT, human resource, purchasing, finance, warehousing.)

5.3.1 Known Organizations

The simplest and probably the first method used to identify potential participants is to list organizations that we know which fulfil the benchmarking requirements. For internal and industry studies this should always yield at least some potential participants. In addition to organizations that the benchmarking team are aware of themselves, they could ask others within the organization, in particular those who are likely to be aware of competitors.

If organizations outside our industry are being considered as participants, suppliers and customers may be able to suggest potential participants, or depending on the subject being benchmarked, themselves be potential participants.

5.3.2 Industry Groups, Journals, Award Winners, Web Searches

Most if not all organizations will work in an industry which has its own professional bodies, meetings, journals, and websites. Accessing these should yield
potential participants. Writing an article in a journal or on a website asking for
contact details of others that may be interested could also yield potential participants. These contacts may be either by industry or function. For example, a bus
company wanting to benchmark vehicle maintenance may search for contacts in
the transportation sector and in the maintenance sector.

Many industry groups offer a service or facility to help organizations who
want to benchmark with each other. This may be as simple as having a benchmarking notice board on the member's website, emailing members, offering a
researcher to identify suitable participants or other services. There are some
organizations, such as Best Practice Club ⇨ 4 Case Study, The British Quality
Foundation and The Benchmarking Network that offer a variety of benchmarking and best practice services.

There are a growing number of quality award programmes and winners of
these awards in particular are well publicized. The fact that they have won an
award may make them suitable benchmarking partners. In the USA the most
famous is the Baldridge National Quality Award, its counterpart in Europe being
the EQFM Excellence Award while in the Middle East the Dubai Quality Award
is well known (⇨ Chapter 2).

Case Study: Benchmarking with a Business Excellence Award Winner

One organization followed this approach and went to visit a winner of a regional
award. Full of anticipation, the management team arrived at the winner's facility and
spent some time investigating and learning what that organization had done, how
they had done it, and the resulting performance levels. Unfortunately, the visitor's
conclusion was that the performance was not as impressive as had been publicized.
The winner of the award had taken a great deal of time and trouble focusing on and
presenting information and results for the award, but their overall performance level
was similar to that of the visitors' organization.

A contributor from the Healthcare industry, UK

As the above example demonstrates, it is always wise to validate that a
potential participant's performance is as good as you think it is before deciding
to learn from them.

Often, a little thought will identify other industries where world class performance is likely to lie. For example, for production line operations and Just in
Time practices, the automotive industry is considered world class. For rapid
setup, Formula1 racing would be hard to beat. To help identify who might be an
appropriate benchmarking participant, a useful question to ask is 'Who in the
world is best at . . .'

5.3.3 Industry Consultant

The above methods should provide an excellent starting point and will probably be enough to progress the study. However, another avenue often worth pursuing is to ask consultants to research potential participants.

5.4 FINALIZING POTENTIAL PARTICIPANTS LIST

In a typical study the next step will be to develop a list of criteria for accepting an organization as a preferred participant. Preferred participants will then be evaluated against these criteria. This may result in a list with more participants than required for the study, in which case the participants are prioritized.

One way of prioritizing preferred participants is to score each participant against each criterion and then add the scores for the participant. The results can be ranked by total score thus resulting in a ranked list of preferred participants. If required, the criteria can themselves be weighted, as shown in the example.

Example: Using Criteria to Rank Preferred Participants

The table below identifies selection criteria on which participants are to be ranked. Each criterion is given a weight reflecting its importance as determined by those carrying out the analysis. In this example participant location is given a weight of 15, and perceived learning potential 20.

The weights are usually allocated to sum to a convenient total such as 100. Each participant is then scored out of convenient total such as 10 (as in this example) or 100, against each criterion.

If participant A is a five minute drive away from our location, we might award 10 for location, but if the perceived learning potential were thought to be low, we might award it a score of only 1. The weighted score for each participant is then calculated by adding up the scores multiplied by the weights. Participant A would score 10 for location multiplied by a weight of 15 = 150 plus a score of 1 for learning potential × a weight of 20 = 20 giving a total of 170, etc.

Criteria	Weight	Potential participant			
		A	B	C	...
Location	15	10	4	10	
Perceived learning potential	20	1	7	3	
Commercial similarity	10	8	4	7	
...					
etc.					
Total	100				

The preferred participants would be those with the highest total score.

In reality, the benefit of such an exercise is not the final scores, but the thought and consideration required to develop the table and assess each potential participant.

For one-to-one benchmarking studies it will then be a case of ranking the potential participants and, starting with the highest ranked, approaching them in turn until one agrees to benchmark with us.

For club benchmarking it is likely that all, or the most highly ranked organizations will be invited to join the club.

It is likely that the team will need to justify their preferred participant list to management or others involved in the study. Using a selection process such as the one described helps to demonstrate objectivity. In addition, summarizing the reason for participant selection in a succinct sentence such as 'We would like to benchmark X because Y' also helps to focus attention on the objective reasons for selection.

SUMMARY

In this chapter we discussed the main criteria for selecting benchmarking participants and found that we need to consider whether we want to benchmark:

✓ Internally.
✓ Within industry: competitor.
✓ Within industry: non-competitor.
✓ Cross-industry.

The attributes of different types of participant are summarized in the table (Figure 5.2).

Other typical criteria for selecting participants include, but are not limited to:

✓ Commercial environment.
✓ Physical environment.
✓ Organization size.

Some organizations use a range of participant classes depending on the needs or each study (Figure 5.3).

Selecting participants may be simple. Where selection is not simple a selection process such as that shown in Figure 5.4 may be employed.

| | Same industry | | | Cross-industry | Location | | |
| | Internal | External | | | Local/national | Zonal | Cross-zonal |
		Competitor	Non-competitor				
Cost/time/ease of communication	1 Very cheap, fast, communication easy	2 Cheap, quite fast, communication quite easy	2 Cheap, quite fast, communication quite easy	3 Most expensive, slow, communication difficult	1	2	3
Ease of data collection & validation	1 Internal political issues may support or hinder	2	2	3	1	1 Assuming submitted by participant	1 Assuming submitted by participant
Learning potential	3	2 Potential to be best of the competition	2 Potential to be best in industry	1 Potential to be world class	1	2	3
Transferability of practices	1 Easy	2	2	3 can be difficult	1	2	3
Commercial concerns	None	Commercially sensitive issues	None	None	3	2	1
Frequently used for	To standardize methods, metrics and definitions. To learn how to benchmark.	Industry specific issues. When non-competitor benchmarking is difficult/impossible.	Industry specific issues. Preferred to Competitor benchmarking because there are no commercially sensitive issues.	Support service benchmarking such as purchasing, IT support, and customer services.	Where speed and cost of study are important. Where the aim is to be best local provider.	Where the aim is to be best zonal provider.	Where the aim is to be world class (either in the industry or cross industry).
But…	May not be possible for small organizations		Non-competitors may be in other countries/commercial environments				

FIGURE 5.2 Summary of attributes of different classes of benchmarking participant. The numbers represent typical ranked preference. 1 = most preferred choice and 3 = least preferred choice.

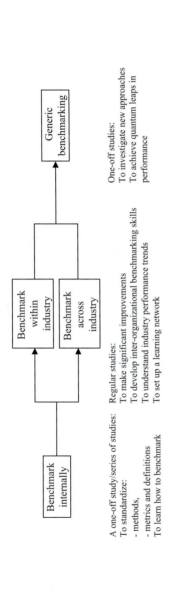

FIGURE 5.3   Uses of different classes of benchmarking participant.

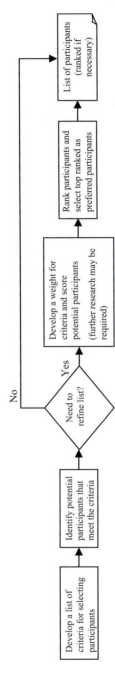

FIGURE 5.4   Process for finalizing a potential participant list.

Y ACTIVITIES

To complete these activities you may need to refer to previous chapters.

? What class of participants are most suitable for your benchmarking project? Why? Why are other classes not so suitable?

? What non-organization criteria should you consider when selecting participants for your study?

? If you do not have a list of potential participants, make a list of methods you will use for identifying them. Use the methods to generate a list of potential participants.

? Identify and use criteria to evaluate each participant and generate a list of preferred participants. A useful way to do this is to draw up a table of criteria versus potential participants.

Chapter 6

Metrics and Data

*There are lies, damn lies and **misuse** of statistics (statistics themselves cannot lie)*

INTRODUCTION

The adage: 'There are lies, damn lies and statistics' is not true simply because statistics cannot lie. However, what frequently does happen is that errors in data collection, poor definitions, inappropriate use of statistics and statistical methods lead to erroneous conclusions.

The issue of metrics can be complex, because there are so many often conflicting issues that need to be considered. For example:

? How many metrics do we need? Too many metrics makes the benchmarking study unmanageable, while too few leads to incorrect conclusions as to relative performance.
? How do we ensure comparability of data? Different organizations report to different standards: perhaps some organizations will not be able to supply all the data, or will report to a different definition.
? How do we balance different aspects of performance such as environmental responsibility and minimizing costs?
? How do we compare disparate facilities such as maintenance of production facilities each with a unique set of equipment or the quality of teaching in schools in different socio-economic environments?

These and many similar questions are not easy to answer. However, there are a variety of tools and methods we can employ to help us, which are discussed in the following chapters.

To add to the confusion, it is an unfortunate fact that all (non-trivial) process inputs, outputs, methods of working, etc. vary over time. Understanding the nature of this variation and how to deal with it is important at the data planning stage as failure to address variation may result in erroneous conclusions later in the benchmarking study.

As the above outline indicates, developing metrics for a benchmarking study can be a very time consuming and, for those not experienced in measurement, difficult aspect of benchmarking fraught with pitfalls for the unwary. Even for those experienced in benchmarking, developing a suite of metrics that will clearly, concisely and accurately compare performance levels of different participants can be a daunting and lengthy task.

In this chapter we outline the role that metrics play in many benchmarking studies and illustrate the importance of collecting consistent data from participants. We also discuss the existence of, and how to deal with variation in data along with the effects of planned process changes and the effect of significant events. Finally we consider specific difficulties when benchmarking projects.

In all but the simplest of analyses we strongly recommend the use of a trained and experienced statistician to at least review, if not be actively involved in, data collection, analysis, and report writing.

6.1 THE IMPORTANCE OF METRICS

In many benchmarking studies metrics are arguably the single most important aspect of the study, as they are the building block for analyses, recommendations, and improvement activities as illustrated in Figure 6.1.

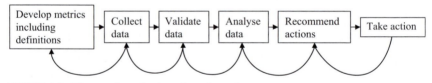

FIGURE 6.1   The use of metrics throughout the study.

The first step is to develop a set of metrics, including clear definitions. Clear definitions are important because without them participants may report to different standards, ⇨ Chapter 10.

Once the metrics are finalized, data will be collected and, having been validated, will be analysed. The analysis will frequently include a gap analysis between each participant and a selected standard (such as 'best in class', top quartile, or average). Recommendations will result from the analysis and some or all of these recommendations will be actioned.

However, as with all processes the requirements of the process work in the opposite direction. From the point of view of benchmarking, the requirement is to take effective appropriate action. The decision on what actions to take will flow from the recommendations in the report, and if the recommendations are inappropriate, inappropriate actions are likely to follow. However, to arrive at appropriate recommendations, the conclusions based on the analysis have to be correct, and these can only be correct if:

1. The data they are based on is correct (which is why validation is important);
2. Appropriate data have been collected to allow the analysis to be carried out; and
3. The appropriate analysis is carried out.

Examples: Importance of Consistent Data – Late Deliveries and Reliability

A benchmarking study includes the metric 'number of late deliveries'. Each participant reports according to their own concept of late deliveries.

- A defines a late delivery as missing the agreed date for delivery
- B specifies a time slot of 2 hours within which a delivery should arrive. It reports late deliveries, but not those that arrive early
- C only records a delivery as late if an item is not available when required because a delivery is overdue. Therefore, a delivery could be days or weeks after the expected delivery date but still not be reported as late.

In the analysis, C is deemed as best in class.

In a manufacturing benchmarking study it is agreed to report production line reliability. Participant A has the poorest reliability, and participant B the highest. B is deemed best in class. However, at the Best Practice Forum B explains that it duplicates the equipment that is most likely to fail so that it can simply switch to the standby unit thereby maintaining high production line reliability. After further data collection and analysis it is discovered that A's equipment is better maintained than B's and it has chosen not to duplicate equipment, thereby suffering slightly worse reliability figures at the benefit of having less equipment (e.g. saving on purchasing, insurance, managing, and maintenance costs).

Finally, the correct data will only be collected if it has been asked for and collected consistently.

6.2 THE PROBLEM OF VARIATION

If we measure any useful aspect of a process over time the measured values will vary. For example:

- The time taken to process invoices, treat a patient, answer an enquiry, etc. will vary from occasion to occasion.
- The number of errors, breakdowns, failures, safety incidents will vary from month to month.
- The performance of a particular person or group will vary over time.

Figure 6.2 shows typical rejects of a process from two participants of a benchmarking study.

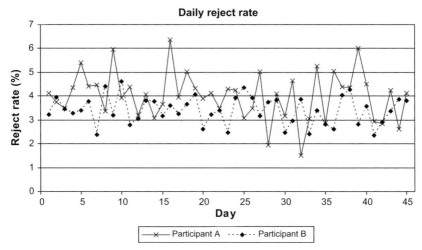

FIGURE 6.2   Daily reject rates for two participants.

If we were to select randomly a day on which to report the rejects the conclusion could be:

1. That participant A has a higher reject rate than participant B (for example on day 1).
2. That participant B has a higher reject rate than participant A (for example on day 2).
3. That there is very little difference in reject rates (for example on day 3).

To overcome this problem it is usual to take average performances over time. In this example the average reject rate for participant A over the 45 days recorded is 4.0% and for participant B is 3.4%.

The same issue occurs even if we report the number of rejects per week or month: the reject rates will vary from week to week and month to month. However, the longer the period over which we measure, the nearer the true process average the sampled value will be. This is because the background or day-to-day variation will have minimal effect on the overall average. Similarly, even moderate aberrations in values will have limited effect on a long-term average. The solution, it would seem, is to report an average figure over as long a period as possible.

This solution is appropriate if and only if the process has not changed within the reporting period, and the longer the time period over which we collect data, the more likely it is that a process change has occurred.

There are two issues that need to be addressed:

1. Has the performance level changed during the period over which data were reported?

2. If the process has changed, should we report the current performance level, the average over the period reported or some other value?

To determine whether the performance level has changed we use a control chart to monitor performance levels. A brief introduction to control charts is given in ⇨ Appendix 1. Organizations will normally be monitoring performance levels and so should be aware of changes that have occurred. As an example, Figure 6.3 shows that the reject rate for participant A (from Figure 6.2) increased after day 45. The current daily reject rate is 6.0%.

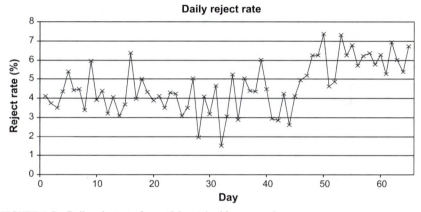

FIGURE 6.3 Daily reject rate for participant A with process change.

There is now a choice of which performance level to use: the previous average, 4%, the average of all the data, 4.6% or the current level, 6%. Usually the current average performance rate is preferred because using a previous rate will justifiably lead to the criticism that things have already changed and the data are out of date. Using a combined figure will lead to confusion since it neither relates to the current reality or previous reality, but a combination of the two. However, some people argue that the correct value to use is the average over all the data as this reflects the actual figure for the period reported.

6.3 DEALING WITH PLANNED CHANGES

Some situations are a little more complex. Many benchmarking studies will report on 'as is' data, e.g. the current manning/expenditure levels, accident/error/failure rates, etc. However, sometimes it is known that the current situation is changing or will change, for example, due to a business review, expansion, contraction, automation, or other reason. A common question from a participant in this situation is should they provide data for current manning levels or the proposed future manning levels. Usually participants will choose to benchmark the planned future manning levels since these will represent normal operations and benchmarking against the current data could be seen as irrelevant. However, the choice does depend on the needs of the participants. One possibility is to submit two sets of data, one representing the current and the other the planned situation. This would allow analysis of the effect of the changes to be seen in the light of relative performance to other participants.

6.4 THE EFFECT OF SIGNIFICANT EVENTS

There is one further source of variation that needs consideration. Sometimes the process may not be changing but is affected by an unusual event, which causes unusually high or low values for the duration of the event.

Example: Variation in Late Deliveries

Late deliveries have been running at 2.5% for many months. A fire at the warehouse causes the late deliveries to soar to 28% one month, drop to 12% the next and 6% the next before returning to 2.5%.

In the example the effect of the fire was clearly reflected in the data. However, such events are not always so dramatic and analysis may be required to identify them. The tool for doing so is the control chart.

Having identified an unusual (more correctly termed an 'assignable' or 'special cause' of variation) event, the data change due to the event is usually omitted for general benchmarking purposes.

Whatever decisions are made regarding the basis of data reporting, anything other than reporting to the agreed definition should be noted so that the analysts can deal with the issue appropriately.

6.5 PROJECT BENCHMARKING

Project benchmarking is different from other types of benchmarking in that the data reported will reflect a complete project rather than a period of time. The concept of variation still applies, but in a different way. The 'variation' of importance with project benchmarking is the difference between projects.

For example, a highways maintenance benchmarking study may benchmark different maintenance projects from different participants. For each participant there are likely to be several different projects that could be benchmarked. In some studies participants may be permitted to submit one or a selection of their projects, in which case the submission is not likely to be representative of their general performance level.

Generally, a better option is that participants submit all projects to the study. This will allow analysis of variability both between projects from each participant and far more accurate conclusions to be drawn concerning the difference in participant performances. A few simple examples will illustrate the issue, but for a more complete explanation ⇨ 'Mastering Statistical Process Control', a handbook for performance improvement, Chapter 13.

In a study, two organizations benchmark similar types of projects. Participant A has 5 potential projects to submit, participant B has 20. Consider the situation where on average both organizations perform at the same level. Variation between project performance will only be due to random variation (in much the same way that the time taken to drive to work every day will vary even if we leave home at the same time, take the same route and drive in the same manner, etc.). The chances are that the 'best' performing project will be one of participant B's 20 projects (in the same way that the fastest time to drive to work over a month will usually be faster than the fastest time recorded in any one week). Therefore, allowing participants to select projects to submit will favour the participant with most projects to choose from.

In some studies participants may be allowed to submit a 'typical' project. In these situations participants will remove the projects with poor performance finding something unusual that went 'wrong'. In the same way, the day that it took the longest to drive to work, we might erroneously say, was because we had to stop at 10 of the 13 traffic lights on the route, and that is not normal.

Reporting the average of all projects is a reasonable solution, except that we are not able to see which participants have consistent performances and which have not: and one thing we would like to learn is how organizations can routinely perform at superior performance levels. For example, if applied to aeroplane arrivals, it would be better to always arrive exactly 5 minutes late than to sometimes arrive very early, sometimes on time and sometimes hours late, even if the average arrival time is only 1 minute late, i.e. often performance consistency is more important than a slightly better average performance with huge variability.

SUMMARY

✓ In studies where metrics are the foundation for making recommendations, it is crucial that the metrics be selected carefully.
✓ Process performance varies over time, and it is important to understand this variation.
✓ It is important to identify whether process performance has changed/will change. If it has, we need to decide whether to report the previous/current/ planned performance level or more than one performance level.
✓ It is important to identify whether the process as been subject to 'special causes' of variation – i.e. explainable as one-off aberrations. If so, we need to decide how to treat that data. Normally the effect of the special cause will be reported separately and the analysts will decide how best to include it in the report.

✓ In **project benchmarking**, variability between different projects for the same operator is a key issue and careful thought is needed as to whether one, a few or all projects should be submitted.

✓ The key tool for identifying process changes and special causes of variation is the control chart.

Before setting out on an intensive data collection and analysis exercise, it is worthwhile seeking the advice of a statistician.

⅄ ACTIVITIES

? If your study is likely to be using process monitoring data (such as failure/ error/production/accidents rates) develop control charts of the performance data to discover whether the underlying average rates are changing (i.e. subject to 'special cause' variation) or not (i.e. the process is in a state of *statistical* control). If you are unsure about how to use control charts, find out how to do so.

? If you think the analysis may call for statistical expertise beyond that of the team, identify a statistician within the organization and ask their advice. If necessary, draft a statistician onto the team.

? Be prepared to address the possibility that participants in the study may not understand the concepts of variation. How will you ascertain:

- What participants need to understand.
- Whether they do understand.

If participants do not understand the concepts variation, what action do you need to take?

Chapter 7

Normalization:
How to Compare Apples with Pears

Normalization is a tool that allows us to compare or combine data that we could not otherwise compare or combine.

INTRODUCTION AND THE NEED TO NORMALIZE

A key aspect of benchmarking is to compare performance levels between participants. In some situations direct comparison of the data are appropriate, while in others it is not. For example, comparing the maintenance cost of different models of cars would be useful when benchmarking the cost of ownership. Similarly, comparing the number of pupils in a school or patients consulting a doctor at the local practice, can provide useful comparative information.

However, directly comparing injuries incurred in different organizations, or the number of pupils passing a certain exam is usually inappropriate. In these situations we need to account for the differences (i.e. variation) in underlying opportunity for injury to occur or the number of pupils sitting the exam. In the former we account for the opportunity for injury by taking the exposure hours into account and reporting, for example, the number of injuries per million hours worked. In the latter example we may report the percentage of pupils who pass the exam. In both cases we normalize the data in order to create an appropriate basis for comparison.

As illustrated above, the problem that normalization addresses, at least partially, is to convert metrics that are not directly comparable into metrics that are. In this chapter we develop six methods of normalizing data:

1. **Per unit**, often thought of as 'rates' or 'frequency', e.g. cost per unit of production, incidents per hundred (or thousand, etc.), percentage of failures.
2. **Categorization**, in which we group data into categories. e.g. number of accidents for women aged 18–25.

The Benchmarking Book: A How-to-Guide to Best Practice for Managers and Practitioners

3. **Selection**, most common in project benchmarking, where we select specific projects for comparison that we deem to be similar.
4. **Weighting Factors**, where packages of work are allocated relative weightings according to the size of package, e.g. when comparing the cost to maintain different chemical plants we can assign a Weighting Factor to the equipment in each plant and compare cost of maintenance per unit weighting.
5. **Modelling**, where performance of individual data items are compared to a mathematical model developed from the data.
6. **Scoring**, where performance of different metrics are weighted and summed, e.g. Business Excellence Models, final marks for academic courses may be weighted and summed across exams, course work and projects.

Per unit, scoring and Weighting Factors are all closely related but are treated separately as each generally has different applications. These methods can be combined, for example as in mortality rate for women in the age range 30–40 with bowel cancer. In this example the category is women aged 30–40 with bowel cancer and the 'per unit' is the mortality rate, e.g. percentage.

Finally, we discuss how to manage data that are subject to occasional abnormally high (or low) values. A typical example of this occurs when benchmarking annual maintenance activities and dealing with activities that occur every few years (such as major overhauls).

7.1 NORMALIZATION AND VARIATION

The key purpose of normalization is to account for certain types of variation. For example, categorization of cancer data into male and female acknowledges that a major variation in cancer data is due to the sex of the patient.

The question arises as to how much detail we should normalize? We could, for example, normalize by categorizing the incidents of heart attack by gender, age, family history, diet, marital status, number of children and so on, but to do so would be of little or no practical use because by over-categorization each group has too little data to draw any meaningful conclusions.

It is necessary to think carefully about what we are trying to achieve by normalization and the usefulness of the resulting data. While in many situations the method of normalization is already established or obvious, in some situations, for example as described in the Weighting Factor ⇨ 7.2.4, it is not.

7.2 METHODS OF NORMALIZATION

7.2.1 Per Unit and Per Hour

The simplest form of normalization, and one that is used so frequently that we hardly recognize that we are normalizing, is the 'per unit' normalization. The aim of this type of normalization is to recognize that the 'volume' varies between participants of a study, and often between time periods within an organization and that we need to account for these changes by comparing the units per volume. There are many examples of this type of normalizing, including:

- Cost per ... unit bought, hour, kilometre transported, passenger, PC supported.
- Hours per ... unit produced, per client.
- Sales per ... unit floor area, salesperson, customer.
- Accidents per million hours exposure.
- Errors/failures/rejects per ... 1000, person, day, form.

One very common 'per unit' measure is percentage. A survival rate after a medical treatment of 78% means that for every 100 people treated 78 survive.

7.2.2 Categorization

In many situations the difference in the activity being benchmarked cannot be accounted for by 'per unit' normalization because the underlying environment is different.

For example, exam results are influenced by the quality teaching, the home background of the pupils, the ethos of the school, etc. Benchmarking exam results purely by looking at the percentage of pupils passing does not tell us anything about the influencing factors. Similarly, comparing the performance of luxury cars with compact cars, or cost per mile for public transport journeys of long and short distances, is seldom appropriate for improvement benchmarking purposes.

In these situations one solution is to categorize the underlying environment into groups or classes, and benchmark within each class. In some situations natural groupings may occur. For example, in the railway industry, journeys may be grouped according to class of travel (commuter, cross country, on versus off peak, etc.) In the Healthcare industry, patient care may be grouped by types of disease, in versus out patient or many other factors.

Categories may be synonymous with products or services, geographical locations, market segments, etc. and can be considered discrete with reasonably concrete boundaries. In other situations, for example projects such as software development or civil construction, group boundaries may have to be determined.

This does not mean that it is not useful to compare across categories. For example, it is useful to know which groups of patients do not do well after surgery, partly to help identify the reasons and also to help select areas for focusing research and improvement. However, for most benchmarking studies the focus is on comparing performance levels within categories.

7.2.3 Normalization by Selection

Normalization by selection can be considered as an extreme example of categorization. The method typically applies to project benchmarking where projects are individually selected for comparison based on criteria that, it is hoped, will result in a group of similar, comparable types of project. This method is most useful where there is a database of projects to select from and those projects most similar are used for comparison. The criteria can be expanded or restricted to capture more or less projects into the comparison as required.

Example: Normalizing by Selection for Oil Well Drilling

In a benchmarking study to compare the drilling of oil wells, it may be required to benchmark a 3000 metres deep well drilled in a swamp (amongst other criteria). A database of wells is scrutinized to find specific wells with which the well can be benchmarked. The first step is to select all swamp wells. If this yields a large selection of wells, the selection can be further reduced by selecting wells between 2800 and 3200 metres. If this yields no wells, the search can be widened to 2500–3500 metres. The criteria are further adjusted until an acceptable number of wells deemed to be similar are identified.

This type of normalization is open to abuse in that the criteria may be determined in such a way as to 'confirm' pre-determined conclusions, for example that the item benchmarked is the 'best'.

7.2.4 Weighting Factors

7.2.4.1 The Application of Weighting Factors

A common question that many benchmarking studies will want to answer is: 'How well do we maintain our equipment?' When defining what we mean by 'How well' a typical response will be:

1. How many hours does it take?
2. How much does it cost?
3. How effective is it?

It is usually straightforward to gather the number of hours expended and the costs incurred to carry out maintenance tasks. Effectiveness is often defined to be mix of items such as reliability, ratio of planned to unplanned downtime, and equipment or system availability. It is often not appropriate to compare maintenance costs directly between participants because of the differences in the type or complexity of equipment being maintained.

Example: Weighting Factor Normalization I

Example: Airline Maintenance I

In a benchmarking study Take-off Airlines maintain a fleet of 40 aircraft of various models. Flying High Aviation maintains a fleet of 42 aircraft, also of various models, but not all the same models as Take-off Airlines. Last year Take-off Airlines required 40 000 hours to maintain its aircraft, whereas Flying High required 38 000 hours to maintain its aircraft. While this might suggest that Take-off Airlines are less competitive than Flying High Aviation, on closer inspection it was discovered that Take-off Airline's aircraft were generally larger and more complex to maintain than those of Flying High Aviation.

Example: Weighting Factor Normalization Gas Processing I

The Big Flame Gas Company processes a high grade, high-pressure natural gas with few impurities. They process 55 million cubic metres of gas per day (mmscmd), and require 110 people to operate and maintain their facility.

Pilot Light Company processes a low quality, low-pressure gas with several impurities. The gas requires significant processing to remove contaminants such as hydrogen sulphide and mercury. After processing the gas they need to compress the gas before passing it on to their customers. Pilot Light Company process 28 mmscmd and require 125 people to operate and maintain their facility.

While comparing the manpower and costs to produce a cubic metre of gas tells us how economically viable each facility is, it does not tell us how well each facility is operated and maintained because Pilot Light Company require significantly more equipment to process and pressurize their feed gas than Big Flame Gas Company require to process their gas.

As the above examples illustrate, comparing raw costs and hours to carry out tasks is not appropriate unless the tasks are similar.

There are several methods that could be used to normalize the data, which would take account of the amount and complexity of the work being done. In the case of the airlines, we could normalize by 'carrying capacity' (for example by number of passengers) and calculate the number of hours to maintain aircraft per 1000 passenger capacity. However, this would assume that, for example, maintaining five aircraft each with a capacity of 200 people requires the same effort as maintaining ten aircraft each with a capacity of 100.

An elegant solution has been developed whereby each 'package' of work is given a Weighting Factor (WF) based on the amount of effort required to complete the task. The Weighting Factors for each package of work are then added to give a Total Weighting Factor (TWF) for all the work being carried out.

Example: Weighting Factor Normalization 2

Example: Airline Maintenance 2

Each different model of aircraft was assigned a WF reflecting the relative effort required to maintain that model. Adding up the WFs assigned to the 40 aircraft maintained by Take-off Airlines gave a TWF of 380. Similarly the WFs of the 42 aircraft maintained by Flying High Aviation resulted in a total of TWF of 320.

The maintenance hours per TWF is then calculated for:

$$\text{Take-off Airlines} = \frac{40\,000}{380} = 105 \text{ hours per unit WF}$$

$$\text{Flying High Aviation} = \frac{38\,000}{320} = 119 \text{ hours per unit WF}$$

This shows that with respect to time, Take-off Airlines is more efficient at maintaining aircraft.

Example: Gas Processing 2

Big Flame Gas Company requires 200 000 hours of effort to operate and maintain its gas processing facilities while Pilot Light Company requires 225 000 hours of effort.

However, when weightings are applied to the different types of equipment at each site the TWF for Big Flame Gas Company is 100 and for Pilot Light Company the TWF is 125. The hours per unit of Weighting Factor are:

$$\text{Big Flame Gas Company} = \frac{200\,000}{100} = 2000 \text{ hours per unit weighting}$$

$$\text{Pilot Light Company} = \frac{225\,000}{125} = 1800 \text{ hours per unit weighting}$$

This indicates that Pilot Light Company is run more efficiently than Big Flame Gas Company.

The same method can be applied to costs.

The steps for calculating and using Weighting Factors are (Figure 7.1):

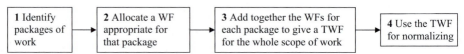

1 Identify packages of work → **2** Allocate a WF appropriate for that package → **3** Add together the WFs for each package to give a TWF for the whole scope of work → **4** Use the TWF for normalizing

FIGURE 7.1   Process for normalizing by Weighting Factor.

1. Identify distinct packages of work within the whole work-scope.
2. Allocate a Weighting Factor appropriate for each package to reflect its relative size/complexity.
3. Add together the WF for each package of work to give a TWF for the whole work-scope.
4. Use the TWF for normalizing.

The examples focus on operations and maintenance of equipment. Similarly, it is possible to apply the technique to any area where packages of work can be identified.

Up to this point the weighting method has only been applied to packages of work. However, it is also possible to include other factors. In the maintenance of aircraft example, the cost to maintain aircraft may be considered higher in remote regions due to high cost of transporting spares and equipment, high wages etc, or may be lower, for example, due to economies of scale where there are greater volumes of work. In all situations the key will be:

1. To identify the key factors that influence the time or cost to complete the work.
2. To decide which of the factors identified should be included in the Weighting Factor and which should be excluded.

7.2.4.2 Cost Per Unit Weighting Versus Cost Per Unit Produced

There are two methods of normalizing the costs and hours to generate an output, such as gas from a processing plant or product from a factory:

1. We can normalize using Weighting Factors, as explained above. This tells us how efficiently equipment is operated and maintained.
2. We can normalize by the amount of product produced, which tells us how economically viable the facility is.

Usually we will only want to compare the hours and costs of carrying out work activities and so only the work scope will be accounted for by the WF.

When we compare companies on the basis of 'cost per unit produced' (e.g. cost per gram of gold extracted in a mine) the lowest cost organizations may be different to when we make the comparison on the basis of 'cost per unit Weighting Factor'. To understand why there is a difference it may be helpful to think in terms of economic viability. The amount and complexity of the equipment required to produce an output (e.g. pure gold) will be largely dependent on the quality of materials (e.g. gold ore) being processed by the equipment. For example, a mine extracting ore with only 5g of gold per tonne will need to process 10 times more ore to produce the same quantity of gold as a mine extracting ore with 50g of gold per tonne. Other things being equal, the mine with lower grade ore is extremely unlikely to be competitive in terms of cost per gram of gold produced, even though it may operate its equipment more efficiently than the mine with higher-grade ore.

This concept is explained by the formula:

$$\text{Economic viability} = \text{Quality of material being processed}$$
$$+ \text{Efficiency of operating and maintaining}$$
$$\text{processing plant}$$

Generally, from the point of view of benchmarking, there may be little we can do about the raw material being processed (e.g. in the extraction industries) and so while the cost per unit produced is of economic interest, all we can influence is the efficiency with which we operate and maintain the equipment required to extract and process it.

7.2.4.3 How to Derive Weighting Factors

Unless we can gain access to a pre-existing set of Weighting Factors, we will need to develop our own. In this section we explain how to develop WFs for costs, but the same methodology can be used for man-hours or anything else.

To derive a set of WFs:

1. Draw up a list of work packages for which the WFs are required.
2. Gather costs for carrying out each work package from several sources (e.g. benchmarking participants).
3. Calculate the average cost for each package: the averages can be used as the WFs (see below for other alternatives).

Example: Deriving a WF for a Hospital

In a hospital benchmarking study the participants want to compare the costs required to treat cancer patients. Each hospital has data on the cost required to treat different types of cancer. For cancer type X Hospital A spends $51k per patient, Hospital B spends $45k and hospital C $48k. The average cost for treating cancer type X is $48k per patient.

Similarly for a second cancer type, Y, the costs from three hospitals are $66k, $77k, $73k, giving an average of $72k per patient.

If we choose a datum such that a WF of 1 to equates to $1k, then the WF for cancer type X = 48 and for cancer type Y = 72.

Using this methodology each cancer type can be assigned a WF.

Notes:

- We do not weight the average by the number of patients at each hospital as to do so would bias the weighting factor towards the hospitals with larger numbers of patients.
- Weightings for other treatment types can be developed in the same manner to build up treatment lists with weightings.

What is important about the WF is not the *absolute* values but the *relative* values. If we were to double all the WFs, the result of the comparisons between participants would not change. This has a number of useful implications: it is not necessary to attach a meaning to a specific WF, but we can choose to make a WF of 1 equate to a useful value. For example, a Weight Factor of 1 could be equated to $1k, 1000 hours of work, or the total workload of a typical participant. For convenience, using a WF = 1 man-year of effort is useful because it is easy to translate immediately from WF to the average number of people required to carry out that amount of work. For example, if we know that a hospital has a Weight Factor of 37, then we expect the equivalent of 37 full time people to be working at the hospital.

Note that ascertaining the weightings for each participant needs to be done consistently. As long as the relative weighting between participants is correct, the absolute values are less important.

☜ This method is subject to bias where there is only a small number of hospitals submitting data on any one treatment (as in the example above). This bias diminishes as the number submitting data increases. The question of bias is beyond the scope of this text and the reader is advised to consult a statistician for more information.

Example: Applying the WF to Hospital Treatments I

The table shows an extract from a list of treatments carried out at two different hospitals, A and B, along with the number of treatments carried out as in-patients and out-patients, and the resulting WFs with respect to hospital staff hours.

For example, Hospital A has 13 patients that underwent treatment 2 each deemed to have a weighting of 1.8 units and 19 patients undergoing the treatment as out patients, each receiving a weighting of 0.4. This gives a total weighting of 31 for this type of treatment ($13 \times 1.8 + 19 \times 0.4 = 31.0$).

Hospital B gives six treatments. Treatments 1, 4, and 5 are the same as hospital A, but with different numbers of patients,

Hospital A

Description of Treatment	In-patient		Out-patient		Total WF = 180.0
	Patients	WF	Patients	WF	
Treatment 1	31	1.0	0		31.0
Treatment 2	13	1.8	19	0.4	31.0
Treatment 3	38	0.5	65	0.1	25.5
Treatment 4	9	3.5	0		31.5
Treatment 5	16	2.2	43	0.6	61.0

Hospital B

Description of Treatment	In-patient		Out-patient		Total WF = 250.0
	Patients	WF	Patients	WF	
Treatment 1	57	1.0	0		57.0
Treatment 4	17	3.5	3	0.4	60.7
Treatment 5	21	2.2	58	0.6	81.0
Treatment 6	4	3.5	0		14.0
Treatment 7	2	6.5	0		13.0
Treatment 8	7	3.0	33	0.1	24.3

The TWF for hospital A is 180, and for hospital B is 250, so we would expect that hospital B would require (250/180) = 1.4 times more hours to provide its treatments than hospital A.

This example illustrates one of the advantages of the WF technique: it allows us to compare the performance of organizations even though they do not carry out the same work-scopes.

Example: Applying the WF to Hospital Treatment 2

In the summary table below, Weighting Factors have been determined for five hospitals, and the associated number of hours for each hospital to provide the treatments is also shown, along with the hours per Weight Factor.

Hospital	WF	Hours to provide treatment	Hours/WF
A	180	9800	54.4
B	250	11750	47.0
C	136	7800	57.4
D	412	18220	44.2
E	312	11220	36.0

Weight Factor summary table.

Comparing the hours per Weight Factor shows that hospital E has the lowest hours to deliver a unit of treatment.

The effect of using the Weight Factor can be clearly seen by comparing Charts 1 and 2. Chart 1 is a bar chart showing the number of hours required by each hospital to deliver its treatments. Hospital C is the lowest, D the highest and A, B, and E are similar.

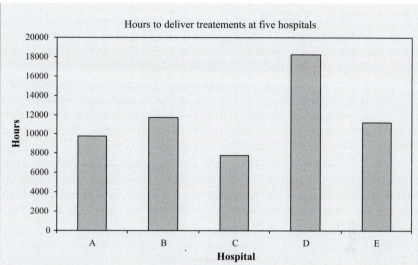

CHART 1   Hours to provide treatments (not normalized).

Chart 2 is a bar chart showing the effect of normalizing for the amount of work at each hospital. Hospital C, which had the lowest hours, is the highest hours per unit weighting, while E is now the lowest.

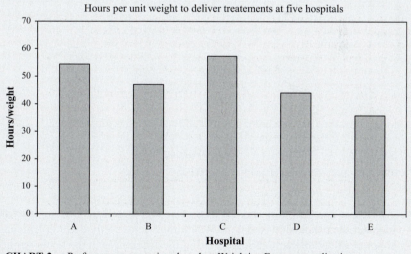

CHART 2   Performance comparison based on Weighting Factor normalization.

7.2.4.4 *Validating a Weighting Factor*

We have explained the theory and application of a Weighting Factor approach to normalizing data. It is important to validate the application and there are a variety of checks that can be used to do so. Validation is discussed in terms of manpower, but similar methods can be used for validating WFs for costs or anything else.

Validation is most easily carried out with reference to a simple scatter diagram of hours versus weighting as shown in Figure 7.2. The interpretation of the chart is straightforward. Each point on the chart shows the manpower hours and TWF of one participant. The added regression line depicts the average performance of all participants, i.e. the study average. If all facilities were equally efficient with regard to hours then all the data points would lie on the regression line. Facilities above the regression line are expending more hours than the study average, while those below the line are expending fewer. Four validation checks that can be made are:

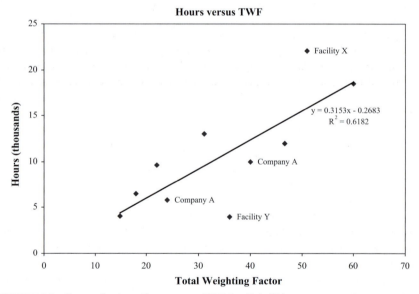

FIGURE 7.2   Scatter diagram of hours versus TWF for 10 facilities.

1. We expect that the hours and TWF increase together. If they do not increase together the conclusion is that the Weighting Factor method as implemented is not an effective method of accounting for differences in the hours worked between participants. The strength of this relationship can be objectively measured by calculating the statistic r^2. The detailed understanding and interpretation of r^2 is beyond the scope of this book and the reader is encouraged to consult a statistician for more information.

2. In general, we expect that doubling the TWF will double the manpower hours. If this is not the case we should be able to explain, or may need to investigate, why this is not so. For example, economies of scale will result in less than doubling the manpower hours. In some operational situations there is a (usually legal) minimum manning level regardless of the amount of equipment being operated. In these situations there is a point below which the man-hours remains the same regardless of TWF.

3. If there are two or more facilities operated by the same organization in the study that organization should be able to confirm that:

   - the relative TWFs are reasonable and
   - the relative performance as indicated by the chart is reasonable.

   For example, in Figure 7.2, Company A has two facilities. They have different TWFs and both appear at approximately the same distance below the regression line. This would be the expected result if, for example, both facilities had similar management styles and operational environments.

4. There may anecdotal evidence either from site visits or participants' own perceptions concerning their relative performance. For example, facility X lies well above the regression line implying that it is a relatively poor performing participant. Similarly, facility Y is below the regression line implying that it is a relatively high performing participant. Personnel at or visitors to these facilities may be able to confirm the general conclusions of the chart.

7.2.4.5 Applications for the Weighting Factor

From a benchmarking point of view the main application of Weighting Factors is to enable the comparison of benchmarking participants carrying out different portfolios of broadly similar types of work.

We have discussed the Weighting Factor in terms of operations, maintenance and providing treatments in a hospital. The same concepts and methodology apply to any benchmarking subject which is made up of basic building blocks of work and where work scope varies widely in size and configuration.

The method can also be applied to work scopes that do not currently exist but are being planned or considered or to answer 'what if' questions *based on actual current data*. For example:

- Once a new production facility, or expansion to an existing facility, has been planned, the TWF for the new operation can be calculated and the cost and hours required to operate and maintain it estimated.
- Two organizations are considering a merger. Knowing the Weighting Factors, and the effects of other considerations such as economies of scale, the effect of the merger on costs and hours to operate the new organization can be estimated.

ⓘ This section is a generalization of the Juran Institute Complexity Factor methodology and has been patented by them for use in the Oil and Gas industry. Further information can be found at their website www.juran.com/benchmarking_approach2.asp.

7.2.5 Modelling: How to Deal with Constantly Varying Metrics

In some situations the metric of interest is not stable; it varies, perhaps continuously and sometimes predictably. A typical example is the journey delays on buses, trains and other forms of public transport. Delays may vary depending on, for example, time of day and distance travelled before measuring the delay. Figure 7.3 is a scatter diagram of the delay on the vertical axis against the scheduled arrival time (to the nearest hour) along the horizontal axis. The data are drawn from industry wide sources and are not overly influenced by any one carrier. Many arrivals are on time and are plotted with delay zero. It can be seen that delays are shortest during the night and longest during the rush hours at the beginning and end of the working day.

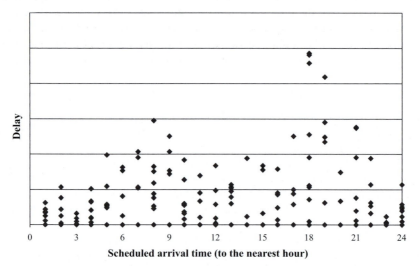

FIGURE 7.3   Transport arrival delays.

Any comparison of delays between carriers is likely to be overshadowed by the scheduled arrival time. It is possible to categorize the data into, for example, night, rush hour and day arrivals and compare delays from different participants for each category, however, the transition from one category to the next is gradual and defining appropriate category boundaries will be difficult.

Fortunately, there is an alternative. It is possible to rank each delay against other delays in the same time period and then combine all the ranks. There are several methods of doing this, and a simple one would be to calculate the delay in terms of the percentile.

Example: Transport Arrival Delays

An arrival is scheduled for 09.40 and is 12 minutes late. Ignoring the minutes, the scheduled arrival time is in the ninth hour. The number of minutes delay experienced by arrivals scheduled in the ninth hour in ascending order were: 0, 0, 0, 0, 2, 5, 7, 12, 16, 19. The 12 minutes late arrival is ranked 8th and so can be assigned a value of 8 (or equivalently, 80% as it was 8th out of 10).

There are alternative comparison methods other than the ranking method used in the example. A statistically better, but more complicated method is to calculate the delay in terms of standard deviations away from the average. Another alternative is to record the scheduled and actual arrival time and develop a formula to estimate the expected delay for a given arrival time, i.e.

$$\text{Delay} = \text{function (scheduled arrival time)}$$

and then calculate the deviation of each arrival from the expected delay in terms of standard deviations.

ⓘ These and other statistical methods are beyond the scope of this book and the reader is recommended to enlist the help of a statistician to ensure that both the normalization and analysis of the normalized data are valid.

7.2.6 Scoring: Business Excellence Models

Scoring is a method of combining disparate types of measure (for example accident rates and profit) to give a score that is comparable between participants. In benchmarking terms this is not very useful because the main aim of benchmarking is to learn from others how they achieve superior performance levels, which implies that the activities being benchmarked are in some way similar. However, managers often look for one overall figure that summarizes how their performance compares with that of other participants.

Two well established highly respected scoring methods are the Baldridge Award in the USA and the European Foundation of Quality Management (EFQM) Excellence model. The methodology behind these and other models is simple. Business activities are split into categories (see the example below). Each category is independently assessed and scored against set criteria. The scores for each criterion are added to give a total score for each category and the scores for each category are added to give a score for the organization as a whole.

The scoring mechanism is the same for all organizations and so two highly disparate organizations can be benchmarked by comparing their total score and/or the score in each of the separate categories.

Example: Scoring Using a Business Excellence Model

Participants A, B, and C are assessed using the Baldridge model. Their results are given in the table, along with the maximum number of points available for each area.

Category	Score			Maximum points available
	A	B	C	
Leadership	85	73	105	120
Strategic planning	62	58	43	85
Customer and market focus	65	60	69	85
Measurement, analysis and knowledge management	70	51	62	90
Human resource focus	41	53	60	85
Process management	53	49	57	85
Results	298	378	312	450
Total Score	**674**	**722**	**708**	**1000**

Analysing the scores for each individual category allows us to benchmark each category, and comparing the total scores allows comparison of the overall performance.

ⓘ For further information on the Baldridge Award, EFQM and Dubai Quality Award Excellence Model ⇨ bibliography and Chapter 2

Most benchmarking studies do not have a scoring mechanism like the Excellence Models. However, it is possible to develop a scoring method for most if not all situations. As an example, consider five participants in a study with three metrics: incidents per million hours exposure, tonnes produced and environmental emissions. The raw values as might be reported in a benchmarking study are given in Figure 7.4, which includes the average for each metric.

	A	B	C	D	E	Average
Safety incidents per million hours exposure	2	6	5	4	8	5
Tonnes produced per day	80	105	95	120	100	100
Emissions per tonne	40	20	30	25	15	26

FIGURE 7.4 Normalization by scoring: table of raw data.

The score for each participant can be calculated as the percentage difference from the average:

$$\text{Score} = \frac{100(\text{average} - \text{value})}{\text{average}}.$$

$$\text{For safety, participant A this is} : \frac{100(5 - 2)}{5} = 60$$

$$\text{For emissions, participant A Scores} = \frac{100(26 - 40)}{26} = -54$$

Similarly we can calculate the score for tonnes produced per day. However, because high values are preferable, the formula used is:

$$\text{Score} = \frac{100(\text{value} - \text{average})}{\text{average}} = \frac{100(80 - 100)}{100} = -20$$

However, the three metrics may be considered not to be of equal importance, and we may wish to assign factors to take this into account. In this example we assign the factors 2, 1, and 0.5 respectively to each of the three metrics. Each participant's score is multiplied by the appropriate factor and added together to give a total. For example, the score for safety incidents for participant A is 60. This is multiplied by the factor 2 to give a total of 120. The full table of results is given in Figure 7.5. The factors used imply that safety is twice as important as tonnes produced which in turn is twice as important as emissions. Note that what is important is the relative size of factors, not the actual numbers. Therefore, we would draw the same conclusions whether the factors were 2, 1, 0.5 or, for example, 100, 50, 25.

		Weighted score				
	Factor	A	B	C	D	E
Safety incidents per million hours exposure	2	120	-40	0	40	-120
Tonnes produced per day	1	-20	5	-5	20	0
Emissions per tonne	0.5	-27	12	-8	2	21
Total		73	-23	-13	62	-99

FIGURE 7.5   Normalization by scoring: weighted scores.

An average performance would result in a score of zero, a positive score results from a better than average performance, and a negative score indicates a poor performance. In Figure 7.5, participant A has a very good incident performance, scoring 120, a poor cost per tonne performance scoring $1(-20) = -20$ and a worse than average environmental emissions score of $0.5(-54) = -27$.

The resulting Total of the factored scores provides an overall comparison between participants. Participant A, with a positive total of $120 - 20 - 27 = 73$ is the best performer, participant E with a negative score of -99 is the worse performer, and participant C, with a small negative score is just worse than average.

There are two main decisions to be made with this type of scoring method.

1. How to calculate a score, and
2. Selecting Factors

7.2.6.1 *Calculating a score*

In the above example, we chose to standardize against the average. Equally it would be possible to standardize against the best or worst performance or some other value, for example a theoretical best performance, such as zero for safety incidents.

Having standardized, we also chose to calculate the score as the percentage difference between the average and each individual performance. This is the simplest, but probably not the most appropriate method. There are a number of possible disadvantages to this method including:

x   It assumes that a difference of 1 unit is of different importance depending on the average value. For example, if the average number of safety incidents is 5, and participant A reports 2 incidents, the score would be calculated as $100(5 - 2)/5 = 60$. If the average were 50 and A incurred 47 incidents, the difference is still 3 incidents, but the score is now $100(50 - 47) = 6$. In some situations this may be appropriate, while in others it may not.

✗   It assumes the units are linear. For example that a difference of 2 units is twice as important as a difference of 1 unit. While this may often be appropriate, in other circumstances it may not. For example, consider pollutants in discharge to sea. It may be legally allowable and environmentally acceptable to discharge up to 5 parts per million (ppm) of a pollutant in water discharge, therefore a difference from 1 to 3 ppm or 2 to 4 ppm is not as important as a difference between 4 and 6 ppm.

Other relatively simple methods of scoring include:

- Calculating the number of standard deviations of each participants' value from the average, and using this as the score.
- Calculating the percentile (or rank) of each participant's value, and using this as the score.

7.2.6.2 *Selecting factors*

The factors may be chosen at will to reflect the relative importance of different metrics in the study. In the example above safety, incidents were considered twice as important as production rates which in turn were considered twice as important as emissions. While this allows less important aspects of the study to have less influence on the overall score, the disadvantage is that the factors can be manipulated to favour any particular participant.

   It is sometimes possible and helpful to consider factors in monetary terms. Using the above example, if a tonne of product is worth $1000, one accident may be deemed to be valued at $100 000 and one unit of emission $50, factors of 1000, 100 000, and 50 may be used.

   ☜ The selection of factors can be difficult and lead to endless discussion amongst participants.

   ⓘ The scoring methodology has been illustrated using very simple examples. A statistician should be consulted for more sophisticated situations.

7.3 WHICH METHOD IS APPROPRIATE: CATEGORIZATION, FACTORS OR MODELLING? EXAMPLE: TREATING CANCER

Of the six normalization methods discussed in this chapter 'per unit', Selection and Scoring have reasonably well defined, specific roles. However, selecting the appropriate method amongst the other three is not always so clear and we need to identify and consider the advantages and disadvantages of each. In yet other situations we may choose to use a combination of these three methods. To illustrate the use of categorization, factors and modelling this section presents a brief simplified medical example and summarizes the attributes of each method.

Example: Choosing Between Categories, Weighting Factors and Modelling: Cancer Treatments

Consider the benchmarking of the survival rates from cancer. Some cancers, such as bladder cancer, have a relatively very good prognosis with over two thirds of patients surviving for at least five years. Conversely, lung cancer has a very poor prognosis with survival rates generally below 20%. It would therefore be inappropriate to compare treatment centres with different mixes of different cancers simply by comparing the survival rates.

We could categorize the cancers into type, but survival rates are also known to vary between men and women, for example, stomach cancer survival rates for men being lower than for women. To account for this we could further categorize into male and female, only to discover that survival rates also depend on the age of the patient at diagnosis, amongst other factors.

Cancer type	Gender	Age range		
		15–39	40–49	etc.
Bladder	Male	98	. . .	
	Female	78	. . .	
Lung	Male	. . .		
	Female	. . .		
etc.	. . .			

Categorized cancer survival rates (percent surviving 5 years).

While it would provide useful information to benchmark categories in this manner, there are practical difficulties:

✗ There may be too little data (i.e. too few patients) in a category to draw meaningful conclusions.

✗ It assumes minimal variation within groups compared to between groups. In this example age is a continuous variable that we force fit into groups. If, for example, an age category is 40–49 years and we assume that the prognosis for woman aged 40 is not significantly different from a woman aged 49. If the survival rates of ovarian cancer were:

55% for a 40–49-year old woman,

44% for a 50–59-year old woman and

32% for a 60–69-year old woman

a women of 51 would have a survival rate of about 48%, whereas as 59-year-old women would have a survival rate of only about 39% (see the chart below), but by categorizing they are both deemed to have a survival rate of 44%.

A simple regressing model estimating average survival rate from age, as in the chart, would overcome the arbitrary categorizing of the continuous variable "age" and provide a more accurate standard to measure against. In addition, because we are no longer categorizing by age the remaining categories will contain more data and so we will be better able to draw statistical conclusions.

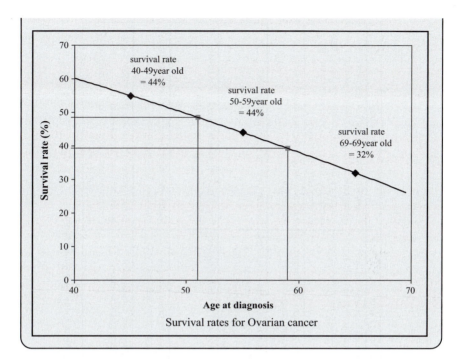

Survival rates for Ovarian cancer

Figure 7.6 summarizes some of the key aspects of using categories, Weight Factors and Models.

7.4 REPORTING INFREQUENT ACTIVITIES

In addition to metrics reflecting day-to-day organizational activities, there may be infrequent regular activities. One such common activity of this nature is major planned maintenance that occurs every few years.

Example: Infrequent Activities: Maintenance

In a maintenance benchmarking study it is required to supply costs and worked hours relating to maintenance activities in the previous year.

In addition to breakdown and on-going planned maintenance, Once In A While Manufacturing Ltd. carried out a major planned maintenance routine during the year taking major packages of equipment off line for 3 weeks. They feel that if they include these costs and hours in their data submission their figures will be higher than usual and unrepresentative of normal operations. They argue that if the data had been submitted in a previous or following year their worked hours and costs would be much lower.

	Categorization	Weighting Factor	Modelling
Application	Good where natural discreet categories exist, AND where there are enough data in each category to make meaningful comparisons.	Useful where benchmarked units consist of "building blocks" such as units of equipment, and we want to benchmark, for example, costs per unit.	Appropriate where the underlying metric is continuous and does not fall into natural categories. Examples include time, age, lengths, numbers of items (e.g. passengers) there the numbers can vary widely.
Ease of use	Easy	Usually difficult. May require an industry specialist and a statistician to develop a Weighting Factor system.	Difficult – usually requires experienced statistician.
Cost	Low	High – both to develop the weight factor system and assess the weights for work scopes. Participants will need to be convinced of the appropriateness of the system.	Usually high – to develop the model.
Risk	There may be too few data in each category to draw meaningful conclusions. In these cases statistical advice may be needed.	Few – as long as the system is well researched and developed.	Few. Modeling should provide an accurate, traceable and objective solution.
Benefits	Quick, cheap, simple to understand.	Allows comparisons to be made between participants with disparate amounts and types of activity.	Makes maximum use of the information in data compared to categorization.
Examples	Product type (e.g. aircraft model), types of disease, mode of transport.	Operation and maintenance of equipment. Construction or other projects, medical treatments.	Public transport delays in arrivals where delays are a function of arrival time. Cost per patient where the cost is a function of disease, patient profile etc.

Figure 7.6 Comparison of categorization, Weighting Factors and modelling methods for normalization.

In some situations there may be a number of items maintained every few years, but on different cycles. For example there may be:

- Production line overhaul carried out every 3 years.
- Compressor maintenance carried out every 4 years.
- Refurbishment every 10 years.

These types of activities need to be treated carefully because if the data are reported for a time period in which these activities were not carried out, the effort expended on them will be excluded from the study. On the other hand if these activities were carried out within the time period being reported, the activity level will be unusually high. A typical profile of organizations with this type of regular but infrequent activity is shown in Figure 7.7.

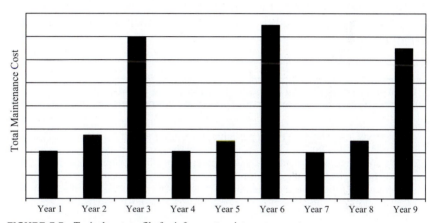

FIGURE 7.7   Typical cost profile for infrequent maintenance events.

In the chart, major overhaul maintenance is carried out every three years resulting in very high maintenance costs in years 3, 6, and 9 and lower costs in the other years. If the benchmarking study were to take place in years 1, 2, 4, 5, 7, or 8 the maintenance cost reported would be much lower than if it were carried out in years 3, 6, or 9.

There are various solutions to this issue, the most common, and usually preferred, being to allocate the costs evenly over the cycle. In the above example, a third of the overhaul maintenance cost would be allocated to each year.

In the above example it would be possible for each participant to report the annualized maintenance as one maintenance item, called for example, 'overhaul maintenance'. Alternatively, the details of each overhaul activity could be reported separately, thus allowing detailed benchmarking of overhaul activities.

In practice there are likely to be many items maintained less frequently than annually and reporting each one separately may result in a very long list. It is likely that while the maintenance details of major equipment are useful, details of low cost minor maintenance activities is not. There are several simple alternative methods of reporting them including:

- Grouping them together and reporting them as one maintenance activity (e.g. an organization maintains 100 valves every 4th year, i.e. all 100 valves are maintained in one year with no planned valve maintenance in the intervening years). This can be reported as one task: '4-yearly valve maintenance', rather than one set of data for each valve.
- Where the maintenance activity is spread more or less evenly every year, (e.g. 100 valves maintained on a 4-year cycle, 25 of which are maintained every year) the data can be included in the normal maintenance activities instead of being reported as a separate item.

Though infrequent activities are probably most prevalent with maintenance, it may occur in other areas and similar methods can be used for reporting them.

Example: Benchmarking Infrequent Maintenance Events

In one study it was noticed that a participant was carrying out fewer overhauls than other participants. During a Best Practice Forum they revealed that they had carried out a controlled experiment on the equipment. They had gradually extended the periods between overhauls from the recommended annual maintenance, carefully examining the effects on the equipment. They finally concluded that the 'annual' maintenance could be carried out every 6 years with minimal negative impact on the equipment.

SUMMARY

In this chapter we have explored the need for normalizing data to ensure that data comparisons are appropriate.

The six methods of normalizing discussed here are:

✓ **Per unit**, often thought of as 'rates' or 'frequency'
  Use: where the metric being reported is directly proportional to the 'volume' of activity

✓ **Categorization**, in which we group data into categories
  Use: Where natural categories exist and the metric being reported is known to vary between categories

✓ **Selection**, most common in project benchmarking, where we select specific projects that we deem to be similar.
   Use: Where items being reported on can be viewed as projects and each project is different.
   Example: benchmarking of IT or construction projects.

✓ **Weighting Factors**, where work packages are allocated a weighting depending on relative work load of each package.
   Use: Where the activities being benchmarked consist of different quantities of common packages.

✓ **Modelling**, where performances of participants are compared to a mathematical model developed from the data.
   Use: Where the independent variable is continuous (e.g. patient age). Where there is a choice, this is a preferred normalization method to categorization.

✓ **Scoring**, where performance in different metrics are weighted and summed.
   Use: To calculate totals for completely disparate metrics.

These methods are often used in combination to ensure that appropriate comparisons are made.

A comparison of categorization, Weighting Factor and modelling is given in Figure 7.6.

ϒ ACTIVITIES

? As metrics are developed for your benchmarking study, consider whether they need normalizing. If they do, which method or methods will be most appropriate?

Chapter 8

How to Identify Benchmarking Metrics

INTRODUCTION

The subject that often causes the most interest with regard to benchmarking is what to measure.

In the previous chapter we discussed methods of normalizing data, and those methods need to be considered when developing a list of potential metrics for a benchmarking study. In this chapter we discuss methods of identifying potential metrics, and in the next we discuss how to review the list and hone it to an effective manageable list of metrics that will meet the objectives of the study.

There are a variety of methods we can use to help us identify appropriate metrics and we discuss the following:

1. The scope and objectives of the study.
2. Review of currently reported metrics.
3. Analysis of organizational structures.
4. "Plan versus Actual" types of metric.
5. Aspirational statements: mission, vision, purpose, values and others.
6. Critical Success Factors and Key Performance Indicators.
7. The Balanced Scorecard.
8. Analysis of customer needs.
9. Analysis of process flow charts.
10. Cause–Effect analysis.
11. SWOT, cost of poor quality, and similar analyses.

Many of these methods are related to one another.

In general, we suggest that combinations of methods be used depending on the nature of the study.

DELIVERABLE

The deliverable of this step of the benchmarking process is a list of potential metrics, which will be pruned and finalized later in the process.

8.1 SCOPE AND OBJECTIVES OF THE STUDY

The project charter, and in particular the scope and objectives of the study, will always be a guiding document for the whole benchmarking project, and the metrics should:

- ✓ Be within the scope of the project
- ✓ Help meet the objectives.

If any metrics do not satisfy these tests we should consider:

- Removing the metric (which is the most likely choice) or
- Changing the objectives or scope or both.

Example: Revising the Scope: Information Technology

An organization was concerned about the cost and performance of its internal IT department. A benchmarking study was proposed with a scope that included all IT activities.

When the project team began to develop draft metrics for the study they realized that the project work carried out by the IT department would be very difficult to measure and benchmark. As the cost and effort expended on these project activities was only 5–10% of the IT budget, they proposed to refine the scope of the study to omit projects.

As in the above example, it is common to exclude some aspects from a benchmarking study, projects being a common example. Modifications to the scope can be made at any stage of the project, though the usual times are when the team reviews the charter, during metric development and while finalizing the details of the project with participants.

Other typical exclusions include:

- Items that the organization will not or cannot alter such as rent and rates.
- Items that are commercially sensitive amongst the participants.

- Items that would contravene law. Depending on the location, this might include such items as contract rates.
- Items where the effort of benchmarking is likely to outweigh the benefits.
- Items which are unique or nearly unique.

Example: Revising the Scope: Oil Well Drilling

In the oil and gas drilling industry it is usual for the drilling rigs to be hired. However, during the metric development stage of a drilling benchmarking study, participants chose not to share rig rates, even though rates are readily accessible and in the public domain.

8.2 REVIEW OF CURRENTLY REPORTED METRICS

Probably the easiest, and certainly an important, method of identifying potential metrics for a benchmarking study is to review currently reported metrics.

All organizations collect data. As a minimum, data will exist for managing the finances of the organization. These metrics are likely to include sales value and quantities, purchases, salaries and other expenses. Most organizations will also collect time, quality, safety and environmental data such as production rates, error rates, losses and accident rates to name but a few. These metrics are a good starting point for collating lists of potential metrics, as in general they are:

✓ Useful (otherwise we should stop collecting them)
✓ Available

These metrics can usually be found by:

- Reviewing reports.
- Looking at notice boards.
- Speaking with people to find out what data they use to help them with their job, why they use it and how they use it.

✍ Developing a list of metrics for benchmarking purposes is a good opportunity to review the metrics that are currently being used to manage the organization. All too often data are being collected and reported because somebody requested it years ago, no doubt for a good reason. The reason may have been forgotten, but the data are still collected. Benchmarking offers a good opportunity to review and improve internal management and operational data needs.

Case Study: Identifying Wasted Effort

While developing the metrics for a benchmarking study a team member decided to review some of the reports being circulated in the organization. She started with a quarterly review report and asked who received copies of it. She then interviewed a number of the recipients with the intention of finding out what people used it for, what data they needed and why.

She discovered that only about a fifth of the people receiving the report used it. Many never read it. Of those that did, most were only interested in a small portion of the report. Some of the information in the report was not referred to by anyone, but others had to carry out their own analysis on data in the report to find out what they wanted to know.

As a result, the size of the report was reduced by about 50%, the circulation reduced by 80% redundant data and information replaced by data and information that would help those receiving it.

Some of the data items were proposed and used in the benchmarking study.

8.3 ANALYSIS OF ORGANIZATIONAL STRUCTURES

The analysis of organizational structures is particularly useful where a whole facility, group, project or similar entity is being benchmarked. Probably the most common examples are organization charts and budget/finance reporting structures. To illustrate the method consider the organizational structure for a typical manufacturing facility as shown in Figure 8.1.

Typically, costs and/or hours or staff numbers can be reported against each element in the structure and sub totals and totals be reported.

Most organizations will use such structures for budgets, safety, and error reporting amongst others. For example, safety incidents may be subdivided into parts of the body injured, cause of incident etc for investigative purposes.

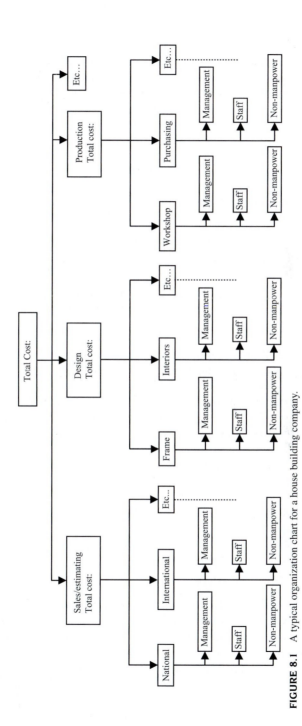

FIGURE 8.1   A typical organization chart for a house building company.

8.4 PLAN VERSUS ACTUAL

One common method of measuring performance is to compare actual performance with plan. Unfortunately it is usually assumed that meeting or beating
the plan or 'target' is 'good', particularly when considering project benchmarking.

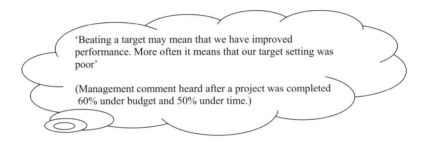

'Beating a target may mean that we have improved
performance. More often it means that our target setting was
poor'

(Management comment heard after a project was completed
60% under budget and 50% under time.)

However, 'beating' the plan, though it does not usually have such a serious
implication as not meeting the plan, is not ideal. The aim should be to develop
good plans that we are able to meet. The main cost of over-budgeting is
that resources are set aside and not used which could have been invested
elsewhere.

The usual metric for comparing actual against plan is:

$$\text{variation} = \text{actual} - \text{plan}$$

which for comparison purposes is often normalized as

$$\text{variation} = \frac{(\text{actual} - \text{plan})}{\text{plan}}$$

and can be multiplied by 100 to turn it into a percentage.

When benchmarking groups of projects from different participants, it is
necessary to carry out more analysis than simply comparing average variation.
In these situations it may be necessary to report plan and actual for each project
(or budget) in order to investigate, for example, whether the percentage variation
increases with budget size, or project type, etc. ⇨ Appendix 1 for an example

8.5 ASPIRATIONAL STATEMENTS: MISSION, VISION, AIM, VALUES AND OTHERS

Many organizations develop statements summarizing their aim, purpose, vision, mission, values, goals, etc. These statements are very useful for identifying both potential benchmarking projects and metrics since they encapsulate key aspirations of the organization. We will consider some typical examples to demonstrate the methodology of developing metrics from these types of statement.

In the following text we have avoided giving specific definitions of these terms, as the intention is to explain how to generate potential metrics.

These and many other types of statement can be used to help develop metrics by defining each statement more fully. The idea behind the method is simple: each actual or implied intention expressed by the statement should be measurable. Not only should it be measurable, but also, if the extent to which these statements are being realized is not being measured and monitored, how does the organization know whether they are being successful in achieving them?

A **Mission** statement:
Is a brief description of the purpose of an organization

Example:
The mission of Merck is to provide society with superior products and services by developing innovations and solutions that improve the quality of life and satisfy customer needs, and to provide employees with meaningful work and advancement opportunities, and investors with a superior rate of return.

A **Vision** statement:
describes where an organization wants to be at some time in the future.

Example:
Amazon, the internet book retailer has a vision statement: 'Our vision is to be earth's most customer centric company; to build a place where people can come to find and discover anything they might want to buy online.'

Value statements:
describe what qualities the organization value.

Example:
Barclay's bank has a number of value statements. The first part of their legal value statement is: "All our activities will be conducted in a manner fully compliant with the relevant law, our licence obligations (including FSA) and any other regulations that may apply. We expect our suppliers to also uphold these standards."

The **Aim** of an organization explains what it is trying to achieve.

Example:
Robert Gordon's College in Aberdeen, Scotland has the following aims:
- To encourage pupils to develop their *individual talents* to the best of their ability both inside and outside the classroom.
- To develop a strong partnership with parents by involving them in and informing them about the life and work of the College.
- To foster a recognition of the merits of hard work to prepare pupils for their future careers.
- etc.

Where these types of statement have not been developed for the subject being benchmarked, it can be very useful for the participants of the study to develop and agree a statement that they feel summarizes the aim of the part of the organization being benchmarked.

The example below demonstrates how these statements can be analysed to develop metrics.

Example: Analysing a Mission Statement to Develop Metrics for a Passenger Railway

A public transportation mission statement could be:

'**To transport people, to schedule, safely, securely, comfortably, at minimum cost**'. This may be analysed in the following way (note that the interpretation is the author's):

- '**Transport**' is defined as the movement of people, i.e. relates to the train journey only and specifically excludes other aspects of the service such as the booking and railway station activities.
- '**People**' implies that freight is excluded from this mission.
- '**To schedule**' means that trains are intended to arrive at stations according to the time schedule.

  There are a number of options for measuring timeliness, including, for example:

  - Percentage of times a train arrives late at its final destination (or all destinations including intermediate destinations where passengers alight or board the train.).
  - Variation in minutes between planned and actual arrival times.
  - Percentage of trains cancelled.

- '**Safely**' means with minimum accidents. This could be measured by counting the number of deaths, hospital admissions, first aid cases, crashes etc. Much of this data will already be available due to legal reporting requirements, and the method of gathering the data is likely to already be established.
- '**Securely**' means free from fear of, or actual, criminal activity. As with safety data, there will usually be a formal record of crimes against passengers. This could be supplemented by customer surveys. In addition to the number of incidents, data could be categorized, for example by:

  - robberies
  - assaults
  - sexual offences
    and each of these may be further subdivided.

- '**Comfortably**' means that the customer finds the journey comfortable and will be defined principally in terms of customer perception which is usually ascertained through customer surveys. 'Comfortably' will need to be defined, for example as 'clean carriages, comfortable seats, sufficient baggage storage, sufficient clean toilet facilities', etc.
- '**Minimum cost**' could refer to the cost to the customer and/or the operator, and is likely to be expressed as cost per kilometre travelled. For the customer, the cost may need to be categorized by type of ticket (e.g. walk on, concession, class of travel, time of travel, length of journey). The operations cost may also need to be categorized (e.g. suburban, long-distance), and we may choose to further subdivide the costs into, for example: train crew, station, rolling stock, maintenance, overheads, etc. These categories can often be deduced from the budgets items.

The key when analysing aspirational statements is keep asking the question 'what do we mean by. . .' until the answer to the question is measurable.

8.6 CRITICAL SUCCESS FACTORS AND KEY PERFORMANCE INDICATORS

> **The Critical Success Factors (CSFs)** are:
> the relatively few **factors**
> in which it is **critical** that an organization **succeed**
> in order to realize its aspirations.

The aspirational statement encapsulates what an organization wants to be or do or achieve. The Critical Success Factors, CSFs, describe the areas in which the organization must perform extremely well in order to fulfil its aspirational statement. CSFs can be developed by asking what must the organization do right in order to survive/succeed.

For example, distribution is critical for industries such as food and clothing or selling in bulk to the public. Similarly, security is critical for financial institutions and the nuclear industry.

While CSFs are useful stepping stones, of themselves they do little more than highlight key areas in which the organization must succeed. The obvious next question is how the organization knows how well it is performing with respect to these factors. To answer this question the CSFs are used to develop what are frequently called Key Performance Indicators (KPIs). The KPIs are metrics that tell us how we are performing against the CSFs and these KPIs are obvious candidates for benchmarking metrics.

> **Key Performance Indicators (KPIs)** are:
> The **key** metrics that
> **indicate** how well an organization is **performing**
> against its Critical Success Factors

The method of development of the KPIs is very similar to the method for developing metrics from aspirational statements. If a Critical Success Factor for the nuclear industry is containment of radioactive material, and we ask how we know how well this is being achieved, one typical metric may be the Number of Reportable Leakages of radioactive material.

8.7 THE BALANCED SCORECARD

Example: The Purpose of the Balanced Scorecard for a Call Centre

The manager of a call centre was required to cut costs by 10%. To do this he cut agent numbers, kept wages low and minimized training and other non-productive costs. He achieved his target.

Some time later his manager noticed that the number of complaints about the call centre seemed to have risen and decided that he should call the manager in and discuss it with him. The result was a target to reduce complaints by 10%.

To achieve this, the manager analysed the complaints data and realized that a large number of complaints were due to callers being kept on hold. To resolve this issue he recruited more staff.

He also noticed that agents often gave incorrect or incomplete answers or had to ask for advice. He resolved this problem primarily by increasing training.
Again he achieved his target but the costs began to rise.

This example illustrates, albeit somewhat simplistically, a management style that is familiar to many of us. In more complex systems the same management style may exist, but the effect is subtle:

Example: The Purpose of the Balanced Scorecard for a Police Force

The government realizes that the electorate are concerned about drug abuse. The poilce are told to focus on drugs crime. The police focus on drug use and drug related crime. The number of drug seizures and arrests increase. The government can demonstrate it is reponding to people's concerns, and its popularity increases.

However, other crime rates increase as police move resources to focus on drugs. In addition, because of seizures, the price of drugs increases which makes the illegal drugs trade more lucrative and encourages more people to become suppliers, once again increasing the supply. However, by now the spotlight has moved on as the police have been told to focus on another type of criminal activity.

These two cases illustrate the result of focusing on one, or even a few, organizational aspects at the expense of others. We need a method of identifying and monitoring all key aspects of an organization at the same time, in much the same way as when driving a car we focus on speed and direction, fuel gauge, traffic lights, road conditions and other road users amongst others.

This problem is not new. In 1996 Kaplan and Norton brought to our attention a method of overcoming the single-issue focus. It is a method that shows us how to identify and monitor all key aspects of our organizations at the same time. They called it the Balanced Scorecard.

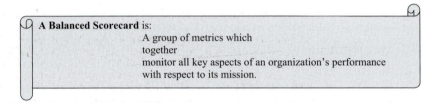

A Balanced Scorecard is:
A group of metrics which
together
monitor all key aspects of an organization's performance
with respect to its mission.

There are other definitions of a Balanced Scorecard, but this captures the key ideas, which are:

- No one metric can monitor an organization's overall performance: we need a group that *together* monitor performance
- We are mainly interested in monitoring *key* aspects of the organization: we would be unable to monitor every aspect – there would be too many metrics.
- The key aspects of an organization are determined by its *mission* (or other statement of intent)

The Balanced Scorecard can be considered as an extension of the CSF and KPI method, with the added requirement that they reflect all aspects of the organization under consideration.

From the point of view of the Balanced Scorecard there are four perspectives that need to be considered:

1. Financial (the investor's/owner's perspective).
2. Customer perspective.
3. Internal business process (manager's and process owner's perspective).
4. Learning and growth (employee's perspective).

Without considering each of these perspectives any conclusion about the organization's performance will be incomplete.

Financial aspect: traditionally many performance metrics focus on measuring financial results. Unfortunately these only summarize the economic consequences of past decisions and give few indications as to future performance. Typical measures include market share, sales value and volumes, profit.

Customer perspective: performance viewed from this perspective reflects the consequences of past decisions (e.g. customer retention) but also helps identify the future direction of the organization. For example, if customers value short lead times, then this will be a requirement for maintaining future customer satisfaction.

Internal business processes: focuses on the processes which enable an organization to deliver the products and services that will retain customers and other stakeholders. This seems similar to other approaches discussed earlier in the chapter; however the scorecard approach may identify new process that are currently not in place but are required to fulfil the organization's vision and strategy. Furthermore, while other approaches focus on current outputs (products/services), the Balanced Scorecard, because it is aligned with the organization's mission, encourages longer term thinking: anticipating and fulfilling future needs as well as meeting current needs. Typical metrics include production and error rates, internal delays, work in progress and those focused on product and service features.

Learning and growth: here performance is analysed in terms of technological and employee capability in order to meet the future needs. Typical measures include employee satisfaction and retention, training availability and take up. Technological metrics could include the timeliness, accuracy and usability of data/information delivered to appropriate people.

In the previous section we saw how to analyse an aspirational statement (mission, vision, goal, etc.) and develop from it a series of metrics. One way of perceiving the Balanced Scorecard is as an extension of that idea. However, the Balanced Scorecard methodology is perhaps more prescriptive in that it specifies four aspects that need to be considered. It is also specifically intended to:

1. Clarify and translate vision and strategy.
2. Communicate and link strategic objectives and measures.
3. Plan, set targets and align strategic initiatives.
4. Enhance strategic feedback and learning.

ⓘ There are many books written on the development and implementation of aspirational statements, Critical Success Factors, key performance indicators and the Balanced Scorecard. Kaplan and Norton's book, The Balanced Scorecard was published in 1996.

8.8 ANALYSIS OF CUSTOMER NEEDS

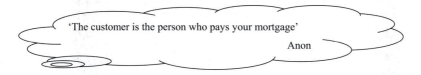

'The customer is the person who pays your mortgage'

Anon

The importance of the customer is difficult to overestimate. Without customers, it is said, there is no work, no job, and no money to pay the mortgage.

Customers may be external or internal to the organization. Some departments, or functional areas such as IT support, may exist primarily to serve internal customers. If the supply of goods and services between groups within an organization is poor it will be difficult for that organization to excel with the external delivery of goods and services.

Therefore, key aspects to benchmark are those of interest to the customer (Figure 8.2). Three important methods of doing so are:

1. Through customer surveys.
2. Through Service Level Agreements.
3. By monitoring customer activity.

FIGURE 8.2   Building up a picture of relationships with customers.

Customer surveys are most commonly used to ascertain the public's perception and experiences of an organization's products and/or services. Organizations often carry out their own surveys, for example hotel groups may contact recent guests to ask for their opinions. However, formal surveys are usually carried out by experienced, professional independent customer survey organizations to ensure independence and unbiased results. Surveys may, for example:

●   Be commissioned by a specific organization to benchmark against their competitors without the competitors' knowledge.
●   Be part of a benchmarking study in which the participants agree to a survey.

- Be carried out by independent consumer groups, magazines and other bodies to report on the customer perception of the subject being benchmarked.
- Be focused on external customers and/or internal customers. A good example of an internal customer survey is one looking at staff satisfaction.

Where an organization supplies a service to other groups (the customers), a Service Level Agreement (SLA) or similar contract is often drawn up between the supplier organization and the customer. The supplier and customer may be two groups within the same organization, or from different organizations. The concept of the SLA is simple: the supplier and the customer agree on the service to be provided and on the metrics to measure the degree of compliance. For example, the SLA between a supplier of IT services and their customer may include a statement that a computer system will be available 99.8% of the time. A useful source of metrics for benchmarking studies are the metrics specified in SLAs.

While measuring against the SLA indicates how well the written commitment is being adhered to, it does not indicate the customers' perception of the service. In many situations, especially where the public are the customer, there is no SLA.

We can measure customer behaviour by measuring such things as:

- Percentage retained customers, and by inference, lost customers.
- Market share.
- Numbers and types of complaints received from customers.

8.9 ANALYSIS OF PROCESS WORKFLOW

Many organizations have developed process flow charts which document their work procedures. These charts are usually accompanied by a textual description of the process and are generally much easier to understand than a textual explanation alone. They are frequently used for training and auditing against, and some organizations also use them for process analysis such as identifying potential sources of error, losses due to re-work and process improvement activities. In addition they can be used to facilitate the development of metrics.

Potential metrics can be deduced for (Figure 8.3):

- Activities or tasks, represented by rectangles:
  - Costs and times (elapsed and manpower effort) to complete the activity.
  - Measures of inputs and/or outputs including, as appropriate, dimensions, quantities, quality.
  - Complaints, errors or other problems occurring at the step.

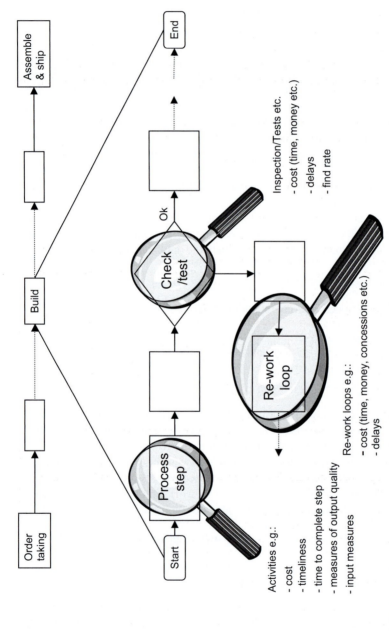

FIGURE 8.3   A flow chart is a good source of ideas for what to measure.

- Inspections/checks/tests, represented by diamonds:

  - Costs and time to carry out the activity.
  - Delays due to checking process including associated costs.
  - Reject rate.

- Rework loops, often represented by work flowing right to left or bottom to top (by convention flow charts are drawn with flow from left to right, and rework from right to left or with flow top to bottom and rework bottom to top):

  - Costs and times (elapsed and manpower effort) to carry out the re-work.
  - Delays and the associated cost of delays.

- Databases, reports or other documents:

  - Accuracy of information.
  - Completeness of information.
  - Time to access information.

Example: Flowchart Metrics for Oil Well Drilling

One petrochemical organization decided to re-design the complete process for constructing oil wells. The process had 12 major steps, each of which was further broken down into a detailed flow chart. In order to measure the effectiveness and efficiency of the process it was decided that each major step should have between 6 and 10 key metrics that would be used to monitor performance and manage the process.

Example: Hospital Records Retrieval

In a hospital, access to patient records is an important aspect of delivering patient care services. Many process flow charts have a step "retrieve patient records". Suitable metrics could be:

- Number of times record is lost.
- Length of time to retrieve record.
- Delay caused because record is unavailable when required.

8.10 CAUSE–EFFECT ANALYSIS

The use of cause–effect analysis to identify metrics is particularly useful when the purpose for the benchmarking study is to improve performance in a specific area. With this type of analysis the problem or effect must be clearly defined.

A list of possible causes is then generated either by brainstorming, analysis of data or any other means. There must be a logical link between the possible cause and the observed effect. For example, if in a hospital the observed effect is incorrect diagnoses, possible causes include poorly trained doctor or incorrect test results (see Figure 8.4). It is possible to go further back along the cause–effect chain and ask why the laboratory results were incorrect. One reason could be that the test equipment was faulty. We could keep asking 'why?' until we get back to a cause that either we can resolve ourselves or is beyond our control at this time. In this example, a budget cut caused maintenance to be reduced, resulting in faulty equipment, incorrect test results and ultimately incorrect diagnosis.

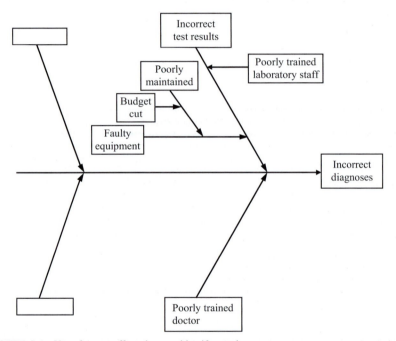

FIGURE 8.4   Use of cause–effect charts to identify metrics.

Having developed the cause–effect relationships and checked their logical connection, we can select as potential metrics any or all of the causes. In the example, these metrics would be:

- Percentage of incorrect diagnoses (perhaps categorized by diagnosis type).
- % of incorrect diagnosis due to incorrect test results.
- % of incorrect test results due to faulty equipment.
- % of incorrect test results due to poorly trained staff.
- % of incorrect test diagnosis due to poorly trained doctor, etc.

This method may lead to a large number of metrics and it may be necessary to benchmark only those that participants believe are the key causes.

8.11 SWOT, COPQ AND SIMILAR ANALYSIS

There are a number of models that organizations may use for analysing their business. Some, like Strengths, Weaknesses, Opportunities, Threats (SWOT) analysis are well established, while at the other extreme, some models may have been developed internally to meet specific organizational needs. As an example of how these tools can be used to generate potential metrics for benchmarking we will consider SWOT analysis.

SWOT is a widely applicable simple technique for analysing many different types of organizational or individual situations. Having identified the SWOT for a particular actual or possible situation the aim is to improve on the weaknesses until they become strengths and to try to eliminate the threats. Benchmarking is often an appropriate tool to use to help turn the weaknesses into strengths. The example below is a typical application of SWOT to business expansion.

Example: SWOT Analysis for Business Expansion

An organization wants to review its status within the industry and decides to carry out a SWOT analysis with the aim of taking appropriate action to improve its position. The team generate a list of attributes that they believe are key to business success, and use these to determine which are strengths, weaknesses, opportunities or threats. They collect information from a wide variety of sources and present the results in the traditional SWOT matrix.

Strengths	Weaknesses
Ownership of good patents	Lack of international presence and associated
Strong marketing expertise	market skills (e.g.:
Strong technical development team	- language, customs, law
Current operations located near good national	- offices, no established distribution network)
and international infrastructure	
Recognized as environmentally friendly,	Poor processes (high re-work, high cost per
safety conscious, and a preferred employer.	unit etc)
Opportunities	Threats
Developing markets (China, India, Latin	Poor brand reputation (high quality but poor
America)	reliability)
Opening up of trade/loosening regulations	Overseas competitors may initiate price war
Merger/JV opportunities	(based on cheap overseas labour)
No established market leader	Possible new legislation

The organization decides to use benchmarking as a tool to address their poor processes and reputation. Metrics will include as a minimum the following areas:

- cost per unit produced (preferable categorized by key factors)
- re-work rate (preferably categorized by cause)
- brand reputation (by developing a customer questionnaire to identify specific weaknesses as perceived by the customer)

Having carried out a comparison of data, the organization wants to learn from others how they can achieve lower costs, less re-work and build customer reputation.

SWOT analysis is applicable in many areas of business such as decision-making (e.g. whether to expand, move into a new market or develop a new product), process management, projects or even for individuals. The key to applying SWOT analysis is that it may be possible to identify the strengths and weaknesses of the current situation and the opportunities and threats for the future.

A Cost of Poor Quality (COPQ) study is carried out by identifying and classifying the losses and costs within an organization or part of an organization. The losses may be due to errors, failures, rejects, rework, contingencies or anything else resulting from processes or practices that are not efficient and effective. Such analyses can identify specific metrics for a study and can also be used to identify benchmarking projects.

Example: Cost of Poor Quality for a Mail Order Company

A mail order company has initiated a benchmarking study to improve the picking and dispatch of items from its warehouse. A COPQ study was carried out to identify specific weaknesses in the system on which the benchmarking team was to focus.

Whether SWOT, COPQ or similar analysis are used to identify specific benchmarking metrics or projects, the same basic process is followed as illustrated in Figure 8.5.

The first step is to use the analysis method to clarify exactly what aspects we want to collect data on and this will naturally lead to identifying the metrics.

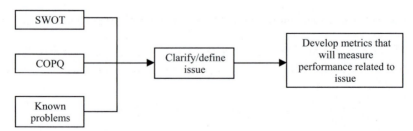

FIGURE 8.5   SWOT, COPQ, and known problems are all potential starting points for identifying specific metrics, or issues on which we may want to focus.

SUMMARY

✓ There are many different, often inter-related, methods of identifying potential metrics for a benchmarking study. In this chapter we discussed a selection
of 11 of the more common methods. Some of the methods are based on
reviewing current metrics, work processes and flows. Others on the
organization's aspirations as captured in, for example, mission and vision
statements. Still others are based on analyses of current situations through,
for example SWOT and cause–effect analysis.

✓ Organizations that have previously analysed their business and developed a
Balanced Scorecard, or identified key performance indicators have a short
cut for developing metrics for a general performance level comparison study.
This is because they will already have identified the key metrics for measuring and managing their organization and it will be some or all of these that
they wish to compare with other organizations. Similarly, those with Service
Level Agreements are focused on measuring customer perception and will
have a list of key metrics they wish to benchmark.

✓ Benchmarking studies often use a mix of methods to identify potential
metrics and those that are chosen often become obvious from the purpose
and objectives of the study.

❡ ACTIVITIES

Considering the subject of your benchmarking study:

? What metrics related to the study are currently being collected? Are the metrics
being used? If not, why not? Should the organization stop collecting the data?

? If appropriate, draw a structure of the subject being benchmarked (e.g. organization chart, financial/budget model). Develop potential metrics based on
the structure(s).

? What are the customer's requirements of the area being benchmarked? How
do you/would you measure them?

? Is the comparison of plan versus actual data appropriate to your study (e.g.
budget versus expenditure, planned time versus actual time)? If so, how could
you measure the variation between the plan and the actual?

? Are there aspirational statements relating to the subject being benchmarked?
If so develop metrics from these that will measure how well these aspirations
are realized.

? If you use KPIs, Balanced Scorecards, CSF or similar methods, what metrics
within these systems could be appropriate for your study?

? If you are benchmarking weaknesses or specific difficulties: What is the
evidence that there is a weakness (e.g. what metrics tell you)? What analyses
have been carried out so far? What (performance) metrics were used? Would
these be suitable for the benchmarking study?

Chapter 9

Reviewing and Finalizing Metrics

INTRODUCTION

In the previous chapter we discussed methods for identifying metrics. As these metrics are being developed there will inevitably be questions in the back of the mind about their suitability such as: can we collect the data? Will it tell us anything? Will others be willing to share these data? However, the purpose and deliverable of the previous step was to develop a list of *potential* metrics. The task discussed here is to review these metrics and produce a list of preferred or recommended metrics that will not only meet the objectives of the benchmarking study but also maximize the amount of useful information that can be gained from the study with a minimum of effort.

In practice, many of the issues raised in this chapter will automatically be considered as the metrics are being identified, as described in the previous chapter. Conversely, as the metrics are reviewed and refined it is likely that the need for additional metrics will become apparent.

When selecting the metrics to be used in the study there are a number of criteria against which we can test the metrics in order to validate their appropriateness. In this section we propose criteria to help ensure that the metrics are complete, collectable, consistent, useable and appropriate:

1. Review against the Charter: will the metric help in meeting our objectives?
2. Availability and consistency of data: are the data available? Will they be consistent?
3. Completeness: do the metrics cover effectiveness, efficiency, utilization and reliability?
4. Ability to test theories and strategies.
5. Inclusion of checking metrics: can the data be validated?
6. Rate of improvement metrics.
7. A suitable mix of metrics.
8. Can/how will the data be analysed?

The Benchmarking Book: A How-to-Guide to Best Practice for Managers and Practitioners

9. The number of metrics: too few metrics and we will not meet our objectives; too many will lead to confusion.
10. Cost data: will participants be willing to share cost data? What are the legal issues? How will we address differences in salary and exchange rates?
11. The issue of non-productive time.

At this stage of the project it is likely that we will be working on our own, without the benefit of comments from other participants. We will need to use our own judgment and perhaps highlight potential issues for discussion with participants at a later date.

DELIVERABLE

A complete validated set of preferred metrics that will meet the objectives of the study.

9.1 REVIEW AGAINST THE STUDY CHARTER: WILL THE METRIC HELP IN MEETING OUR OBJECTIVES?

A process begins with inputs, processes them and produces outputs. When designing processes we work backwards: we start with the desired output and plan what tasks need to be completed in order to arrive at the output. When applied to travel, for example, we plan by starting with our required arrival time and working back to determine a departure time.

When developing the metrics for a benchmarking study the same concept applies. In order to meet the objectives of the study we need to arrive at certain findings or conclusions (e.g. how fast our costs are rising compared to others in the industry); to arrive at these findings we need to carry out certain analyses; and in order to carry out the analysis we need certain data (Figure 9.1).

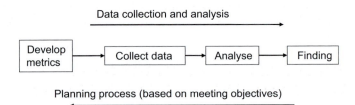

FIGURE 9.1   Processes and requirements flow in opposite directions.

The Project Charter is our guide throughout the benchmarking study and when deciding what metrics to use we need to ensure that the metrics, along with suitable analysis, will allow us to meet the terms of the Charter.

▼ Activity

Questions we need to consider are:

? If we collect the metrics, and carry out specific analyses, will we meet the terms of the Charter?
? Are we collecting more metrics than we need to meet the terms of the Charter? If so, do we want to change the Charter or remove the metrics?

These questions imply and encourage us to consider the analysis that we will carry out in order to meet the objectives of the study. In some situations the required analysis and associated charts may be quite obvious and straightforward, however, in other situations complex analysis may be required. Whatever the situation, it is important to be aware of how the data will be analysed and charted so that we can be sure that the metrics will lead us to meeting the objectives of the study. ⇨ Section 9.8

9.2 AVAILABILITY AND CONSISTENCY OF DATA: ARE THE DATA AVAILABLE? WILL THEY BE CONSISTENT?

It does not matter how well the planned metrics would help us to meet the terms of the Charter if it is not possible to collect the data.

▼ Activity

To ensure that the data are collectable and will be consistent between participants consider:

? Can we develop and agree on definitions for the required data? (⇨ Chapter 10)
? Will participants be able to report data according to the definitions?

👍 Data that are collected to industry or legally recognized standards, for example environmental emissions, safety and financial reporting will normally be available and, where there is a choice, are therefore preferred to those that are not.

9.3 COMPLETENESS: DO THE METRICS COVER EFFECTIVENESS, EFFICIENCY, UTILIZATION AND RELIABILITY?

It is important to ensure that all types of metric are considered for inclusion in the benchmarking study. Costs, hours, failure/reject rates and similar metrics are likely to be identified from analysing the objectives of the study or other metric identification methods. In this section we discuss three groups of metrics that may not have been identified earlier.

9.3.1 Completeness: Effectiveness and Efficiency

Two aspects of metrics that are frequently considered together are effectiveness and efficiency. Effectiveness can be considered as the degree to which we achieve our aim (e.g. our plan or target), and efficiency is how quickly or cheaply we do it. Peter Drucker put it succinctly when he wrote that effectiveness is 'doing the right things' and efficiency 'doing things right'.

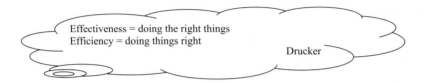

Effectiveness = doing the right things
Efficiency = doing things right
                                    Drucker

Both effectiveness and efficiency are important, though their relative importance can vary. For example, compared to the cost of flying (efficiency), the safe arrival (effectiveness) is far more important.

Unfortunately effectiveness and efficiency measures are often in conflict. We can speed up many activities (efficiency), frequently at the cost of making more errors or compromising on the output (effectiveness). Often it is efficiency that gets measured (cost of maintenance, cost of production, cost of a patient bed night in a hospital) whereas the effectiveness (unplanned downtime, returns, lost customers, quality of patient care) often go unmeasured or are not linked with the causes of effectiveness.

Example: Efficiency and Effectiveness: Hiring Personnel

One of the most common metrics of the process of hiring people is the length of time to fill a vacant post. Another typical measure would be the cost to recruit. These are efficiency measures because they measure how quickly or cheaply we do something.'

However, it is more important to hire the right person for the job than to hire them quickly or cheaply. Unfortunately, because this effectiveness measure is very difficult to define or measure it is often ignored.

The difficult truth we face is that we measure what we are able to measure and manage our organizations based on these metrics. Sometimes these are the most important aspects of the organization but unfortunately it is often the case that the most important aspects are not measured and therefore often are not managed.

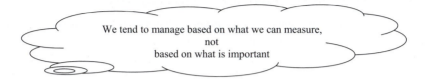

We tend to manage based on what we can measure,
not
based on what is important

ϒ Activity

Do we need to consider effectiveness and efficiency in our study? If so:

? Have we considered both efficiency and effectiveness metrics?
? How are effectiveness and efficiency likely to be related?
? What is the relative importance of effectiveness and efficiency?

9.3.2 Completeness: Efficiency and Economic Viability

It is an understandable mistake to believe that if we run our organizations effectively and efficiently then we will succeed economically, but this is not necessarily the case. We may, for example, make the best gas lamps in the world, and at a low cost; however, if there is no market for gas lamps we will go out of business.

Example: Economic Viability in the Extraction Industry

Economic viability is briefly described with reference to gold mining ⇨ Chapter 7.2.4.2

ⓘ When benchmarking with competitors, it is unlikely that it will be appropriate, or that participants will agree, to benchmark economic competitiveness.

ϒ Activity

? Do we need to consider efficiency and economic viability in our study? If so, how will we do it?

9.3.3 Completeness: Reliability, Availability, Utilization, etc.

There are several terms and concepts aimed at capturing information about use of resources. The most popular are:

Resource usage definitions
Planned downtime: planned time when a resource is not available for use.
Unplanned downtime: unplanned time when a resource is unavailable for use.
Availability (uptime): time when a resource is available for use
Unavailability: time when a resource is not available for use
Availability = total time – downtime
Unavailability = downtime
Idle time: time that a resource is available for use but not in use.
Utilization: time during which a resource is being used
Utilization = Total time – availability - idle time
Loss due to unavailability of resource: may occur when equipment is not available for use. However there may also be losses due to other reasons such as incorrect set up of equipment, lack of training, lack of inputs. These are usually accounted for separately.
Frequency of failures, e.g. of equipment, incorrectly completed forms

These terms may be applied, for example to:

- A specific resource such as a generator,
- Groups of resources that constitute a homogeneous unit, such as production lines or processing trains,
- A complete facility.

The usage of equipment and facilities can be important. For example, low utilization compared to other organizations may result in high maintenance costs due to maintaining excess equipment. Conversely, high utilization may lead to late delivery whenever unplanned downtime occurs.

Example: Availability Metrics: Port-Loading Facility

A port has five berths, labelled A–E, each with associated loading facilities.

For a benchmarking study reporting on operations over 1 year:

Planned downtime:

Each berth has planned downtime for maintenance of 100 hours per year.

Maintenance for each berth is planned at different times and in such a manner that the remaining four berths can manage the workload.

Planned downtime for each berth = 100 hours,

Availability = 8760 − 100 = 8660 (8760 hours in a year)

Dredging (planned) operations were carried out for 150 hours at times when the port had no scheduled activity:

Planned maintenance = 150 hours,

Availability of port = 8760 − 150 = 8610 hours.

A crane on berth B failed, and was replaced by a spare. The failed crane was repaired and became available for use. No downtime or loss was incurred.

Ancillary equipment failure on berth C resulted in 200 hours of unplanned downtime and one delayed berthing resulting in a loss of US$ 24 k.

Unplanned downtime = 200 hours, availability = 8460 hours

(remembering there was 100 hours of planned downtime),

loss = US$ 24 k

A problem with berth D was noticed and maintenance planned to be carried out in 10 weeks. How this is reported will depend on specific definitions for the study.

ϒ Activity

? Do we need to consider reliability, availability, utilization in our study? If so, have we included appropriate metrics?

9.4 ABILITY TO TEST THEORIES AND DETERMINE STRATEGIES

In many benchmarking studies there are a number of participants. This gives us an excellent opportunity to go beyond simple comparison of performance levels, and gather data on and test operational theories and relationships between metrics. This helps us to begin to investigate why organizations' performance are at certain levels, investigate possible operational strategies and identify more complex industry norms.

Example: Testing Theories: Maintenance

There is a belief amongst participants that the more preventive maintenance is carried out, the less corrective maintenance (e.g. due to breakdowns) is necessary, and that as a result total maintenance costs will be minimized. To test this theory we need a minimum of around 10 participants to report both preventive and corrective maintenance. The theory could then be tested by drawing a scatter diagram of total maintenance versus percentage of preventive maintenance, as shown in the chart.

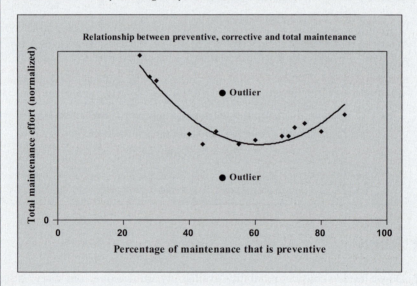

The vertical axis shows the total amount of maintenance being carried out, normalized to take into account any differences in, for example, amount or type of equipment. The horizontal axis shows the percentage of maintenance that is preventive and is calculated as:

$$\text{Percentage} = \frac{100 \times \text{preventive maintenance hours}}{\text{Total maintenance hours}}$$

A regression line is often fitted to highlight any relationship between the plotted variables.
In this case the chart shows that:

1. As the percentage of preventive maintenance increases from 20%, so the total amount of maintenance decreases.
2. As the percentage of planned maintenance increases from 60%, so the total amount of maintenance increases.

The Chart has Other Uses, for Example:

• It is possible to estimate the expected total amount of maintenance required for any given mix of preventive and corrective maintenance.
• It is possible to identify performances that do not follow the industry pattern, i.e. outliers. For example, a participant with 50% preventive maintenance that is significantly above the trend line could suggest that the participant should be able to reduce the total amount of maintenance. Similarly, the outlier that is significantly below the trend line could suggest that the participant has learned a different and better way of maintaining their equipment than the other participants could learn from.

There are other theories that could be tested with some additional data. For example, if the maintenance data were from a production facility, we could hypothesize that facilities running at high utilization would not want to risk incurring unplanned shutdowns and so would follow a strategy of minimizing corrective maintenance by increasing preventive maintenance. Conversely, facilities with low utilization may aim to minimize total maintenance costs by carrying out less preventive maintenance and suffering higher breakdowns. This theory could be investigated by charting utilization versus percentage of total maintenance that is preventive.

ϒ Activity

? What philosophies, theories or ideas could we test with the data we are collecting?
? Are there additional philosophies, theories or ideas that we want to and could test by collecting extra data/information? Would it be worthwhile collecting it?

9.5 INCLUSION OF CHECKING METRICS: CAN THE DATA BE VALIDATED?

Once the data have been collected, one of the first tasks will be to validate it to ensure that the data are correct, complete and in the correct units. At the metrics review step it is useful to consider how the data will be validated. In particular, it may be useful to add some metrics to help with data validation. For example, budgeted figures, or plans, can be collected as well as actual figures. This allows a variation analysis of actual figures against budgeted or planned figures, and it also allows a consistency check between the two. If collecting budgets/planned figures for every item is perceived to be onerous, just the total budget could be collected and compared to the total actual figure. Similarly, if manpower figures are being collected, these can be checked against an organization chart.

ᛪ Activity

? How will we validate data received from participants?
? Do we need to collect extra data for validation purposes?

9.6 RATE OF IMPROVEMENT METRICS

One of the advantages of taking part in regular benchmarking studies is that it is possible to monitor changes in performance metrics over time.

Most benchmarking studies are a snapshot of past, or at best current, performance. In Figure 9.2 organization A may be performing better than organization B today, but if B is improving faster than A, B may soon overtake them.

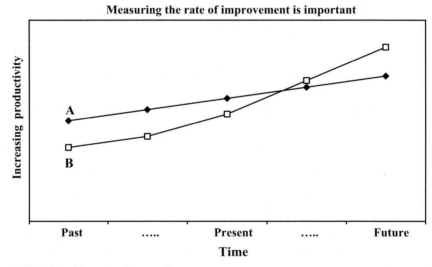

Measuring the rate of improvement is important

FIGURE 9.2   Measuring the rate of improvement.

One of the other advantages of ascertaining improvement rates is that it helps us to predict the future: information which organizations can use to help with medium and long term planning.

ᛪ Activity

? Do we plan to benchmark regularly? If so:
? What extra data (if any) do we want to collect to allow analysis of improvement rates?

9.7 A MIX OF METRICS

9.7.1 Upstream Versus Downstream; Lead Versus Lag

In the terms of process flow, we can consider metrics as being upstream or downstream. The ultimate common downstream metrics are concerned with customer satisfaction, while the upstream metrics would measure order taking, suppliers and design, depending on the industry.

Downstream metrics tell us the consequences of historic decisions while upstream metrics indicate future performance, and to this extent these two types of metrics can be considered as lag and lead indicators. Lag metrics are also often referred to as outcomes.

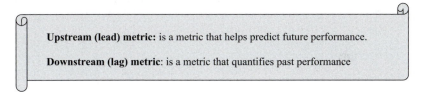

Upstream (lead) metric: is a metric that helps predict future performance.

Downstream (lag) metric: is a metric that quantifies past performance

Most organizations will use a mix of both upstream and downstream metrics, and we may choose to include a mix in a benchmarking study. For example, equipment breakdowns (downstream metric) may be caused (at least in part) by lack of preventive maintenance (upstream metric) (Figure 9.3).

One frequently asked question is how to link upstream and downstream metrics. To link them we need to consider what causes the outcome. Where the relationship is not simple, and usually it will not be, it can be useful to illustrate the link with a cause–effect diagram (⇨ Chapter 8). As most outcomes depend on a number of upstream variables, we need to consider which of them should be included in the benchmarking study.

While the concept of lead and lag metrics is useful, as Figure 9.4 shows, illustrating the relationship as a flow diagram indicates that many metrics are circular in nature. The figure shows that late deliveries influence customer satisfaction. If satisfaction is high, orders will increase, leading to more activity, which will influence the number of late deliveries, and ultimately customer satisfaction.

In some situations it may seem more natural to consider the downstream outcome as the result of a formula. For example, customer satisfaction may be considered as a function of:

- Timeliness of delivery
- Cost
- Quality of product
- After-sales service

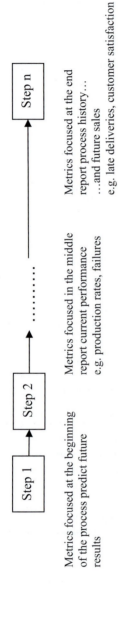

FIGURE 9.3 Generalization of lead and lag indicators.

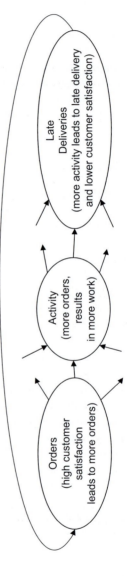

FIGURE 9.4 One of many paths showing cause and effect relationships for a typical manufacturing organization.

Whichever approaches are used, the result should be the same, i.e. to identify relationships between variables and ensure an appropriate mix for your particular study.

ⓘ For further information see The Fifth Discipline, Peter Senge.

ᵀ Activity

? What mix of lead and lag metrics do we want?
? What are the relationships between our lead and lag metrics?
? Do we want to test the relationship between lead and lag metrics in the benchmarking study? If so how will we do it?

9.7.2 Short-Term Versus Long-Term

It is well known that it is often possible to boost short-term performance by sacrificing long-term performance. This can be done, for example by:

- Reducing preventive maintenance, which will cut costs now but is likely to lead to more breakdowns later.
- Reducing research and development, resulting in fewer new products and services in future.
- Cutting back on staffing levels, resulting in inability to meet future workload or expand.
- Cutting back on training resulting in lower work quality and possibly loss of staff.

Such short-termism can lead to difficulties in a benchmarking study as participants with lower costs, for example, may be seen as the performance leaders, and for several years may not suffer from the effects of the cutbacks. For this reason it is preferable to have a mix of both short-term and long-term metrics in the study. It is also a key reason for seeing benchmarking as a regular activity rather than as a one-off exercise. In time, if cutbacks do lead to a drop in performance this will be seen. If it does not lead to a drop in performance, there may be an opportunity for the other participants to learn.

Case Study: Effect of Short-Term Maintenance Cutback

One participant in a benchmarking study submitted the lowest maintenance costs for three years in a row. Then a major failure at the facility resulted in a complete review of maintenance strategy, after which they submitted higher maintenance costs, which were more in line with the other participants in the study.

👍 One reason for benchmarking on a regular basis is to check that performance levels are sustainable.

❢ Activity

? What mix of long/short-term metrics do we want?
? How will we check that a good performance from a participant is sustainable?

9.7.3 Internal Versus External

> **Internal metric:** is a metric by which people inside an organization or process measures its performance.
>
> **External metric**: is a metric by which people outside a process or organisation measure its performance.

Many internal activities have a direct effect on external metrics, and where possible we would like to include a mix. For example, reducing costs by reducing call centre staff may result in lower customer satisfaction. Reducing costs by eroding the staff remuneration package may result in the better people leaving for higher rewards elsewhere and fewer people applying to join the company.

> **Case Study: Internal and External Metrics for Purchasing**
>
> A clerk became very frustrated that her printer often jammed. After several attempts to fix it, and much wasted paper, she finally phoned the IT Help Desk. They informed her that this problem was being experienced by a number of people within the organization. The problem was not with the printer but with the last delivery of paper. The purchasing department had been told to cut costs, and to do so they had bought the last batch of printer paper from a different supplier. While the cost to the purchasing department went down, the cost to the organization as a whole increased through wasted time and wasted paper.

❢ Activity

? What mix of internal/external metrics do we want?
? How will we check that a good performance in one area is not resulting in a poor performance elsewhere?

9.7.4 Global Versus Local

For many organizations public image is perceived as important and so while it is important for them to focus on corporate issues (global metrics) it may also be important to examine local effects. For example, a supermarket may benchmark both sales per square metre of floor area and involvement in local issues and local perceptions.

ϒ Activity

? What mix of global/local metrics do we want?

9.7.5 Soft and Hard Metrics

In many organizations there is a tendency to focus on what are known as hard metrics such as costs, sales, production and profit. These issues are easy to measure and important. Soft issues such as employee satisfaction, training effectiveness, marketing and research are often given less prominence because they are much more difficult to measure. The soft issues are just as important as hard metrics because, as we have seen elsewhere, some are predictors of future hard metrics.

ϒ Activity

? What mix of soft/hard metrics do we want?
? If we do not know how to measure soft issues, how will we find out?

9.7.6 Customer Satisfaction

No customer = no business

Over the last 20 years there has been a growth in the realization that the customer is central to the success of the organization. Many people believe that customer satisfaction is based on quality, delivery and price. While this may or may not be a complete representation of customer satisfaction, what is important is to ensure that appropriate measures of customer satisfaction are included in the metrics.

ϒ Activity

? What, if any, customer satisfaction metrics do we want?
? Will we need to employ an external organization to measure customer satisfaction?

9.8 HOW WILL THE DATA BE ANALYSED?

Data analysis is addressed in ⇨ Chapter 14; however, when ensuring that the objectives of the study can be met, we discussed the need to consider how the data will be analysed. The same need has been implied in other sections.

However, as the list of metrics continues to change, the final check will usually be to confirm that if we collect the metrics as proposed, and carry out certain analyses, will we meet the objectives of the study.

Case Study: Environment

One of the objectives of a benchmarking study was to compare the environmental impact of participant organizations. Each participant produced different quantities of various pollutants and each pollutant caused different effects on the environment, and so it was not appropriate to compare them directly.

Before deciding to collect the metrics, research was carried out into discover if a pollutant index had been or could be developed that would take into account impact of each pollutant.

Statisticians involved in benchmarking, and indeed other consulting statisticians, have many experiences of being asked to analyse data to test theories or determine relationships between variables only to find that the data collected is not appropriate for the required analysis. In order to save much frustration later in the study it is well worthwhile to invest some time ensuring that appropriate analyses can be carried out on the data to meet the objectives of the study.

One method of ensuring that the analyses and metrics will be appropriate is to carry out a test analysis using hypothetical data similar to that which we think might be submitted by participants.

☝ The resulting charts and analyses can be used to help with participant recruitment.

⟁ Activity

? Review the list of metrics and objectives of the study. Determine what analyses and charts will be required to meet the objectives.
? If appropriate, trial the charts and analysis by developing test data.
? If in any doubt as to the suitability of the metrics or analysis, consult a benchmarking statistician.

9.9 HOW MANY METRICS DO WE NEED?

A common question people ask is how many metrics should there be in a benchmarking study? Unfortunately there is no easy answer. There are however, some simple guidelines.

- The more metrics there are, the more complicated every aspect of the study becomes. If there are more than about 50 metrics in the study, it will take considerable effort to carry out any in-depth analysis. For this reason, studies with this many metrics are only likely to be manageable if they are split into distinct unrelated groups. Each group can then be treated as a substudy with its own analysis.
- With too few metrics there is a risk that the objectives of the Charter will not be met.

  If there are more metrics than we think we can manage it will be necessary to reduce them. There are various methods for doing so including:

  - Perhaps the simplest solution is to restrict the scope of the study thereby cutting out metrics pertaining to the excluded scope. For example, when benchmarking operations and maintenance, we may choose to exclude capital projects.
  - Another simple solution is to only benchmark metrics at a high level. For example, rather than collect cost and man-hours data on each of the purchasing department's functions such as order placement, ware-housing, supplier selection and monitoring, we would collect cost and hours for the department as a whole. This is particularly appropriate where the department as a whole forms only a small element of the whole study.
  - Follow the Pareto principle ⇨ Appendix 'Data Analysis Tools'. The Pareto principle, frequently thought of as the 80/20 rule, proposes that 80% of an effect is due to 20% of the causes. For example, 80% of costs will be due to 20% of cost items, 80% of problems will be due to 20% of the causes. We can choose to benchmark in detail the 20%, known as the 'vital few' items of interest, and the remainder, known as the 'useful many', can be excluded or combined in an 'other' category. For example, in a safety benchmarking study if we incurred 100 injurious incidents one of which resulted from a spilled chemical, even if we found a method of eliminating chemical spills it would only reduce incidents by 1%. If there were 15 incidents involving lifting, it would be more beneficial to focus effort on reducing lifting incidents.
  - Consider the likely outcome of relatively poor performance of the metric. This idea is closely related to the Pareto principle, but approaches the issue from the potential benefits view. For example, resolving the problem of

late deliveries may result in an estimated saving of $10 000 per year, while resolving the problem of deliveries of damaged goods may result in a saving of $100 000 per year. We would prefer to focus on the areas of highest potential return and either ignore or combine areas with lower potential return.

ϒ Activity

? Are we collecting data in too much detail?

- If the performance level were to be shown to be poor would the potential improvements be significant in terms of other potential improvement areas in the study?

? Is the number of metrics manageable:

- Will we be able to explore relationships between metrics?
- How long do we think it will take to analyse the data and understand the results?
- If there are a lot of metrics, can they be categorized into separate independent groups?
- Would the data/metrics in each group be manageable?

9.10 OVERCOMING DIFFICULTIES OF COST METRICS

An issue that is likely to arise in all except internal benchmarking studies is that of benchmarking costs. Typical issues and potential solutions are presented in Figure 9.5.

ϒ Activity

? If costs are included in your benchmarking study, what confidentiality or legal issues may arise? How do you propose to cope with them?

Issue	Potential solutions
Legal restrictions ⇨ Chapter 21	Laws such as The Treaty of Rome in Europe and Anti-Trust law in the USA prohibit some forms of cost sharing for fear of anti-competitive activities. However, most costs can usually be shared in some manner (see below). Check with local laws.
Commercial sensitivity ⇨ Chapter 21	Organizations may choose not to share some commercially sensitive data. However, there are some methods of reducing or removing objections (see below)
Cost-sharing	For both legal restrictions and commercial sensitivity there are four possible methods of benchmarking financial data: • Sometimes costs may be shared openly. • With more than two participants it is often possible to report costs anonymously. • With more than two participants is it possible to report each participant's costs against summary costs such as average, highest or lowest. • Rather than sharing actual values, participants can be given ranges of values and report against the appropriate range. The use of independent facilitators helps to anonymise / mask the data and so lessen or remove objections.
Intrinsic differences in costs (e.g. salaries in Scandinavia are higher than in Africa)	Comparison of costs may be of limited use even if a "country factor" is developed to at least partially account for differences in costs between countries.
Accounting practice differences	It may be possible to extract required costs even if accounting practices are different. Alternatively, it may be possible to quantify the differences. For example, if one organization adds in overheads to manpower costs whilst another does not, it may be possible to estimate the overheads as a percentage of the manpower costs.
Exchange rate volatility	Short term variability can usually be catered for by using an average exchange rate over an appropriate period (e.g. a year). ⓘ Historic currency conversion tables are available online from sites such as www.oanda.com
Inflation (when benchmarking costs over long periods)	Inflation can usually be accounted for by simply inflating or deflating costs.

FIGURE 9.5   Typical cost benchmarking issues and potential solutions.

9.11 ☞ THE RISK OF NON-PRODUCTIVE TIME AS A METRIC

A common measure of equipment utilization is non-productive time (NPT), i.e. the time during which equipment is not being used. In many situations this is an appropriate measure.

However, in some situations it is possible to reduce NPT by running equipment slower. This may result in less NPT it will also take longer to complete the task. It may be preferable to work equipment harder, complete the job sooner and incur greater NPT.

> **Example: The Risk of Non-Productive Time as a Metric**
>
> ---
>
> The Easy-Does-It organization has a stated goal to minimize NPT and runs its equipment accordingly. One particular task takes 100 hours, which includes 90 hours of productive time 10 hours of NPT.
>
> Get-It-Finished Ltd. has a stated aim of completing the tasks as quickly as possible. It completes the same task as Easy-Does-It in 80 hours, which includes 68 hours of productive time and 12 hours of NPT.
>
> Easy-Does-It incurs only 10 hours (10%) NPT compared to the 12 hours (15%) NPT of Get-It-Finished Ltd., however, Get-It-Finished Ltd. completes the task faster and is able to move on to the next task sooner than Easy-Does-It.

While it is useful to collect NPT data for analysis of NPT causes and improvement, comparing NPT figures can be misleading and needs to be used with care.

SUMMARY

In this chapter we have reviewed a number of methods of appraising potential metrics and refining them into a list of preferred metrics that will ensure that they:

- ✓ Are collectable.
- ✓ Conform to legal and commercial requirements.
- ✓ Will allow us to meet the objectives of the study.
- ✓ Will maximize the information gleaned from the study with a minimum of effort.

ϒ Activities

? Each section of this chapter has recommended activities. Choose and carry out those that are most applicable to your study.

Chapter 10

The Importance of Operational Definitions and How to Create Them

Differences in performance levels between participants in a benchmarking study are often due to differences in definitions against which their data were reported, NOT differences in performances

INTRODUCTION

Definitions are probably the least interesting aspect of benchmarking. But, as the following example shows, definitions are important.

Example: The Importance of Definitions

In his acclaimed book Out of the Crisis, Deming quotes an example of a factory where there was no definition of a defective product. As the level of rejects was high, 11%, it was decided to focus on reducing them. The first step was to agree a definition of a reject. The result was that rejects fell to 5% – less than half of what they had been. It took seven weeks to develop the definition and explain to those involved what a reject was. Then the true benefits began to emerge: by reducing reject rates the production rate, capacity, profit, customer and employee satisfaction increased. The only cost was that people took the time to define what they wanted and tell the workforce.

Out of the Crisis, Dr W.E. Deming, p. 6

In benchmarking we are more likely to be puzzling over definitions for the number of hours spent on tasks, or the cost of maintaining a building, or the number of accidents or errors. The result of participants using different definitions may be insignificant or they may not. We will not know, because we will not know what definitions they have used.

In this chapter we discuss the importance of definitions and how to develop definitions to help ensure that data are reported consistently by all participants in the study.

Defining the scope of the study has already been discussed ⇨ Chapter 4. In this chapter we focus on definitions for metrics.

DELIVERABLE

The deliverable of this step is to produce definitions:

✓ That are clearly understood by all participants and
✓ Against which participants can submit data.

10.1 THE IMPORTANCE OF OPERATIONAL DEFINITIONS

The example in the introduction demonstrated the importance of definitions in day-to-day activities. The case study below demonstrates the importance of definitions in the context of benchmarking.

Case Study: The Importance of Definitions: Maintenance

Two very similar organizations want to compare maintenance effort. They decide to report the number of people in the maintenance department:

Failures Are Us Inc. (FrUS) looks at its organization chart and adds up the number of people: 32.

Let It Rust Pty. (LIR) looks at their organization chart and report: 38.
That is nearly a 20% difference.

When they visit each other they sit down to discuss how FrUS manages with so few people they discover:

- The effectiveness of the maintenance is similar.
- LIR has included the person on long-term sick leave. FrUS does not have anyone on long-term sick leave.
- LIR has included their two trainees. FrUS has not included trainees because they are not considered productive.

So the numbers are refined to FrUS = 32 people and LIR = 35 people.

Over lunch the topic of working conditions comes up. It transpires that at FrUS people work 40 hours a week while at LIR people work 36 hours per week. They decide to calculate the number of hours worked rather than the number of people as this will better represent the actual effort expended.

$$FrUS = 32 \text{ people} \times 40 \text{ hours} = 1280 \text{ hours per week}$$

$$LIR = 35 \text{ people} \times 36 \text{ hours} = 1260 \text{ hours per week}$$

The difference in hours worked on maintenance between the two organizations is now insignificant.

During the afternoon they discuss the responsibilities and tasks for which each maintenance group is responsible:

LIR explain how the maintenance crew purchase most of what they need themselves. They believe that this is more efficient as the maintenance crew know exactly what is required and are better able to discuss requirements and resolve problems with the suppliers.

FrUS explain they have a purchasing department that does all their purchasing. They believe that it is more efficient to have purchasing experts carrying out all purchasing activities.

LIR explains that the maintenance team leaders plan their maintenance because they believe that the skills required to best plan maintenance activities are within the maintenance crew.

FrUS explains that they have a planning team which does all production and maintenance planning. This is the best way, they believe, to ensure that the needs of every group are taken into account.

LIR has included the maintenance stores area activities which the maintenance teams manage themselves.

FrUS has excluded stores management as it is the responsibility of the purchasing department.

They estimate the effects of these differences on the hours worked per week and conclude that if FrUS were to include the activities that LIR includes they would require about 1370 hours per week.

That is a difference of about 8%, but this time in favour of LIR.

That is a turnaround from 20% in favour of FrUS to nearly 10% in favour of LIR and it is all due to each participant using different definitions.

Adapted from a maintenance benchmarking study

Differences in performance levels due different interpretations of definitions are common, albeit differences of 30% are unusual.

The clear, concise development of definitions is:

1. Not easy,
2. Important.

Remember that the people using the definitions will be those collecting the data. Where possible these people should be involved in developing the metrics and definitions to help ensure that they are collectable, consistent and appropriate.

10.2 DEVELOPING DEFINITIONS

Developing definitions for benchmarking is usually an iterative process. They are usually addressed:

1. As the metrics are developed and reviewed during the internal phase of the study. The reason for developing them at this stage is to provide a framework for future refinement and to clarify what the metrics are trying to achieve.
2. When planning with participants the definitions will be reviewed and refined to ensure that they are both clear and that participants can report against them. ⇨ Chapter 12.
3. When those collecting data are unsure of what is required and ask for help ⇨ Chapter 13.
4. At future benchmarking kick off meetings where the metrics are likely to evolve as the participants work to clarify the definitions in the light of having used them.

We discuss several methods for developing clear concise definitions; however, not every method will be appropriate in all cases.

✓ **Ensure that the name of the data item to be defined is as specific and focused as possible**.

The first thing that the person supplying the data will see is the name of the data item required. Careful naming to reflect as accurately as possible the data to be collected is important if only because many people do not read the associated definition.

Example: Clarity of Metric Name

When reporting safety statistics, accidents could include events varying in severity from death to a small bruise. It would be more helpful to refer to the metric as, for example, 'number of accidents resulting in time away from normal duties'.

✓ **Provide a general definition**.

It often helps to begin a definition with a description of the general context. In addition, if this is short it may be possible to include it alongside the name of the data item.

Example: Use of General Definition

Allocated cost: Costs imposed by others over which you have no control.

✓ **Itemize what is included and excluded**.

One of the most useful methods of clarifying a definition, particularly for costs and hours for activities, is to list typical items that should be included and excluded.

Example: Use of Inclusions and Exclusions

Non-productive time due to contractor:

Includes: Downtime due to failure of contractor's equipment, lack of personnel, operational errors ...
Excludes: causes outside the control of the contractor such as regulatory or customer inspections and tests ...

✓ **Provide examples**.

Examples are similar to specifying what is included or excluded and help to provide a context. The main difference is that while itemizing what is included or excluded aims to provide an exhaustive list, providing examples will not be an exhaustive list.

Example: Use of Examples

HSE costs:

Includes: HSE materials costs (e.g. fire extinguishers, CCTV, breathing apparatus, personal protective equipment (PPE), foam for fires).

✓ **Specify context**.

In more complex situations if often helps to explain how data are to be divided, and then define each division.

Example: Specifying Context

In this example the total manpower costs and hours are to be reported.
Total man-hours are split into three different categories: Management, Staff, and Contractors.

1. Management includes ...
2. Staff includes ...
3. Contractors include ....

✓ **Specify units of measurement**.

Ensuring that the units of measurement are specified is important. For example, whether costs are supplied in Euros or US dollars may be difficult to ascertain from looking at the data (because at the time of writing the exchange is close to 1), but could result in important differences in the analysis.

There are three common approaches to this issue:

1. Allow the participant to specify the units. This works well with currency units as the participant will be used to working in one specific unit and for the analyst it is easy to look up exchange rates on the internet and apply them. ⓘ Current and historic exchange rates can be found at www.oanda.com.
2. Specify the units that the participant is to use.
3. Allow the participant to select from a restricted group of units. This choice is useful if it is known that there is a limited number of units used by participants and the analysts are able to convert all of them to a common unit for analysis purposes.

✓ **Refer to standardized definitions wherever possible**.

In many situations there will be accepted definitions. These may be due to statutory reporting requirements or industry wide standards.

Example: Use of Widely Accepted Definitions

In many countries it is mandatory to report safety and environmental incidents. The definition of a safety incident will already be defined, and, because it is required to report to the authorities, readily available. Two such standards for safety incidents are OSHA (Occupational Safety and Health Administration) and RIDDOR (Reporting of Injuries, Diseases and Dangerous Occurrences Regulations).

✓ **Sometimes definitions are not possible**.

Occasionally it will not be possible to develop a definition that all participants can agree upon. In these situations we will usually not collect the data. However, occasionally such data may be useful even though it is inconsistent between participants.

Example: Situations where There are No Definitions

Safety statistics may be divided into several categories such as:

- Serious injury;
- Minor injury;
- Accidents requiring first aid;
- Damage only accidents;
- Near misses.

Many people believe that the frequencies of different types of injury are related. For example, Bird's Accident Triangle (1960) proposed that for each major injury there would be 10 minor injuries, 30 damage only incidents and 600 no injury or damage (i.e. a near miss).

However, it is likely that while serious injury, minor injury, first aid cases and damage only accidents are well documented and defined, organizations will vary greatly in their definition of near misses. Even if a definition were agreed it is unlikely that each organization would capture the data with the same rigour. One may only capture near misses that are reported and be lax in doing so while another may include potential accidents as near misses (i.e. there was no incident but a situation is such that an incident could occur).

While consistent reporting of near misses will not be possible, reporting near misses according to each participant's definition will allow comparisons to be made between the different categories.

✓ **Check availability**.

However accurate and clear a definition may be, if the data are not available, or only available with great difficulty, it will be of little use. It will sometimes be necessary to accept that some participants may not be able to supply data to the exact definition. This often occurs because job specifications vary. A workable solution is to ask participants to estimate the effort.

Example: Use of Estimates

When reporting IT support effort, Participant A has a dedicated IT support desk which resolves all queries. However, the support desk not only supports the area being benchmarked, but also other IT users excluded from the study. In this situation the participant will have to estimate the time the IT support desk spend supporting the area being benchmarked.

Participant B also has a dedicated IT help desk support team. However, many of the simpler queries are resolved informally within the departments by whoever seems to have adopted the role of 'local expert'. The support hours of this informal support effort are not known.

Participant C has a dedicated IT support person in each department. However, their job includes non IT support activities such as telecoms support and minor software development projects.

In Chapter 4 we discussed methods of helping to define projects including the use of statements such as 'the process begins/ends when . . .', process maps, ground layouts and context diagrams. Some of these methods may also be useful when defining metrics.

SUMMARY

In this section we have:

✓ Demonstrated how failure to provide clear definitions can lead to inconsistent data between participants and incorrect conclusions regarding comparative performance levels.
✓ Provided guidelines on developing useable definitions including using clear naming, providing a general definition, examples, a context, units of measure and widely accepted standards.

ⵖ Activity

? Develop draft operational definitions for the metrics in your study.

Chapter 11

Finalizing the Project Plan and Gaining Management Support

INTRODUCTION

At this point in the study:

- The Project Charter, which includes the objectives and scope of the study will have been carefully reviewed and amended as necessary.
- The list of preferred participants, with the reasons for their selection will have been drawn up.
- A list of planned metrics will have been developed, along with any appropriate metric structures and definitions.

In many benchmarking studies the work up to this point will have been carried out within the organization initiating the project. As the next step will be to invite external organizations to join the study many organizations will want to hold a review at this point before proceeding with the project.

In this book we have taken the approach that this review is carried out by means of a presentation by the project team to management and other interested internal parties, with the aim of gaining approval and release of resources. However, there are other alternatives. For example, an organization experienced in benchmarking may have decided at the outset that the project will proceed and at this stage there may only be, for example, a review by the project team, a progress presentation, report to appropriate people, or peer review, depending on the organization's norms.

To prepare for the presentation the team still needs to develop a detailed project plan including resource requirements, tasks, timescales, and project deliverables.

In this chapter we discuss how to finalize the project plan and prepare for the presentation.

DELIVERABLE

The deliverable of this step is a persuasive proposal to management to support the benchmarking project and release required resources.

11.1 THE NEED FOR A PLAN

Successful benchmarking projects, like all other projects, require good project planning and management. The tools and methodology for project planning and management are well known and documented, and organizations will use their own planning methods. The purpose of this short section is to briefly address benchmarking planning issues.

When considering any process, a key aspect is to identify the internal and external customers of the process and their needs. The process can then be designed to meet these needs.

For a benchmarking project this list of typical key customers is also used to help identify what each will want to get from the project plan. Typically:

- **Management**: who will want to know:

    - The cost and manpower resource requirements and
    - The deliverables and schedule of the project.

    The fact that a plan has been developed is evidence that the team has considered and planned resource requirements.
- **Potential participants**: have similar needs as the initiating organization's own management. They will want evidence that the project they are being invited to join has been carefully thought through and planned, not only with respect to scope, objectives metrics, benchmarking process, etc. but also timescales and resource requirements.
- **Implementation team**: for this team, likely to consist of participants from different organizations, the plan is the guiding document of their benchmarking project. They will need to know what the deliverables are and the dates by which they are due.

    A detailed project plan including resource requirements, tasks, timescales, and project deliverables is a key tool in meeting these needs.

11.2 DETERMINING RESOURCE REQUIREMENTS

A good starting point for developing a project plan is to draw up a list of tasks that need to be completed. The tasks are derived from the scope and objectives of the study and some of them will already have been identified. Many of the remaining tasks are discussed in the following chapters of the book.

For each task it is helpful to estimate the resource requirements, probably mainly manpower, as well as deliverables and timescale. The resource requirement and timescale will depend on many factors such as:

- **Scope of the study**: the wider the scope, the more effort will be required to benchmarking it.
- **The planned number of participants**: the greater the number of participant the slower the decision processes and the more difficult it is to accommodate the needs of all participants.
- **The ease with which participants will be persuaded to join the study**: the more difficult it is the more resources will be required to recruit them.
- **Expected proximity of participants** (i.e. local, national, international): the wider spread the participants, the greater the travel time, and, for international studies, the more difficult it is to cater for differences in such things as culture, cost structures, and exchange rates.
- **The complexity of data requirements and analysis**: the more complex the data requirement the more care is required with data definitions, development of data collect packs, analysis, and reporting.
- **How cost data will be handled**: anonymizing costs will require more effort than sharing costs openly.
- **Team members' experience**: the greater the experience the less time should be required to complete the study.
- **Need for external resources** such as consultants or facilitators: use of external resources may bring valuable experience, but is likely to be expensive.

The planned timescale for invited participants will be similar to the plan for the instigating organization except that it will exclude the activities prior to the participants becoming involved in the project, and any internal follow up activities such as presentations and improvements.

The manpower resource requirement will also vary between initiating the organization and participants depending on whether participants will share in the work load, for example, organizing, facilitating and minuting meetings, preparing data collection documents, analysing results, and report writing.

Similarly the financial contribution to the study will depend on how participants agree to share costs. At one extreme the initiating organization may finance the whole project, at the other, participants may share costs.

Typical planning tools of the study include:

- ✓ A Gannt chart including planned dates for meetings and deliverables.
- ✓ Personal/participant task list with expected resource requirements (i.e. number of people, tasks, skill requirement).

Individual organizations may require the use of other project planning/management tools.

11.3 BENCHMARKING PROJECT REVIEW

Once the details of the project plan have been finalized, the team can prepare for the presentation.

Each organization will have its own standards for presentation and we present here a typical one.

The starting point is likely to be a review of the Project Charter which defines the terms of reference for the team. It typically includes:

1. What is to be benchmarked (the scope).
2. Why it was chosen for benchmarking (the justification for the project).
3. Expected benefits.
4. The objectives of the project.
5. Team members.

The team will have reviewed the charter and may, with agreement, have changed it. In particular, it is likely that the team will have reviewed, clarified, or added more detail to the expected benefits. Any changes, comments, or further details will be incorporated into the charter as appropriate. For example, if a more detailed analysis of the expected benefits has been carried out, this may be attached as a separate document, while if the scope has changed it is likely that the charter will have been, or will need to be, re-issued.

The next part of the presentation will include details of:

- The proposed metrics, presented as a model where appropriate.
- Preferred participants and how and why they were chosen.
- The project plan, including, for example, timescales, deliverables, resource requirements, use of external facilitators/consultants.

11.4 DEALING WITH OBJECTIONS

With thorough preparation for the presentation there should be few if any objections to the proposal. Preparation for a successful presentation will include:

✓ Thorough and careful consideration of the Project Charter and the resulting research and proposal for metrics and participants.
✓ Frequent eliciting of comments and concerns from those who are likely to be impacted by the project, including the decision makers at the presentation. Where appropriate, any concerns should have been addressed before the presentation. Where the team feels the concerns are not valid, they should develop a robust response.

Where possible, the people who will be at the meeting should be contacted beforehand to gauge their support or otherwise of the proposal. For those who do not support it, or have objections or concerns there are several possible courses of action. The team may:

1. If the objection is valid, address the issue and resolve it before the meeting.
2. If there is a lot of support from other opinion leaders it may be possible to ask them to discuss the issue with the objector beforehand with the aim of changing the objector's view.
3. If the issue cannot be resolved before the meeting, consider raising the issue during the meeting yourself and addressing it before the objector has a chance to raise it. This has the advantage of demonstrating that you have considered the issue and have an appropriate response.

Objections are likely to fall into two categories:

- Those that reject the concept of benchmarking.
- Those that accept the concept of benchmarking but have concerns about specific aspects of the project.

Unfortunately it is not always easy to distinguish between these two types of objections. Those who object to benchmarking will often couch their objection in terms of a detail of the specific project, such as the impossibility of finding a truly appropriate participant with whom to benchmark.

Often the objector's stated concern is a smoke screen for their real concern. For example, a manager may think that his group's performance will be poor compared to others in the study and is concerned about possible ramifications if this turns out to be the case. However, the manager is unlikely to admit that concern and may object, for example, on the grounds the resources could be better used elsewhere. In these situations, once the presenting issue has been resolved, the objector will often then find another objection and repeat the delaying process in the hopes of derailing the project. One way of dealing with this issue is to ask the objector for all of their objections and an agreement that if these are resolved he or she will be happy to continue with the project.

Changing the opinion of those who genuinely object to benchmarking may be a little easier and some methods of changing their opinion are discussed in ⇨ Chapter 20.

Those who genuinely object to specific details of the study should be the easiest to persuade as this type of objection usually emanates from people who accept the concept of benchmarking and want to make it work. These objections need to be dealt with on their merit.

SUMMARY

As with all projects, the development of a benchmarking project plan is important for project management. The project plan is likely to be used:

✓ To gain management support for the project.
✓ Recruit organizations to join the study.
✓ Communicate details of the project to those impacted by the project.
✓ As a guide to the team.

A typical project plan will include resource requirements, tasks, timescales, and project deliverables.

Project planning for benchmarking is similar to project planning for other types of project and organizations may have their own planning methods and software.

It is often useful, if not necessary, to review progress before inviting external participants to join the study. In this book we assume that the review takes the form or a presentation to management to gain project approval and release of resources.

The review presentation will typically include the scope, objectives, draft metrics, participants, project schedule, and other details of the project. The presentation should also seek the support and release the necessary resources to continue the project.

Common types of objection to benchmarking include:

• Genuine concern regarding a specific project or aspect of a project. This type of objection needs to be dealt with on its merits.
• A smoke screen, usually masking a fear that the objector may in some way be disadvantaged by benchmarking. This type of objector can often be identified by their continually raising objections as each previous one is resolved.

ϒ Activities

? Prepare a project plan for your benchmarking project. Use your own organization's methods and standards as appropriate, bearing in mind that the plan, or a version of it, will be used to help recruit participants onto the study.
? Identify the objectives and method of review that is appropriate for your organization and study.
? Develop the materials necessary for the review (e.g. presentation, report, proposal).
? Elicit opinions, ideas, and objections of key opinion leaders and those impacted by the project and include these in the review as appropriate.
? Plan how to deal with objections and deal with them.

Chapter 12

Inviting and Working with Participants

INTRODUCTION

As in all aspects of this book we address the more complex situations and encourage the reader to simplify or even omit specific tasks as appropriate for their study. Inviting participants is one of the more obvious tasks that will need to be tailored to the specific study. At one extreme, an internal study where the members of the project team themselves represent the participants, this section will hardly seem appropriate. At the other extreme, inviting participants into a benchmarking club that is likely to run annually may require significant effort over many years.

If the intention is to benchmark with one or a few participants, either all together or sequentially, then the list of preferred participants will need to be ranked as described in Chapter 5 and organizations approached one after another until a suitable number of partners have agreed to take part.

If the intention is to create a club for a 'one-off' study, a number of organizations will need to be invited in a short space of time to encourage them to attend an inaugural (often called kick-off) meeting for potential participants.

If the intention is to create a club that grows and develops, then it may be more appropriate to start by building up the membership locally/nationally and extending the geographical catchment area in later years.

For the purpose of this chapter, we consider the situation in which a number of participants are being approached to join a new club. The first step is to gain agreement to hold a meeting of interested organizations to take the study forward.

The objectives of this step of the process are

1. To identify the appropriate person/department to approach within the target organization.

2. Present information about the study to the appropriate person. This may be via a one-to-one phone call, meeting, email, website forum or, more likely, a combination of methods.
3. Gather feedback and suggestions.
4. Gain commitment to attend an inaugural meeting and/or to participate in the study.

The inaugural meeting is, if the project is supported by enough potential participants, probably only the first of a number of meetings that will be held in order to finalize and implement the project plan developed by the initiating organization. The number and length of participant meetings required will depend on such factors as:

- The type and scope of the study: the greater the scope and the more complicated the subject, the longer it will take.
- The geographical location: participants geographically close will find it easier to meet and may prefer a greater number of shorter meetings. Participants travelling around the world to a meeting will prefer fewer meetings that last longer.
- Benchmarking experience of the participants: participants new to benchmarking will generally need longer to identify and work through issues compared to those familiar with benchmarking.

However this part of the study is run, i.e. the mix of meetings, sub-groups, use of email and websites, the basic objective remains the same: to finalize the benchmarking project plan. For studies that run regularly, typically annually, the amount of effort required to kick off subsequent studies will be significantly less because in future years the details of the study will generally only require modification and clarification.

DELIVERABLES

The deliverable for inviting participants is to gain agreement from the targeted organization to either:

- Join the study (preferably) or
- Attend an inaugural meeting or
- Agree to be kept informed of progress.

The deliverables of the meeting(s) are

- An agreed detailed project plan including metrics, timescale, and confidentiality agreement.
- Committed participants.

12.1 IDENTIFYING THE CONTACT

The first step in contacting a potential participant is to identify the right person to contact. There are a number of methods of doing this including:

- **Personal introduction**. Personal introduction is the preferred method of contact and will be through someone in the organization having a contact in the target organization. The aim is to gain some form of personal introduction to the person/department which is most likely to be interested in the proposed study.

- **Interest groups**. Usually a personal introduction will not be possible, especially if the target organization is in another country. In this situation the next most likely approach will be through an interest group. Many organizations, or individuals within the organization, belong to professional, trade, or other body. These bodies often host meetings, conferences, or other events to promote contact amongst members. Such events are ideal for making and developing contacts for benchmarking projects. Many also have notice boards on their websites, or will contact members on your behalf or in some other way facilitate contact between appropriate people.

- **Benchmarking Service Organizations**. Benchmarking Service Organizations are a special type of interest group dedicated to meeting the needs of organizations that are interested in learning from each other. Each will have its own rules and services, but most will be able to help identify from their own membership organizations that may be interested in benchmarking (⇨ Part 4 'Best Practice Club').

- **Contact the appropriate Manager**. The above methods are almost always preferred for approaching organizations. However, if all else fails, it is possible to contact the appropriate department head or benchmarking contact in the target company. When using this approach it is advisable to find out the contacts' name and, if possible a little about the target organization in general and the contact's department specifically. Their website, advertising materials, and/or a phone call to the switchboard should provide the required information.

12.1.1 Presenting Information

The initial phone call may be the only chance of contact and it is important to get across the key information clearly and concisely and be able to answer any questions and objections effectively.

The key information is likely to include the scope, objectives, benefits to the target organization, outline of data/information requirements, deliverables and timescales, costs, confidentiality and names other confirmed or invited participants.

It is unlikely that the invitee will immediately agree to join the study, or to attend an inaugural meeting. More usually the outcome of the first phone call is to gain agreement to send further information: an invitation pack.

12.2 SENDING AN INVITATION PACK

Having gained agreement to send further information to the potential participant, the next step is to do so. The purpose of the invitation pack is to lead the recipient to take the next step. This may be to commit to the study, or to commit to attend an inaugural meeting, or agreement to a visit to discuss the study in person.

The pack will typically consist of:

✓ A covering email/letter thanking the recipient for agreeing to review the benchmarking proposal and outlining the key purpose and objectives of the study. The email/letter can also be used to address any specific issues that were raised during the telephone conversation.
✓ The proposed objectives, metrics, timescales, costs, and other resource requirements.
✓ An invitation to a meeting of interested organizations. It is important to clarify that attending the meeting does not signify a commitment to joining the study.
✓ Any other information that may be of use.
✓ A statement that you will call to discuss the information in the next few days.

It could also be useful to include an automated presentation of the key issues of the study and/or links to the benchmarking study website, and other pertinent information such as a list of frequently asked questions (FAQs).

12.3 FOLLOW-UP ACTIVITIES

About a week after the information is expected to arrive, it is helpful to call the recipient to discuss their thoughts and ideas about the study. The objective of the phone call is to receive feedback on the proposal and ascertain how best to progress the organization to make a decision about the next step.

Assuming the recipient is positive about the study, the next steps are either for the recipient to agree to attend an inaugural meeting or a visit to be arranged to discuss the study.

A visit at this time is very useful for developing a personal relationship with people in the target organization and will be particularly important if those people are not able to attend the inaugural meeting. This site visit may be before the inaugural meeting or afterwards. The key advantage of visiting before the meeting is that the target organization's views can be taken into account before the meeting; it may be too late to take action on them afterwards.

'It seems that with some organizations, however much I talk to them over the phone, it is not until I visit them that they join the study.'

　　　　　　Benchmarking marketing manager

Although presented here as single activities, i.e. one follow-up phone call and one meeting, there are likely to be a number of phone calls, emails, and possibly meetings during this process.

The ultimate aim of the follow-up activities is to get a favourable decision regarding joining the benchmarking study or at least for the invitee to maintain genuine interest.

12.4 DEALING WITH OBJECTIONS

It is likely that at various times in the benchmarking process people will raise objections. Objections at the project review stage were discussed in Chapter 11. Here we discuss the other step where objections are most likely to be raised, i.e. when inviting participants to join the study.

Some objections may be valid, some may be due to misunderstanding while others are excuses to hide the real objection. Some objections and possible responses are provided in Figure 12.1.

Objection	Possible responses
Lack of time/manpower availability to collect the required data.	• Ensure that the effort required to take part is clear. • Offer to supply personnel to help (perhaps for a fee).
The costs of participating are too high.	• Review the potential benefits with the aim of determining whether the benefits would outweigh the costs. • Compare the cost of the study with, for example, a saving of 1% in the area being benchmarked.
No-one else does what we do, and so there can be no comparable data.	• Some parts of the process may be comparable. • All participants are unique, and we have methods of accounting for these differences (e.g. through normalization). • Other participants have raised the same issue, but believe that it is still possible to benchmark.
There is no need to benchmark because we are the best.	• How do they know they are "the best"? • External benchmarking will provide independent evidence confirm the extent to which that this is true.
The organization has a policy of not sharing data externally and/or the data is too sensitive/confidential.	Ask if sharing data anonymously would be acceptable ⇨ Chapter 15 for examples of how to do this with costs.
How do we know that other organizations will submit accurate data?	It is true that other participants may enter inaccurate data. However: • Data validation will help to ensure data is accurate. • Site visits by participants and/or facilitators will help ascertain whether data is accurate. • Better performers will be asked to explain how they achieve their performance levels.

FIGURE 12.1   Some objections to joining a benchmarking study and possible replies.

12.5 THE FIRST MEETING

Hopefully a number of organizations will have committed to at least attending an inaugural meeting, if not to joining the study. There are now likely to be a series of meetings or other forms of contact aimed at finalizing and then implementing the project plan.

Meetings operate at different levels. The obvious agendered objectives are generally task oriented. However, there is also a need to build relationships, trust, a sense of belonging and ownership of the project. These 'soft' issues are important because people are unlikely to commit to a group to whom they feel alienated even if they agree with the study.

12.5.1 The Situation at the First Meeting

The first meeting is likely to be very different, both in terms of goals and ambiance, to later meetings. People will be coming together for the first time who will probably not know each other. There are likely to be people with very differing project objectives, differing views as to how the project should progress, differing commitment levels, and differing levels of confidence and abilities.

12.5.2 Meeting Preparation

In addition to the usual preparations for a meeting such as preparing an agenda, organizing breaks, catering, etc., before the meeting, facilitators should aim to discover each attendee's attitudes, aspirations, and concerns regarding the study in general and the first meeting in particular. The following issues are relevant to all meetings, but are especially pertinent to the first:

✓ It is useful to know whether participants have experience of other bench-marking studies; if they do, we may be able to learn from their experiences and we will certainly be able to learn about their expectations of benchmark-ing in general and benchmarking meetings in particular.

✓ Understanding how participants plan to use the study and whether they have any specific objectives or issues that they want to address.

✓ It is likely that the delegates at the meeting will represent a mix of stakeholders in the study. Some may be managers, others may be from the workforce, or responsible for gathering the data, and so on. Understanding their roles will help identify their needs and skills that they bring to the meeting.

✓ Encouraging delegates to submit agenda items helps them to feel ownership for the study, though care and diplomacy may be required to deal with contentious or inappropriate issues.

✓ It is likely that at times opinion will be divided on an issue. For these situations it is helpful to decide beforehand how decisions will be made on contentious issues without alienating delegates.

✓ There may be participants with specific needs or views who are unable to attend the meeting. It is important to devise a method of taking their views into account during the meeting.

✓ It may be appropriate to employ an independent consultant to advise on technical aspects of the study (and to advise the facilitators during the study) (⇨ Chapter 21).

Addressing these issues will help the facilitators to steer the meeting using the knowledge, experience, and strengths of those present in the meeting, and to prepare for any negative issues that may arise.

Case study: A Domineering Participant

One 'kick-off' meeting was derailed by the dominance of one particular person who wanted to discuss one particular issue. It became obvious that his position on that one issue was at odds with every other participant. However, he would not accept that their viewpoint was valid, let alone acquiesce to their wishes or accept that his organization could be excluded from that one aspect of the study if he wished. Despite the facilitators best efforts he would not allow the discussion to move on.

By the end of the meeting only a third of the agenda had been covered as three hours had been spent labouring over one item that all but one participant felt was a waste of time.

Had the facilitators been aware of the issue in advance they could have prepared a better response to the situation.

As with all meetings, it is helpful to issue an agenda prior to the meeting so that people know what to expect and can prepare appropriately for it. It also allows those who are unable to attend the meeting to submit their comments via the facilitators.

12.5.3 Focus of the First Meeting

The early meetings, and particularly the first meeting, will focus on building relationships, learning to work together, reviewing common ground, and mapping the general course of the project.

If the participants work together and develop a team spirit and cohesion they will naturally want the team to succeed and will be more prepared to compromise their own needs and wishes to encourage others to join and remain in the study. While the level of commitment to join the study should always be in the facilitators mind, agreement to join the study can only realistically be confirmed once the participants take (emotional) ownership of the study. This is far more likely to occur if participants feel they are working with people they know, trust, and with whom they have a common interest.

In later meetings the balance will shift away from relationship building towards finalizing the details, implementing, and managing the study.

12.5.4 Facilitation of the First Meeting

In the early days of working together, the role of the facilitator(s) of the project is primarily to build group cohesion and secondly to progress the task oriented aspects of the study.

For this reason, the first meeting(s) are likely to set aside more time for team building activities, both formal and informal. How each specific benchmarking study will set about achieving this will depend on a number of factors including the number of people at the meeting, cultural norms and time available. An evening meal before the meeting, working groups and asking participants to give an introduction to themselves and their organization are all common useful team building tools.

Case Study: Presuming on Commitment

One can never be sure of a participant's level of commitment. In one study a participant successfully persuaded the others to make a significant conceptual change to the metric structure. It was a surprise to the facilitators when he withdrew from the study some weeks later.

Commitment to the study will come at different times for different people, and may wax and wane. The facilitator needs to be aware of how the participants are feeling about the study and aim to help the participant make the best decision for their organization, and then help them adhere to it. In general, for people to want to continue working in the study they will need to believe that both they and their organization will benefit and that they can work with the other people involved in the study.

12.5.5 Typical Agenda for the First Meeting

A typical agenda for a one-day first meeting would include:

- Participants invited to an evening meal the night before the meeting, allowing people to meet informally.
- Review of the agenda.
- Introductions, possibly including a short presentation from each participant.
- Presentation of the proposal (e.g. scope, objectives, metrics, timetable, resources). This session reviews the progress to date and sets out the current status. There should be plenty of time allowed for discussion on these points.

- Next steps. The facilitators should have formulated a plan, perhaps modified by events of the meeting, as to the next steps. These need to be discussed and agreed. The next steps are likely to include at least three aspects:
  - location, date, and rough agenda for the next meeting/contact. It is not uncommon for participants to take it in turn to host meetings, and if appropriate, provide a guided tour of their facility.
  - agreement of actions to be taken by participants (with timelines).
  - setting up of sub-groups (if necessary/appropriate).
- Review of the day.
- Close.

12.6 FOLLOW-ON MEETINGS

The follow-on meetings will clearly be based on agreement reached at the first meeting. In general terms they will review and develop all aspects of the study including:

- Finalizing the metrics, definitions, metrics model(s), and any other information to be reported (⇨ Chapters 7–10).
- Arranging a Help Desk facility (⇨ Chapter 13).
- Agreement on legal issues, including confidentiality (⇨ Chapter 21).
- Agreement on financial arrangement for funding the study.
- Agreement on use and role of external facilitators (if required) (⇨ Chapter 21).
- Agreement on report format (⇨ Chapter 15).
- Agreement on methods of sharing best practices (⇨ Chapter 17).
- Agreement on meeting location.

12.7 BETWEEN-MEETING ACTIVITIES

As soon as possible after the meeting the facilitators should contact attendees to get further feedback, comments, and suggestions. While it may not be possible to contact everyone, select those who may have been quiet during the meeting, those whom you suspect may be feeling negative about the study and opinion leaders. It is also good practice to thank those who contributed significantly or presented to the meeting.

As with all minuted meetings, it is important to issue minutes promptly, within 24 or 48 hours is ideal. Sometimes it may even be appropriate to have the minutes written up during the day and issue them at the end of the meeting. However, the advantages of waiting a day or so include:

1. It allows opportunity to review the minutes before they are issued.
2. A useful tactic is to include items either *not* discussed during the meeting or about which the participants or facilitators have further comments that were

not aired during the meeting. These items must, of course, be highlighted as such, but it does leave scope for tidying up and communicating unresolved issues.

3. Actions that are completed immediately after the meeting can be reported as closed out or at least progressed.

Finally, the facilitator(s) should review how they can improve for the next meeting.

At the meetings it is likely that tasks were assigned to the facilitators and/or participants. Experience has shown that often these tasks are overlooked because when participants return to their normal work routine they get engrossed in more immediate and ongoing issues. The result is that the enthusiasm and good intentions engendered at the meeting sometimes erode.

The facilitators have an important role to play in maintaining the momentum of the project between meetings. In addition to completing their own tasks, they can encourage others to do the same by remaining in contact with them and supporting them in whatever way they can.

Sub-groups may have been formed to address specific issues, such as developing definitions for the metrics, reviewing legal issues, or drawing up a benchmarking protocol. If there appears to be little progress, facilitators may offer to help progress them with a view to getting the tasks completed before the next meeting.

In addition to moving the project along in this way, the facilitators should keep participants aware of progress, and engaged in the study. There are a number of ways of doing this including:

- Ensuring that progress meetings, as outlined above are held regularly.
- Issuing of minutes.
- Issuing of progress and status reports.
- Developing a website facility, for example, to post notices and track progress.
- Personal contact with participants.

SUMMARY

The aim of inviting participants to join the study is to gain their commitment to doing so.

Inviting organizations to join the study usually consists of three steps:

- ✓ Identify the right person to contact.
- ✓ Gain agreement to give them information about the study.
- ✓ Discussing aspects of the study in terms that are of interest to the potential participant until they make their decision.

Gaining agreement is likely to require a number of emails, phone calls, and/or meetings.

Working with participants to agree the details of the study can begin without all the participants having agreed to join.

Meetings are important not only to agree the details of the study but also to build up relationships and cooperation between participants and ultimately to foster a feeling of ownership and commitment to the study.

Meetings need to be carefully planned, facilitated, and minuted, especially the first meeting, which is likely to be the first time many of the participants have met each other.

Between meetings facilitators need to continue working to progress the study and ensure that participants complete their tasks as agreed.

𝕐 Activity

Develop a plan for contacting potential participants.

? How will you identify the correct person to talk to?
? How will you contact them?
? What will you say/write?
? What information will you give them about the study?
? What do you want them to do?
? What objections may they have?
? How will you deal with objections?
? What is your schedule for inviting people?

Planning participant meetings:

? What tasks need to be completed before the end of the study?
? What work can be done between meetings?
? What tasks can be done by sub-groups?
? How many meetings do you think will be required?
? How will the participants get to know each other?
? How will you encourage relationship building, especially during the first meeting?
? How will you make participants feel welcome and committed to the study?
? Draw up an agenda for the first meeting.
? How will you ensure that any between-meeting actions are completed?

The Benchmarking Process –
Benchmarking Performances

Chapter 13

Effective Acquisition of Complete Accurate Data

INTRODUCTION

By this stage in the benchmarking study the required data should have been agreed by participants and clearly defined. The next step is to acquire complete accurate data from participants prior to beginning data analysis and in this chapter we consider three key aspects of doing so:

1. Development of a Data Collection Pack that will facilitate the submission of complete correct data.
2. Provision of a Help Desk whose role includes answering participant queries, but also acts as a key source of information for analysis, monitoring progress, and improving the study.
3. Validation of received data to help ensure that the conclusions from the analysis are correct.

DELIVERABLES

The key deliverables of this step are

✓ A complete set of reliable data received on time.
✓ Explanations of unusual data values.

13.1 DEVELOPMENT OF DATA COLLECTION PACKS

The aim of the Data Collection Pack is to help elicit the required data, information, documents, and anything else that may be required from the participant as quickly and easily as possible.

The more onerous supplying data and information are, the more participants are likely to delay and the less accurate the data are likely to be. For these reasons the pack needs to be clear, concise, and easy to complete and return; and, if there are any queries, provide details of how the participant can obtain support.

> The **Data Collection Pack** is everything issued by the facilitators of the study to the participants in order to elicit the agreed data and information from them.

Clearly, the Data Collection Pack will vary depending on the requirements and agreements of the study. It typically includes items such as:

1. Guidelines for completing the data entry.
2. Data entry and/or questionnaire forms.
3. Metrics structure, where appropriate.
4. Definitions.
5. Contact details for the Help Desk (in case of queries).
6. Confidentiality agreement for signature (though this may be dealt with separately).
7. Request for agreed documents, e.g. organization charts and workplace layouts.

If the pack is to be used in countries with different native languages it may be necessary to translate part or all of it. For most international studies the language of business is English, but the people supplying data may not speak enough English to understand technical language. It is therefore worthwhile investigating whether and how much of the pack needs to be translated.

13.1.1 Selecting a Suitable Format for the Pack

There are three basic formats that the pack is likely to take: electronic spreadsheet, web based, and paper.

Electronic spreadsheet based packs are perhaps the most common for bespoke benchmarking studies. They have the advantages of being relatively easy to prepare, can be emailed or downloaded from websites and then emailed internally to people collating the data. Guidance on data completion as well as data checks and validation can be built into the spreadsheet, and a copy kept by the participant before returning the completed file to the facilitators. Subsequent re-submissions (quite common as data validation queries the submitted data) are simple and, with relatively simple controls, all parties can have access to the latest version of the documents. Spreadsheets also allow a limited amount of textual information to be entered. Any required documents that are not already electronically held can be scanned.

Web based data entry is more common where data entry is relatively simple both regarding the amount of data and where errors are unlikely. However, creating such a website can be costly and time consuming compared to electronic spreadsheets for all but the simplest types of data submission. Security checking is an added difficulty, both from the point of view of allowing only authorized people to access the site and, for public sites, ensuring that people do not enter data more than once. There are many examples of this type of data entry on the internet such as ⇨ Cronar salary review at www.croner.co.uk.

Paper based packs can be posted, handed out, or downloaded electronically (e.g. using pdf format) and are most commonly used for customer and opinion based surveys. They are, perhaps, the simplest to administer because there are no software compatibility issues, and the recipients do not need to have access to computers. Unlike electronic and web entry systems, data will probably have to be manually transcribed from the completed form into the computer, although with multiple choice questions it can be done electronically.

13.1.2 Guidelines for Data Entry

People completing data entry may not have been involved earlier in the benchmarking process and so may have little idea of the context in which the data they are to supply are being reported.

To help ensure a complete and accurate data/information submission it is useful to include guidelines for the person completing the submission. These guidelines will depend on the specific requirements of the study, however, typical guidelines include:

- ✓ Name and contact details of the person who will answer queries about the submitted data.
- ✓ Date by which the data should be submitted.
- ✓ To where to send the submission.
- ✓ Overall guidance as to what items should be included and excluded from the submission. For example, for a benchmarking study comparing the costs of running a manufacturing facility, capital expenditure items, insurance costs, rent, research and development, etc. may be excluded.
- ✓ An explanation of where to find data definitions.
- ✓ For data submissions using spreadsheets, an explanation of how validation and other built-in self check mechanisms can be used to validate the data.
- ✓ How to treat data that are unavailable, not applicable or zero.

There are three common situations that may cause difficulties when interpreting submitted data.

1. The true value is zero.
2. The true value is not known.

This might occur, for example, where equipment failures occur but are not recorded.

3. The data item is not relevant to the participant. For example, costs of regulatory visits will not be applicable if there are no regulatory visits.

These three situations are commonly dealt with in three different ways. The participant may:

1. Leave the cell blank,
2. Enter 'na' (which could imply not available or not applicable),
3. Enter zero.

This results in ambiguity because the analyst does not know what a blank cell or 'na' signify. A blank cell could imply, for example, that the value should be zero or that the participant forgot to enter data. Similarly 'na' could signify that the value is not applicable, or is not available.

A solution is to specify that:

1. No items should be left blank.
2. Zero should be entered where the value is zero.
3. 'nk' should be used where the value is non zero and not known.
4. 'nr' should be used where a value is not relevant, and an explanation provided.
5. 'na' should not be used because its meaning is ambiguous.

13.1.3 Designing Effective Data Entry/Questionnaire Forms

The main purpose of the Data Collection Pack is to extract the required information from the participant. In many benchmarking studies this will consist mainly of data, but may also include textual answers and documents. Spreadsheet packages are ideal for data submission as they facilitate data validation and can be designed to help participants move around the spreadsheet easily and identify where data need to be entered.

Some of the many useful attributes of spreadsheets are that it is possible to include:

● **Data definitions**. It is useful to provide a list of definitions in one place in a printable format so that people can easily refer to them while filling in the data or pass them on to others who are collating data. It is also helpful to place a definition as near as possible to the point of data entry, albeit that for long definitions it may be preferable to supply a shortened version of it.

- **Basic validation**. With electronic data entry it is often possible to provide at least basic data entry validation, for example, checking that numeric data lie within certain ranges. Drop-down lists for item selection is a useful method of ensuring that only pre-determined valid responses are provided.

- **Basic checking**. Moving beyond single number validation, basic checking allows groups of data items to be checked and if the results are suspect, or data are incomplete, messages can be displayed. For example, if both the number of children and the number of teachers at a school are entered, the pupil to teacher ratio can be calculated and if outside certain limits, a message displayed asking for confirmation.

- **Textual information**. Limited textual information can usually be entered along with data. Longer items or documents could be submitted as separate attachments.

- **The Metrics Structure**. Where a metrics structure has been developed, it is very useful to include it in the collection pack for two main reasons:

  1. It helps the participant understand how the data fit together. While it might be obvious to those who have spent days developing the model and data structure, those supplying the data may have little idea as to how their data fits in with all the data being reported.
  2. The structure can be populated with data entered into the spreadsheet to add a summary checking facility.

  Figure 13.1 shows a simplified extract of a typical manpower and cost model for a manufacturing facility benchmarking study. For process benchmarking a process flow chart may be more appropriate.

- **Contact Details for the Help Desk**. However carefully the data collection documents are prepared there are likely to be queries from participants. Generally the queries will relate to data reporting issues, but may also include project management issues such as submission dates, status of the project, and requests for late submission.

  The existence of a well run responsive Help Desk goes a long way towards answering these and similar queries and the contact details, usually email and telephone, should be clearly and prominently displayed in the data collection documents. The roles and management of the Help Desk are discussed in more detail below.

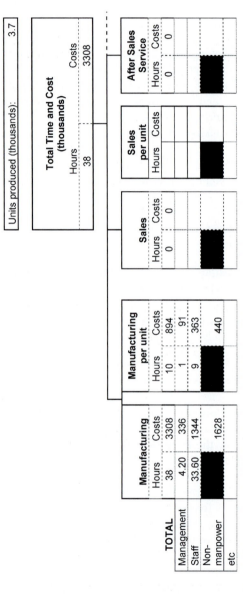

FIGURE 13.1   Simplified extract of a part completed metric model.

- **Confidentiality agreement and other documents**. Depending on how member administration is managed, there may be a confidentiality agreement included in the data collection documents. This may be for information only, or may need to be signed and returned as part of the data submission (⇨ Chapters 17, 21). In addition to data, there may be other information required for the study. This may include, for example, organization charts, process flow diagrams, or floor plans. This information will probably need to be supplied as separate documents, but the requirement for and description of such documents can be included in the pack.

13.2 THE ROLE OF THE HELP DESK

Once the Data Collection Pack has been issued, it would be nice to be able to sit back and await submissions of complete, accurate, and on-time data. Unfortunately, experience suggests that there is likely to be a long wait!

We begin by looking at the central role that the Help Desk plays not only in helping ensure that supplied data are correct and on time but also how it acts as a key depository for information to drive improvements for future studies.

At first glance the role of the Help Desk may be seen as a necessary but relatively unimportant aspect of the benchmarking study. However, the reality is that the Help Desk can play an extremely useful role far beyond answering queries: it can play a pivotal role in monitoring progress and gathering information for use throughout the study. These roles, as depicted in Figure 13.2, include:

1. Resolving participant queries relating to data submissions.
2. Capturing queries as a source of potential improvements for future studies.
3. Monitoring study progress and expediting data.
4. Awareness of participant operations used for data validation, analysis, and reporting.

As the Help Desk is the centre of some of these tasks and can provide useful information for other tasks, the people involved in running the Help Desk should be involved in or have knowledge of the other tasks also.

1. **Resolving Participant Queries**. As the name implies, the main purpose of the Help Desk is to respond to any queries from participants regarding data submission. However carefully definitions have been discussed and agreed they will not be perfect. There may, for example, be inconsistencies, omissions, or poor wording. In addition, some participants will have data items that do not fit neatly with the definitions or items that have not been included in any definition. The role of the Help Desk is to resolve queries about data and information supply and, where appropriate, inform all participants of any changes or clarifications.

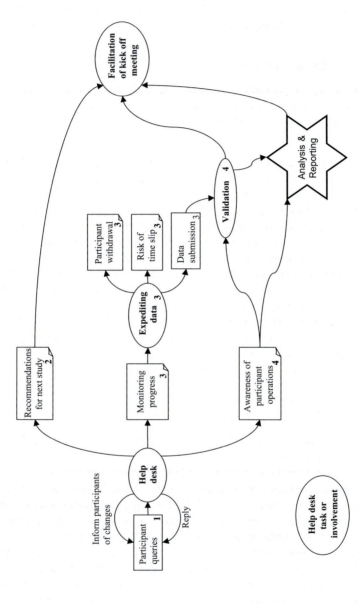

FIGURE 13.2   Roles of the Help Desk.

Help Desk personnel should develop a system for capturing, logging, and responding to queries efficiently and effectively. For studies with many participants or that may run regularly these can be used to build up a database of frequently asked questions (FAQs) which can be disseminated to participants. The advantage of such as system is that responses to similar queries will be consistent.

Typical Help Desk Queries

The variety of possible queries for the Help Desk is almost limitless. However, many queries will be of a similar nature. For example, a review of queries over a number of studies showed that most queries fall into the follow types:

- **Scope of the study**. For example, in a benchmarking study looking at food stores and home delivery, one store also delivers local produce overseas; is that included?
- **Data definitions**. For example, in a study looking at transport in different cities the listed modes of transport include taxis and bicycles. One city operates a fleet of bicycle–rickshaws; are these classified as bicycles or taxis or something else?
- **Special circumstances**. Manufacturing process times are severely affected when a supplier's warehouse is damaged by a freak storm. How should this a-typical data be recorded?
- **Unavailable/confidential data**. For example, a participant works in a joint venture with two other organizations. One of the venturers objects to financial data being submitted to the study.
- **Benchmarking process queries**. These cover all aspects of how the benchmarking study is or should progress such as data submission deadline, requests to submit data late, status of the project, analyses that will be carried out, and how special situations will be addressed.

2. **Capturing Improvements for Following Studies**. Queries usually arise because some aspect of the required data is not clear. Having resolved the query, the information should be filed for review at the next kick-off meeting as a potential improvement to the study. In addition to unclear definitions, data items, titles and requirements, resolving queries may lead to other ideas of how the study could be improved and these should also be logged for discussion at the next kick-off meeting.

3. **Monitoring Study Progress**. The Help Desk is ideally situated to monitor data collection progress. They will be in contact with those people who have queries and can use this contact to gauge when they are likely to submit data. They can also contact those participants from whom they have not heard and expedite data, offer help, and will be aware of any participant who has

decided to withdraw from the study. Being aware of progress they will be able advise of the risk of time slippage.

4. **Gathering information on participant operations**. In order to answer queries correctly it is sometimes helpful for the participant to explain pertinent aspects of their organization's operations. This information can be logged and is often useful both for validation and analysis. In the examples listed above, awareness that a participant's supplier's warehouse fire caused a manufacturing problem could be an important factor during the analysis.

As can be seen, the Help Desk can be a rich source of information for validation, analysis, and reporting. It is therefore often very useful if those who are involved in the Help Desk are also involved in these other activities.

13.3 FACILITATING ON-TIME DATA SUBMISSIONS

13.3.1 Why Late Data are a Problem

Late data submission is a common difficulty that benchmarking facilitators have to contend with. Deliverables deadlines will have been agreed with participants and, as participants may use the results of benchmarking studies as inputs to other activities such as budgeting, delays in reporting can result in frustration. However, the facilitators also suffer because they will have scheduled their workload according to the planned timetable freeing up resources as required by the plan. Delays for them not only result in unsatisfied participants but also staff idle time as they await submissions and/or a shortage of people/resources when the data do eventually arrive (Figure 13.3).

13.3.2 Common Causes for Delays

Those involved in agreeing the terms of the benchmarking study want to see the study succeed and bring benefits to their organization. Frequently these are not the people who are tasked with gathering and submitting the data. Herein lays perhaps the biggest difficulty in getting data submitted on time: the person retrieving and submitting the data may not have been involved in any aspect of the study and perceives this as an extra task, and often an unwelcome interruption by his manager. The chances are that the data supplier has a full time job and is already working hard just to keep on top of his normal activities. To add to the difficulties, data required will often not be in a format readily accessible to the supplier and they may have to do considerable work to calculate or gather the data themselves or try to obtain it from others.

Another common cause for delays is unavailability of data. This applies particularly to cost data which may not be available until well after the close of the period to which it relates.

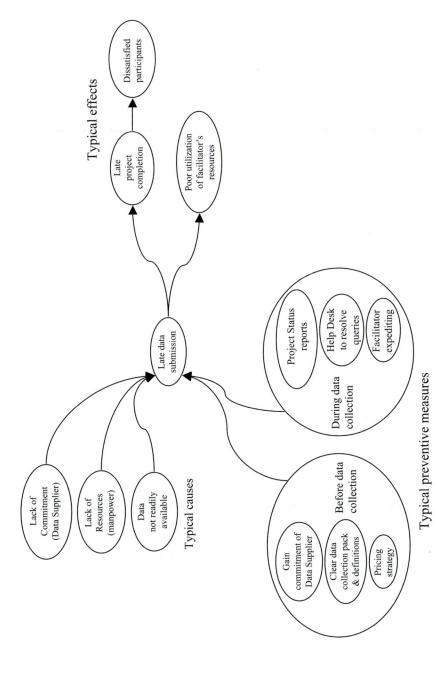

FIGURE 13.3   Causes, preventive measures, and effects of late submissions.

13.3.3 Delay Prevention Action

The participant's management can do much to facilitate timely data collection. The simplest method is to encourage buy-in from the key people in the study, i.e. those whose cooperation is vital to the smooth running of the project. There are a number of ways of doing this and each organization will have its own methods depending on the culture. Firstly, it is important that the data supplier be aware that the benchmarking project has priority. This may be achieved simply by the appropriate manager taking an active interest in progress. Where data collection is a particularly onerous task, it can be further facilitated by assigning someone full time to the project.

The benchmarking study team can also help in the following ways:

✓ By providing a user-friendly data collection facility with clear unambiguous data definitions.

✓ By building a supportive working relationship with the data suppliers in the participating organization.

✓ Where the study is run by consultants, they may agree a fee for participation with a discount for on-time submissions. This sometimes gives leverage to those within the participating organization to persuade management to release resources to submit the data on time. However, the opposite, issuing a penalty for late submission, is likely to be seen as punitive and create antagonism.

✓ Issuing regular project updates of the status of the project including what data have been received, what has been validated, and what is still outstanding. This 'naming and shaming' policy needs to be implemented with care, and should preferably be agreed with all participants beforehand.

✓ In some studies it is usual that the participant's access to results of the study is only granted once certain conditions have been met. For example, the participant has submitted a full data submission and payment has been made.

✓ New participants in particular may struggle to understand the data requirements despite support from a Help Desk. One solution is to provide on-site assistance with data collection.

✓ Despite everyone's best efforts the data submission may be delayed past the deadline. At some point in time the facilitators and participants may have to accept that the project must progress to the analysis stage even if some participants have not submitted data. In this situation there are several possible methods of mitigating the effects:

1. The analysis can be revised once all the data have been received and validated. The extent of the revision (e.g. whether only updated charts will be issued or whether the revision will include partial or complete re-analysis) will depend on issues such as the terms of the study and confidentiality.

2. Participant(s) who cannot supply the data in a timely fashion can withdraw from the study.

3. The 'late' participants(s) can ask for a separate review of their data, with the agreement of the other participants.

One of the advantages of a rolling study (i.e. updated results are issued on a regular basis or as soon as data are available, as opposed to an annual project) is that the issue of late submissions, although still important, need not delay the study – participants only have access to results once they have fulfilled agreed conditions (⇨ Part 4 'Drilling Performance Review').

13.4 DATA VALIDATION

We cannot assume that submitted data are correct, and often it will not be. Therefore, the last step in data acquisition is validating received data. This important step in the benchmarking process is often ignored completely or inadequately addressed.

Unfortunately, it is all too easy for the facilitator to absolve himself of the responsibility to validate the data by hiding behind the fig leaf that it is the participant's responsibility to ensure that the data are correct before it is submitted. While it is undoubtedly true that the facilitator cannot take responsibility for the accuracy of data, it is also true that participants should expect facilitators to take all reasonable steps to ensure that data are correct. In this section we discuss:

1. The importance of accurate data.
2. Typical validation checks and how validating data can help identify useful information for analysis and reporting.
3. Tools for validating data.
4. Tips on querying the data and how querying data can help identify useful operational practices.
5. Automated validation for situations where the volume or complexity of data make manual validation difficult or too time consuming.

13.4.1 The Importance of Accurate and Complete Data

Accurate data are vital for a successful benchmarking study. Inaccurate data may lead to data analysts drawing wrong conclusions and participants making inappropriate decisions regarding improvement or other activities. ⇨ case study p. 182

ⓘ Exactly what is meant by 'accurate' data are debatable. For example, if sales volumes are in the order of 10 000, we are unlikely to need to know the actual volume to the nearest unit.

In most benchmarking studies we are seeking to identify large differences in performance – often of 50% or more. In this context, it is unlikely that errors in data of up to 10% would significantly alter the major conclusions. However, where the benchmarking study includes ranking amongst many participants, it is likely that errors of 10% will change the relative ranks of participants.

In addition to the main aim of ensuring correct data, another advantage of effective data validation is querying suspect data can lead to and provide useful information for the analysis and reporting phase of the project (Figure 13.4).

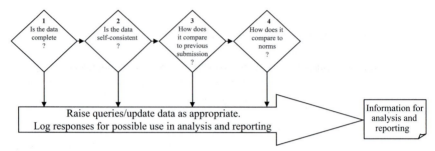

FIGURE 13.4   Validation not only aims to ensure that the data are correct; it often uncovers useful information for use in analysis and reporting phases.

13.4.2 Typical Validation Checks

Despite best efforts of facilitators with regard to a well organized Data Collection Document and Help Desk, on the one hand and diligent participants aiming to enter correct data on the other, errors will creep in.

The facilitator will need to develop validation methods to suit the study. However, in broad terms, validation falls into four sections, which may overlap:

1. Checking that the data are complete.
2. Checking that the data are self-consistent.
3. Comparing data with submissions from previous studies (where available).
4. Comparing the data with that of other participants as well as to norms/ expected values.

1. Checking that the data are complete should be straightforward. As explained above blank cells (or non-response) should not generally be acceptable, and should be queried.
2. Checking for self consistency within the data submission is more difficult, and the checks will clearly be data dependent. However, typical checks include:

   • Has the participant used the correct units (e.g. if costs should be in thousands, have they been submitted in millions, what currency was used?) This common error is usually easy to identify, but should be confirmed with the participant.
   • Are the figures of the right magnitude? The facilitator will usually have some information about the participants, and may have visited the site.

This information can be used to gather an impression of likely data values, for example, number of personnel, level of activity, workloads, and rates.

- Where groups of figures are related, are they consistent? Typical examples include

    - Manpower costs and hours (calculate cost per hour),
    - Process throughputs/production rates (e.g. units produced per man hour, cost per unit produced),
    - Incident rates (e.g. absenteeism, injury, failure, rejection),
    - Ratios (e.g. there are rules of thumb/industry averages for the ratio between maintenance, operations, and overheads).

3. Where the participant has submitted data in earlier studies, how does the current data compare with previous data submissions? Where there are significant differences, they should be queried. If the change in data reflects a process change information about the change may be useful to include in the analysis and/or report.

4. The final set of checks is carried out in comparison with data submissions from other participants. The idea is to compare the data with that of other participants and query extreme values. If the data are confirmed as being correct then there is a learning opportunity. If the data value represents a good performance, there will be an opportunity to explore the reasons for the good performance and share the learning amongst other participants. On the other hand, where the performance is poor, it represents an opportunity for the participant to learn from others in the study and therefore improve performance.

Occasionally there may be other methods of validating data. For example, if there are two or more participants from the same parent organization, the parent organization may have some knowledge of relative performance levels. These can be checked against the submitted data. Some data items may already be in the public domain and can be checked against other sources.

One final method of validation is to ask participants to review the data themselves to identify any concerns that they may have. This works well in benchmarking studies where participants have some knowledge of each others' performance levels. Two methods of doing this are to hold a data review meeting at which data, possibly with a selection of charts and other information are shown and discussed, or to circulate them electronically. Alternatively this may be done as part of a review at the reporting stage ⇨ Chapter 15 (though the result of finding data errors at that stage may be significant re-work and delay of the deliverables). Such reviews not only serve as a final peer-review of data but will often lead to participants requesting certain analysis or explanations from one another, all of which may be included in the final report.

Case Study: Learning from Zero

One participant returned zero for maintenance planning costs and hours. When queried he explained that there was no maintenance planning department or group. The maintenance engineers planned the maintenance themselves.

The same participant also returned zero for maintenance purchasing activities. The explanation was again that the engineer responsible for a particular piece of maintenance would carry out purchasing activities.

These pieces of information were passed through to the analysis phase and the effect of this unusual method of working was analysed and highlighted in the report. It gave all participants the opportunity to consider whether changing from the more traditional functional to a project based approach would be right for them.

13.4.3 Tools for Validating Data

A certain amount of data validation can be done simply by looking at the data and considering whether the value provided is reasonable. Unfortunately the errors identified by this method will usually be the obvious ones: missing values, wrong units used, etc.

For less obvious errors we can use charts to help. One of the key tools for data validation is the bar chart. In the example (Figure 13.5), from a call centre benchmarking study, the average seconds per call are plotted for each participant.

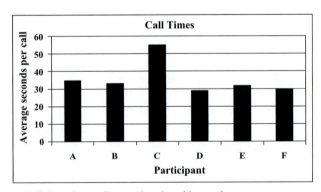

FIGURE 13.5   Call times for a call centre benchmarking study.

From the chart we see that participant C has a very high average call time compared to the other participants and should be queried.

Example: Data Validation: Use of a Scatter Chart

In the chart below, planned and unplanned maintenance are compared for 16 participants of a maintenance benchmarking study. The scatter chart shows that in general as planned maintenance increases, unplanned maintenance decreases. The solid line is a regression line, which can easily be added by most spreadsheet packages. The advantage of showing the regression line is that it draws the eye to the relationship between the variables plotted and allows rough estimates of one variable from the other. In the chart there are two outlying values, drawn as squares and labelled A and B.

Participant A is carrying out about 160 (thousand) hours of planned maintenance, and we would expect that the amount of unplanned maintenance should be about 60 hours. However, they are reporting about 90 hours, a very large deviation from the regression line compared to other participants.

Similarly, participant B carried out about 70 hours of planned maintenance, and from the chart we see that we would expect B to carry out about 90 hours of unplanned maintenance, whereas it is only carrying out less than 40 hours. If B's data are correct, other participants, particularly those with higher unplanned main-tenance, may have an opportunity to improve their maintenance activities. There may, of course, be other explanations and these will be explored in the Analysis section.

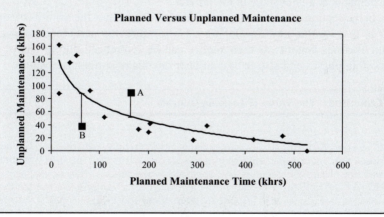

There are other tools that can be used for data validation, but those presented here are some of the most common.

13.4.4 Querying the Data

The analyst should never change a participant's data without their agreement. While this might seem an obvious statement, it is always tempting to assume that a blank cell (a non-response) should be zero, or that the data are wrong by a factor of a hundred, thousand, or million.

Case Study: Validating Unexpected Values: Stores Area

One benchmarking study requested the value of goods in the stores area. A participant entered $215. Assuming the value should have been $215 000, which, while low, would have been of a similar order of magnitude to other participants, the facilitator changed the value. On confirming the suspected error with the participant, the participant confirmed that the value was indeed $215, not $215 000 and explained how they managed the stores so that they required such a small amount.

Within a study, some of the queries may follow the same format. An appropriate standard wording for each type of query can be developed both for speed of raising the queries and, over time, to improve the wording so that it is unambiguous. To help the participant, and minimize the risk of misunderstanding it is always advisable to pinpoint where the data item is and, if values have been supplied, quote the data giving a clear reasoning for concern.

Queries about submitted data values may be raised by phone or email. However, if discussed over the phone, it is useful to confirm in writing what was agreed as there may be related queries either during the analysis or from the participants at a later stage in the project.

In the section on resolving participant queries we saw the benefits of logging them. In a similar way, developing a log for querying participants about their data also has benefits as their replies can be referenced during analysis and reporting phases and also in case of later queries regarding data values.

Case Study: The Value of Logging Queries

In one benchmarking study a participant wanted to refer to data they had submitted over a number of years. While they had kept some of the data they had submitted it was not always the same as had appeared in the reports. Having kept the data queries and participant replies from previous years it was possible to identify the changes they had made in response to queries and quote the reasons they had given for making those changes.

13.4.5 Automated Validation

Where the data are particularly complex, or there is a huge quantity of data, one viable option is to automate the validation by writing validation software. The specification, development, and testing of such software is likely to be beyond the scope of all but IT professionals, and, because of the likely financial investment in developing the software, is likely to be viable for only studies carried out on a regular basis.

Even with automated validation it is likely that not all errors can be identified, but it can provide a very quick method of checking against a large number of rules. A further advantage is that specifying and developing the software ensures that complex data relationships are thoroughly understood.

In addition to checking for suspect data, the software can be further developed to produce error messages, prepare wording for queries to send to participants, and even prepare emails. However, it is still vital that the suspected errors be reviewed before issuing (⇨ Part 4 'Drilling Performance Review').

SUMMARY

Complete, accurate reliable data are a requirement for a successful study.

Achieving reliable data is the result of painstaking work from metric selection right through to data validation. Figure 13.6 shows the steps that have been discussed up to this point in the book to ensure that the data received from participants are reliable.

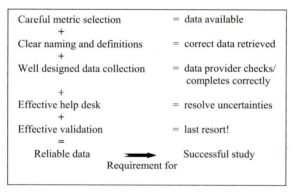

Careful metric selection	= data available
+	
Clear naming and definitions	= correct data retrieved
+	
Well designed data collection	= data provider checks/ completes correctly
+	
Effective help desk	= resolve uncertainties
+	
Effective validation	= last resort!
=	
Reliable data ⟶	Successful study
Requirement for	

FIGURE 13.6   Steps for ensuring reliable data.

The purpose of the Data Collection Pack is to elicit correct and complete data and information from participants. This is typically achieved by issuing a formatted spreadsheet or word processing document that participants complete.

In order to help participants complete the Data Collection Pack it is useful to include guides such as:

✓ Clear instructions on what data to enter and how to enter it.
✓ Metrics models to explain the structure of the metrics.
✓ Definitions of data easily accessed at the point of data entry and in a printable format.
✓ Validation and checks to facilitate checking of data before it is submitted.
✓ Help Desk contact details.

A Help Desk plays a key role in supporting participants through the data collection phase of the study. It also has a key role to play in monitoring the progress of the study as well as collecting information that may be useful during the analysis phase of the project.

Effective validation is the last scheduled chance of correcting data errors before the analysis and reporting begin.

ϒ Activities: Data Collection Packs

1. Identify what types of information will be required for your benchmarking study (e.g. data, floor plans, confidentiality agreements).
2. Keeping metrics models or other data summaries in mind, design an appropriate Data Collection Format. Consider:

   ? How will you clearly indicate what data are required?
   ? Where will be the best place to put definitions and help messages?
   ? What data entry validation can be built in?
   ? What data checking facilities can be built in?
   ? What overall instructions will be required, where will they be located?

3. Ask someone to review the data collection documents. Are they:

   ? Clear?
   ? Concise?
   ? Are items located in logical places?
   ? Are help facilities helpful? Easy to access?
   ? How easy is it to check that the right data have been entered?
   ? Are error/warning messages helpful?

ϒ Activities: Help Desk

1. Identify and define/detail the needs of your study regarding aspects such as:

   - Resolving queries from participants.
   - Monitoring data submission progress.
   - Expediting data/information.
   - Data validation.
   - Data analysis/reporting.

2. Decide how best to meet each of these needs.
3. Are there subsidiary benefits that the solutions in item 2 could bring to the study – e.g. fostering relationships between participants, between participants and facilitators, identifying possible topics for Best Practice sharing? If so, how will they be realized?

4. How will you evaluate how well the above needs were met and how to improve for future studies?

⅄ Activities: Validation

? What data validation checks can be carried out by the facilitators?
? How will the check be done?
? What other information can be gathered during data validation and how can it be used?
? Who are the appropriate people to validate the data and work with participants to resolve queries?

Chapter 14

Turning Data into Recommendations: Data Analysis

INTRODUCTION

In this phase of the project the data and information submitted by the participants along with information gleaned during site visits, Help Desk activities, and data validation are analysed. The results of the analysis will be presented as conclusions which in turn will lead to recommendations and be used to drive improvements within participants' organizations.

In this book we consider the situation where both analysis and reporting is carried out by study facilitators who are independent of the participant organizations. The advantages of the facilitators carrying out the analysis include:

✓ They have some familiarity with *all* participants' organizations, processes and data, and are therefore better able to account for mitigating circumstances, understand causes of differences in performance, etc.
✓ They have experience of analysis and reporting.
✓ They will be independent and impartial.
✓ It is more cost effective for the facilitator to carry out the analysis once than for each participant to carry out similar analyses.

It is important that the participants review the analyst's findings and perhaps supplement them with their own analyses, as discussed in ⇨ Chapter 16.

Whoever carries out the analysis, the methodology is likely to follow a similar pattern, and while the specific objectives of the analysis will vary from study they will typically include:

• Performance Gap Analysis (i.e. quantifying the gap between each participant and a target value).
• Analysis of the causes of the gap.

- Analysis to identify what actions are likely to improve performance, and by how much.
- Investigation of managerial and operational theories and philosophies (e.g. does outsourcing improve performance).
- Industry and/or individual participant performance trends.

The results of the analysis will be carried over into the report writing phase of the project.

In this chapter we focus on the process of the analysis. Here we discuss:

1. Typical sources of data and information for carrying out analysis.
2. Typical benchmarking analysis from comparison of participant performance levels through to recommendations.
3. The appropriate use of a statistician.

In most benchmarking studies the charting and analyses methods require only basic statistical methods, albeit they need to be correctly applied and interpreted. We present a number of charting tools that may be of use in ⇨ Appendix 1. Quartiles are commonly used in benchmarking and ⇨ Appendix 2 discusses some of the pitfalls of using them. For examples of charts used in reporting ⇨ Chapter 15. Charts used in data validation (⇨ Chapter 13), are also likely to be of use during the analysis.

DELIVERABLE

The deliverable of this step of the study is a list of findings with supporting charts, diagrams, data, practices, and other information as appropriate to be used as the basis of the report.

14.1 DATA AND INFORMATION AVAILABLE FOR ANALYSIS

The aim of the data analysis is to glean nuggets of information from the data, called findings. In the context of benchmarking, analysis includes not only statistical analysis of data but also the use of information to explain the data. From this viewpoint there are a number of potential sources of data and information available for analysis as shown in Figure 14.1.

The most obvious data source is the data submitted by the participant to the study. However, in addition useful information may be available from other sources (some of which have been discussed earlier and which are repeated here with different examples):

- **Participant Queries**

  During the data collection phase the participant may have asked questions about how to enter certain data items. While answering these queries, the

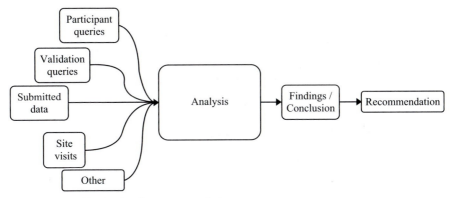

FIGURE 14.1   Turning data into recommendations.

Help Desk often gleans useful nuggets of information about the participant's organization or data.

Case Study: Learning from a Participant Query

One participant wanted to know how to apportion security costs at the facility. It was explained that while some costs were incurred totally by the facility, the neighbouring organizations had cooperated to reduce security costs by employing the same supplier to take care of external security such as the exterior wall/fence, creating one access point for all facilities. This novel approach was later copied by other participants.

● **Validation Queries**

During the validation phase, any unexpectedly high or low values will normally be queried. Where the data are correct, the facilitator will often be told or be able to find out why a value is high or low. This information is often useful to explain performance gaps.

Case Study: Mitigating Circumstances

One participant in a benchmarking study of construction projects had made very slow progress in a couple of projects. When queried, the participant explained that the work had been affected by a labour dispute in one sector of the workforce.

● **Site visits**

Site visits are very useful for gaining a general impression as to how well an organization is managed. In addition, simply by watching what is happening, reading notice boards, and chatting with the people who work there, the visitor is likely to glean useful information.

Case Study: Learning from Site Visits

Chatting over lunch one day at a participant's site a facilitator happened to be talking to a consulting engineer. When he discovered that the facilitator was involved in benchmarking, the engineer suggested the facilitator take a look at the shift changeover activities. Fortunately, the facilitator had the opportunity to follow this lead during his visit and did indeed discover that a significant source of waste in the system lay in the shift changeover process.

● **Other sources**

There may be other sources of information available to the facilitators such as safety procedures/guidelines, internal magazines, organization charts, and company websites. These may all contain useful information for explaining performance levels.

14.2 TYPICAL BENCHMARKING ANALYSES

While the process of analysing the benchmarking data may at first appear straightforward, there are many aspects to consider. Figure 14.2 outlines a typical analysis process showing the relationships between data/information, analyses, and results.

14.2.1 Performance Gap Analysis

The first major step in most benchmarking analyses is the Performance Gap Analysis, outlined in Figure 14.3, the first step of which is to compare performance levels of participants.

This is typically achieved by drawing a series of bar charts, which, by convention, are ranked with the superior performances on the left as shown in Figure 14.4. In this chart higher values suggest better performances as would be typical for yield, percentage of people seen by a doctor with a specified length of time, etc. Equally, low values could indicate superior performances for metrics such as accidents, error rates or costs, in which case the low values would appear on the left of the chart. (A common

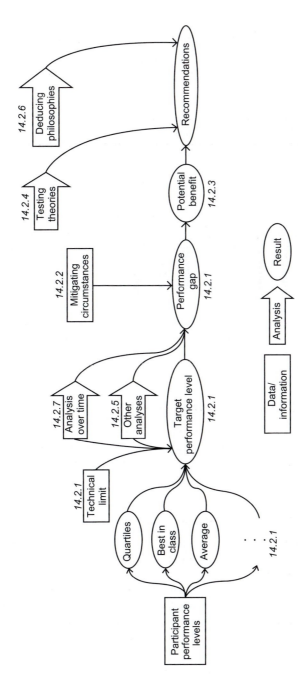

FIGURE 14.2   Key relationships of the data analysis (relevant section numbers in italics).

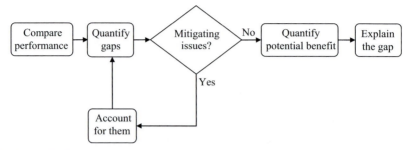

FIGURE 14.3 Performance Gap Analysis process (14.2.1–14.2.3).

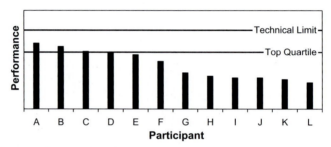

FIGURE 14.4 Ranked bar chart showing performance levels. Superior performances are on the left.

alternative is to keep the order of the bars the same so that each participant's bar is always in the same place. Either format can be used both in the analysis and reporting phase.)

In addition to the participants' performance levels, lines may be added to show various performance levels such as average, top quartile, best in class, technical limit, or some other level. The number and type of lines that are included on the chart is limited only by what is useful and will not clutter the chart. One of these values is often selected as a target performance level, more rarely the analyst will use such values to derive a target performance level. Typical values drawn on charts include:

- **Technical Limit**: This is a (usually) theoretically determined value which is seldom if ever attained. It could reflect, for example, the performance level that would be achieved if there were no failures, delays, interruptions, etc. The advantage of using a technical limit as a target is that it is objective and not dependent on the participants. Sometimes technical limits need to be derived, e.g. minimum effort to operate a production line, while in other situations the limit is clear, such as accidents, errors, and failures where the limit is zero.

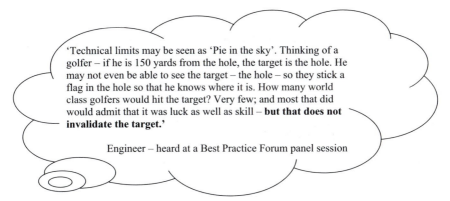

'Technical limits may be seen as 'Pie in the sky'. Thinking of a golfer – if he is 150 yards from the hole, the target is the hole. He may not even be able to see the target – the hole – so they stick a flag in the hole so that he knows where it is. How many world class golfers would hit the target? Very few; and most that did would admit that it was luck as well as skill – **but that does not invalidate the target.'**

Engineer – heard at a Best Practice Forum panel session

- **Top quartile**: Many organizations like to use the idea of being 'top quartile', i.e. being amongst the top 25% of performers. Though dependent on individual participant performances, quartiles are relatively robust to individual participant performances (see Figure 14.8) and membership changes. They can be considered to estimate overall industry performance. Top quartiles should, however, be used with caution as explained in ⇨ Appendix 2.
- **Average**: Some participants like to know whether they are in the top half of performance levels. The average is more robust than the top quartile, however, setting a target to be average is unlikely to be appropriate for most participants. Despite this, many people like to include the average on these types of chart.
- **Best in class**: This is the best performance level in the study. In Figure 14.4 the best in class is participant A. Using the best in class performance as a target is open to justified criticism. While it can be defended as being an achievable performance level, it suffers from justified criticism because:

1 The performance may have been achieved to the detriment of another metric (e.g. low maintenance cost but high unavailability).
2 The performance is believed to be unattainable by other participants for a reason beyond their control.
3 There is some doubt about the accuracy of the best performance data.
4 The 'Best in class' value is highly dependent on the current membership of the study and so will vary from one study to the next as membership of the study changes (see Figure 14.8).

Target Performance:

A performance level believed to be achievable by a typical participant.

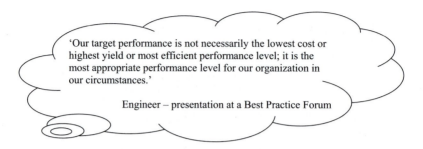

'Our target performance is not necessarily the lowest cost or highest yield or most efficient performance level; it is the most appropriate performance level for our organization in our circumstances.'

Engineer – presentation at a Best Practice Forum

Example: Relating Targets and Technical Limits

If applied to golf, the **technical limit** would be a hole in one.

A **target** could be par (the pre-determined number of strokes that a golfer should require to get the ball from the tee into the hole).

However, for a beginner, the **target** for a round of golf could be their previous least number of hits to complete the course.

With respect to issues such as safety, errors, failures, etc. it is often argued that the only appropriate target is zero.

☞ **Considering metrics in isolation can lead to erroneous conclusions**. It is not always appropriate to consider one metric in isolation and determine which performance is 'best'. For example, low costs are often easy to achieve by reducing maintenance, training, research and development, customer service, etc. Although cutting such costs may give short term benefits, in the longer term they may lead to unforeseen problems. Often 'good' or 'best' performances can only be identified by analysing groups of metrics, sometimes over long periods of time. Regarding costs, as a typical example, it is not always appropriate to minimize costs, it is usually more appropriate to target expenditure in such a way as to achieve the desired results at minimum cost. This is an important and difficult reality to address in a benchmarking study and all too often is ignored. The organization that is deemed to have the best overall performance for a group of related metrics is often referred to as the Pace Setter.

Pace Setter:

The PaceSetter is that participant that is deemed to have the best overall performance, including consistency over time.

14.2.2 Account for Mitigating Circumstances

Having developed the performance comparison bar charts, and determined an appropriate target to compare participants against, it is natural if not unavoidable to notice the differences in performance levels (the 'gaps') between participants and the target. The next step is to analyse the causes for the gaps.

One method of showing the causes of the performance gap graphically is shown in the bridge bar chart shown in Figure 14.5.

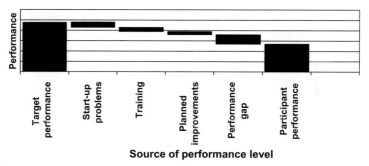

FIGURE 14.5   Bridge bar chart showing causes of gap between participant's actual performance and target performance.

In the example, the target performance is shown on the far left and the participant's actual performance on the far right. The sources of the gap in this example were identified as:

- Start-up problems, not expected to recur.
- Intensive training programme for new staff.
- Improvements that have already been identified and will be/have been implemented.
- Performance gap – the remaining difference after all explained differences have been taken into account. This is the potential improvement available to the participant.

This chart has been drawn as a stacked bar chart with each part of the stack separated, it could equally well have been drawn as a two column bar chart with the bar on the left representing the target and a stacked bar next to it.

Some of the causes for the gap are mitigating circumstances, i.e. causes that are outside the control of the participant. Typical mitigating circumstances include legal requirements, location and, as in this example, start-up problems that are not expected to recur and a heavy initial training programme.

> A **mitigating circumstance** is a known cause of relative poor (or superior) performance usually outside the control of the participant.

Mitigating circumstances can themselves be grouped appropriately for the analysis, for example:

- Causes outside the control of the participant such as legal requirement and physical location.
- Causes due to policy such as keeping high spares in order to minimize unplanned downtime.
- Causes due to one-off events such as extreme weather, commissioning of new facilities, and expansion.
- Known causes that are scheduled for action.

The remaining gap is the performance gap due to both known and unknown causes.

If, at the planning stage, the *likely* or *possible* causes of the gap are known, data can be collected with a view to:

1. Identifying whether these are indeed causes of gaps and
2. Estimating the effect of the cause.

One of the advantages of running a benchmarking study on a regular basis is that as the participants and facilitators gain experience they can begin to pinpoint more accurately likely causes of gaps and so collect appropriate data and information in subsequent studies in order to investigate further.

14.2.3 Quantifying the Potential Benefit and Explaining the Performance Gap

Having identified and quantified the mitigating circumstances, the remaining performance gap is the potential benefit available to the participant. This gap will include any unidentified mitigating circumstances.

In studies which include appropriate information or data it may be possible for the analyst to deduce causes of the performance gap and so help the participant identify where to focus improvement activities.

Figure 14.6 analyses the performance gap identified in Figure 14.5 and shows how this could be presented to the participant.

The chart shows the performance gap as 100% on the left and constituent parts of the gap to the right. In the example, the percentage that each cause contributes to the performance gap is shown.

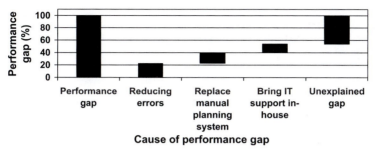

FIGURE 14.6   Bridge bar chart analysing the performance gap.

If the participant reduced error rates to those of the target performance, the performance gap would be closed by just over 20%. Similarly, the analyst estimated that by replacing the manual planning system by an automated system, the gap would be closed by just under 20%, and by bringing IT support in house the gap would close by just over 10%. This leaves about 45% of the gap that the analyst is unable to account for.

In this example the performance gap has been charted as a percentage, but it could equally well have been analysed in terms of the appropriate metric, e.g. costs, time, failures, etc. It would also be possible to draw the chart as shown and annotate the cost (or other) implication on the bars.

14.2.4 Testing Operational Theories

In addition to the more common gap analysis we may have data to allow us to test operational theories. Testing these theories may be reasonably straightforward or may require some in-depth analysis. For example, investigating whether older equipment requires more maintenance can be investigated by drawing charts of maintenance effort against equipment age. Often, especially with regard to soft issues such as people management, the analysis will not be so simple or obvious.

14.2.5 Further Analysis

While the routine tasks of drawing charts and quantifying gaps can be done by most numerate people, it often requires someone more used to analysing data to spot the not-so-obvious nuggets of information that are hiding within the data, as shown, for example, in Figure 14.7.

The chart in Figure 14.7 shows the same data as discussed in a validation example in ⇨ Chapter 13. (Example: Data validation: use of a scatter chart.) The analyst noticed that there seemed to be a set of five data points lying below the trend line and, through working with the data, was aware that these were different participants from the same parent organization. The chart was redrawn with these points highlighted as squares.

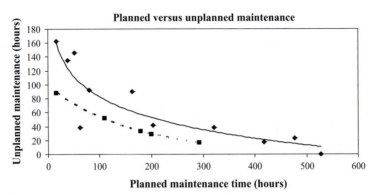

FIGURE 14.7   Planned versus unplanned maintenance.

Identifying this separate set of data is important for several reasons:

- A single outlying superior participant performance is always suspect and the most likely explanation is incorrect data. However, if several participants from the same parent company all return superior performances, and assuming they have interpreted the data definitions correctly, it is likely that the data are correct and that there is some other effect at work.
- It confirms that performances below the line (i.e. better than average performances in this chart) are consistently possible and are not just due to random variation between participants or one-off aberrations.

   This line would probably be adopted as the 'best in class' or 'target' performance curve in preference to the single very low value which could be due to data error or a special situation.
- It suggests that the participant with five facilities probably has a well thought out and implemented maintenance process, though this needs to be confirmed.
- Other participants are more likely to accept a target value that is being achieved consistently by different participants from the same organization than by a single facility.

(In passing we note that other simple explanations for the very low data value include the delaying of planned maintenance into the following financial year, which in reality may only be a delay of a few weeks; equipment being commissioned and so not being within the maintenance umbrella; equipment about to be decommissioned; planned maintenance being brought forward into the previous year, perhaps to take advantage of a breakdown or lull in demand. Of course, the analyst would hope to find and take these explanations into account, but without the participant's help it is unlikely that all the reasons would be identified.)

The analyst should always look for possible groupings such as participants from the same parent organization, geographical area, etc. Whether or not there is a link, something will be learned. For example, it is likely to be as valuable to an organization with several participants in the study to be aware that performance levels between them are different as it is that they are the same.

These charts are not only of use to the analyst, but if included in the report allow participants to consider the likely effect of changing, in this example, their mix of planned and unplanned maintenance.

14.2.6 Deducing Philosophies

Another type of analysis that can often prove very useful for understanding why participants have different performance levels is to deduce or investigate likely management philosophies.

Case Study: Deducing Management Philosophies

In one study which included production stoppages, the analyst noticed that one participant incurred significantly less stoppages than others. The analyst realized that one possible explanation was that this participant had spare equipment that could be brought on-line when a component failed thus allowing production to continue while maintenance was carried out. The analyst reviewed data relating to maintenance effort, which confirmed that the participant was carrying out similar amounts of maintenance to other participants, and was able to confirm that his theory had been correct. It was now possible to begin to investigate the relationship between spare equipment and stoppages which became a feature of the next benchmarking study.

Identifying philosophies such as these is important to both analyst and participant as it is only appropriate to compare some aspects of performance once the operational situation is taken into account.

14.2.7 The Importance of Historic Performance Analysis

It is often recommended that benchmarking be carried out on a regular basis for a variety of reasons including:

- It sometimes occurs that a participant has a good performance level for one or even more rounds of benchmarking but that the level of performance is not sustainable. For example, an organization with low marketing costs may not see the effect in lower market share for one or more years. Similarly, most organizations could reduce, for example, training, maintenance research and development resulting in short term performance improvement at the detriment of long term performance.

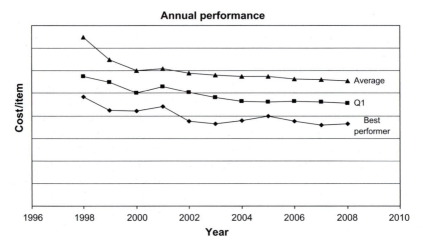

FIGURE 14.8   Run chart of annual cost levels.

- It is possible to predict future performance levels. Figure 14.8 gives an example of a key performance metric reported over the last eight years of an annual study.

The chart is interesting because:

- The Best Performer (Best in Class) performance is more erratic than the average line. This is because it depends on the reported performance of one particular participant and is susceptible to participants leaving the study.
- The quartile (Q1) and average values move smoothly from year to year and do not jump. The consistency from year to year is due to the values being based on all the data and not overly influenced by one or a small number of participants. For this reason, metrics like the top quartile are often preferred as an indicator of good performance.
- There was a steady decrease in costs for the first few years, followed by a prolonged time of slowly declining values. In the industry, this relates to significant step changes made in the management and manning structures in the first few years, followed by a gradual improvement in processes.

14.3 USE OF A STATISTICIAN

This chapter has outlined some of the simpler methods of carrying out basic analysis of benchmarking data. Much of what has been described can be carried out by non-statisticians. However, unless the analysis is of a very basic nature, there are many pitfalls that the non-statistician may fall into.

   The availability of many statistical tools within spreadsheet packages is a mixed blessing. On the one hand, it does allow many analyses to be carried out quickly without having to transfer data from one application to another. The serious disadvantage is that even simple statistics such as the correlation coefficient and regression analyses are frequently used and quoted without understanding the assumptions and whether they are appropriate or what information they really impart. For these reasons it is strongly recommended that a statistician be involved in the analysis phase of benchmarking studies.

The use of a trained statistician throughout a study with all but simple data requirements is important as statisticians can ensure that analyses are sound, meet the requirements of the study and make the maximum use of the data. They are also adept at interpreting data and are likely to identify findings that others may miss. Conversely, being aware of analysis pitfalls, they are less likely to be misled by aberrations thrown up by the data. Finally, for more advanced analysis such as hypothesis or significance testing, the use of a trained statistician is vital to ensure the integrity of the conclusions.

14.4 DEALING WITH MISSING VALUES

One of the more difficult aspects of dealing with data is missing values. Despite earlier best efforts it may be that some participants are unable to provide some of the data required. There are three simple solutions in this situation including:

1. Ignore the missing value and continue the affected part of the analysis with fewer participants. This may be most appropriate where the missing value is not required to calculate totals.
2. Estimate the missing value either from previous data supplied by the participant or by reference to other participants believed to be operating in a similar environment.
3. Estimate the missing value from industry averages.

Any estimated or missing values need to be highlighted in the report and treated with caution.

SUMMARY

In this chapter we:

✓ Showed how information gathered during previous steps of the benchmarking process can be helpful in analysing and explaining participant performance levels.
✓ Discussed a typical analysis process which include:

- Deducing a target performance level.
- Determining the performance gap between the participant's actual and the target performance levels.

- Identifying and accounting for mitigating circumstances.
- Estimating the potential benefit to the participant for operating at the level of the target performance.
- Recommending follow-on and improvement actions.
- Testing theories and deducing philosophies.

Although we carry out analyses to reach conclusions we need to design the analysis starting with the required results and determine what data will be required to carry out the required analyses to lead to these findings, conclusions, and recommendations.

⑂ ACTIVITIES

In the design phase of the benchmarking study:

? Identify what types of recommendation (if any), conclusions, or other findings you want to research in the analysis.
? Deduce what analysis will be required to reach these conclusions, findings and make these recommendations.
? Deduce what data and information will be required to carry out the analysis.

Chapter 15

Finalizing the Report and Other Deliverables

INTRODUCTION

From the facilitator's point of view the report is the culmination of the pains-taking meticulous work that has gone before. For the participants it is a key document that should help them select and drive through improvement projects resulting in the benefits promised by benchmarking.

The existence and purpose of the report will be determined at the planning stage of the project, and usually serves several purposes which typically include:

- Informs each participant of their relative performance.
- Identifies performance gaps, perhaps with proven or possible causes for the gap.
- Acts as a driver for improvement.
- Provides information regarding relationships between metrics.
- Provides conclusions regarding theories.
- Shows how analysts arrived at their conclusions.
- Provides source information for participants to carry out their own analyses.
- Provides material for participants to develop presentations for internal use.

The content and style of the report will be specific to the needs of the study. In this chapter we look at the content and layout of a typical report and explore some useful presentation tools. We also discuss anonymizing data, providing information other than a report, and when to issue summary information and hold a 'report review' meeting.

DELIVERABLE

A clear concise report meeting the requirements of participants and helping them drive through improvements within their organization.

The Benchmarking Book: A How-to-Guide to Best Practice for Managers and Practitioners

15.1 REPORT CONTENT

The requirements and content of the report should have been discussed as part of the planning process, but in all probability this will only have been done in general terms. While it is difficult to know what should be included in the report when the data have not even been gathered, let alone analysed, planning the report in advance helps identify what data will be needed.

The detailed report contents will be specific to each study and will be based on meeting the objectives of the study, in-house formats and cultural considerations. As with all reports it is important to identify the key customers of the report and their needs. There are likely to be two main types of customers:

1. **Managers**

    Managers will usually want a concise overview of the study highlighting the main attributes such as participants, comparative performance levels, areas to be targeted for improvement, and recommendations.

2. **Workforce**

    The workforce consists of those people who will be using the information in the report to improve their aspect of the organization. They need information to help them identify exactly where improvement is required, what the likely causes of relatively poor performance are, likely improvement actions, which participants may be able to help them improve, etc.

    The report should be structured to meet both sets of needs.

While it is not appropriate to lay down a content or structure list for all situations, many studies will include sections such as:

1. Introduction to the study: includes summary information for all participants.
2. Participant Specific Report: includes participant specific information such as management summary/key findings and detailed comments and recommendations.
3. Additional analysis such as best practices, results of testing theories, and industry trends.
4. Supporting charts, tables, analysis, and comments: contains supportive information not included elsewhere.
5. Additional project information:
    - Metrics and definitions.
    - Organization charts, process maps, and facility layouts.
    - Participant contact details.
    - Agreement, restrictions on report use.

    As well as a contents page, list of charts, description of the report layout and list of other items.

15.1.1 Introduction to the Study

Many readers of the report will not be familiar with the details and progress of the study and the Introduction provides a common text for all readers explaining the key features of, and reasons for the study. It will typically include all the background information and may include a summary of the key findings along with summary charts. The table (Figure 15.1) summarizes a typical benchmarking report introduction.

Introduction	
Item	Description
Disclaimer	A statement by the facilitators disclaiming responsibility for accuracy of data and other information supplied by the participants.
Summary of participant agreement	Highlights key aspects of the agreement (usually including confidentiality). It is important to highlight the key aspects of the agreement prominently in the report, especially where the contents are of a sensitive nature. (The complete agreement needs to be included somewhere, perhaps as an appendix.)
Purpose, objectives and scope	Succinctly summarizes the study.
List of participants and summary of participants' operations	Provides a quick familiarization of key participant attributes.
Metric model	Provides an overview of the metrics and their inter-relationships.
Brief log of events	Provides a timescale of the study's activities and, if there are future planned activities such as Forums, the schedule for these events.
History of the study	The history of the study, if there is one, possibly including realized benefits and testimonials. It may also include summary statistics such as changes in key performance metrics over time.
Summary charts, conclusions and recommendations	Provides main non-participant specific findings, conclusions, recommendations and key performance levels.

FIGURE 15.1   Typical contents for the introduction of a benchmarking report.

There are three common options for providing an overview of the charts and findings. If there are only a few charts, it may be appropriate to include them all in the Introduction. If there are a larger number of charts, tables and/or findings, it may be more appropriate to include them as a common separate Management Summary section after the Introduction and before the Participant Specific section. The third alternative is to include them in the Participant Specific section of the report, in which case it will probably be augmented by the summary findings, recommendations, and charts specific to that participant.

Suggestions for summary charts are included in the Participant Specific Report section.

15.1.2 Participant Specific Report

In most studies each participant will only see participant specific recommendations and detailed comments relating to themselves. The Participant Specific Report contains two main parts: the management summary/key findings, and the detailed comments and recommendations pertaining to the specific participant.

The management summary is likely to consist of some summary performance comparisons. The spider diagram is a very useful tool for doing this, and one use of the chart is demonstrated in Figure 15.2.

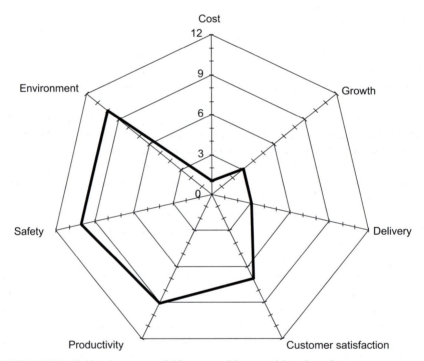

FIGURE 15.2   Spider charts are useful for summarizing a participant's performance.

This chart summarizes the key aspects of a benchmarking study. In this example there were 12 participants in the study and the ranked position of this particular participant is shown on the chart for each of the seven areas of the study. For example, for 'cost' the participant scores 1, i.e. they were the lowest cost (and hence best performer) of the 12, whereas for safety they scored 10, i.e. there were only two participants with a worse safety record.

Values plotted near the centre of the chart indicate higher ranked performances, while those plotted towards the edge indicate lower ranked performances. From the chart it can be readily seen that on the measures of cost, growth, and delivery the participant performs well, while on safety, environment, and productivity they perform poorly, and these are areas that could be selected for improvement. The result for customer satisfaction is close to average.

Spider charts such as these are very useful for summarizing performance. In Figure 15.2 the charted values were ranks, but where there is a common unit of measure such as cost for different activities, the cost rather than rank can be plotted. Further details and options for the chart are discussed in ⇨ Appendix 'Data Analysis Tools'.

The summary charts show relative performance levels and quantify potential gains. The accompanying text and/or tables should highlight key areas for improvement and recommendations. One simple method of arranging the text is as a table summarizing the strengths, weaknesses, mitigating circumstances, recommendations, and comments for each area of the study.

Where an organization has more than one participant in the study it may be appropriate to compare and contrast performances amongst these participants, identifying where their best performances and practices lie.

The detailed part of the Participant Specific Report will expand on the management summary. It will include detailed charts, explanation of the summary comments, and the analysis that led to them.

While the spider diagram is useful for displaying the performance of one particular participant, a bar chart is useful for displaying results of all participants, thereby allowing comparison between participants.

Figure 15.3 is a stacked bar chart showing total costs for a typical study.

There are many variations on drawing this type of chart. The chart in Figure 15.3 shows the total cost per unit produced for participants of a study. The chart also shows the breakdown of the costs into its constituent parts. This allows the reader to compare not only the total cost per unit but to gain some basic insight into how the costs are spread across the constituent parts. In this somewhat simple example, the participants with lower maintenance cost per unit spend proportionally more on equipment management. While this finding would need to be explored in more detail later in the report, at a summary level the observation is worth making with a reference to where in the report the issue is analysed in more detail.

In addition to the chart as shown it is possible to add many other features such as an additional stack representing the target performance, average or some other value; or the total cost per unit above each stack. In addition to the stacks representing the current values, each participant stacks from a previous study could be added to allow comparison with previous

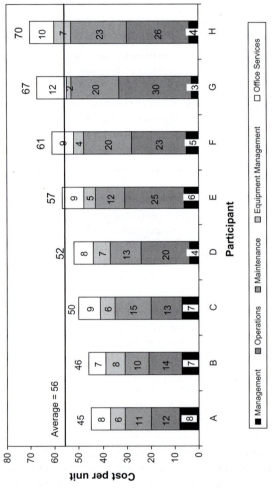

FIGURE 15.3   Stacked bar charts are useful for summarizing data.

years, albeit that adding too much information will make the chart look cluttered.

Summary charts showing how performance has changed over time can also be useful. An example is given in ⇨ Chapter 14, Figure 14.8.

Scatter diagrams can also be used to advantage in the summary. These can be especially helpful where participants differ greatly in terms of size, production/ throughput, complexity, or similar attribute. In Figure 15.4 the relationship between the number of clients, along the horizontal axis and the cost of managing and developing the client relationship, on the vertical axis, is not linear, as can be seen by the fitted line. Therefore, to compare the cost per client across all participants is not an appropriate measure of how well that part of the business is run: participants with more clients will always be at an advantage compared to those with fewer clients. However, it is appropriate to compare participants' data points with the fitted line since this is based on all the data and takes into account the number of clients.

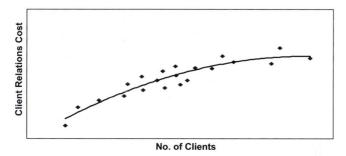

FIGURE 15.4   Scatter diagram highlighting the non-linear relationship between two variables.

15.1.3 Additional Analysis

There may have been a requirement to carry out non-comparative analysis such as the testing of theories, identifying current practices, and trends. One of the many methods of displaying such learning is shown in Figure 15.5.

The chart in Figure 15.5 shows the performance level on the vertical axis and different management philosophy areas on the horizontal axis. For example, organizations that have a philosophy of providing minimum of training report low performance levels, while those that provide job-based training report better performance levels, while those that provide person-based training report the highest performance levels. While charts like these provide succinct summaries, there will clearly be a need to supply detailed analyses elsewhere in the report showing how these conclusions were reached.

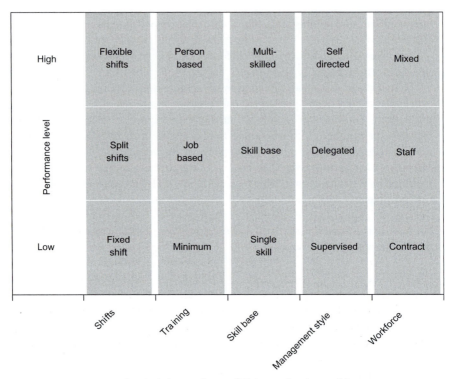

FIGURE 15.5   Results of analysis into attributes of higher performing participants.

15.1.4 Supporting Charts

When reviewing the report participants are likely to want to carry out their own investigations and/or make presentations to management or others in the organization. To facilitate this, participants should either be given a complete suite of charts and/or electronic copy of the database so that they can draw their own charts and carry out their own analyses.

15.1.5 Additional Information

There will probably be a certain amount of additional information that is not appropriate for the main part of the report. This is likely to include:

- A complete list of metrics and definitions, and, if it is not included elsewhere, the metric model along with a description of the model.
- Participant information such as organization charts, process maps, facility layouts, policy documents, questionnaire survey results, and submitted data.
- A list of participant contact details.

- A copy of the participants' agreement, restrictions on report use.
- Details of any calculations, for example, of Weight Factors.
- An explanation of how to interpret more complex charts and analysis.

15.2 ANONYMIZING CHARTS AND DATA

There are three common situations regarding the anonymity of data and the report:

1. All the data are fully open – all participants gain access to all data (in which case the submitted data may be part of the deliverables from the facilitators).
2. All the data are to be anonymized.
3. Part of the data is open and part is to be anonymized.

Fully open studies are the easiest to manage and provide the most information for participants to work with. The key advantage is that participants can be supplied with the validated data and carry out their own analysis. Each participant can also directly contact other participants for clarification or any other reason (if this is allowed in the agreement). Fully open studies may also result in a lesser need for the facilitators to analyse data and supply a comprehensive set of charts. The difficulty arises, especially with regard to cost data, that sharing cost data may contravene legal requirements or, regardless of the legal situation, participants may be reluctant to share cost data openly.

Fully anonymized studies are also easy to run. Probably the best method of producing charts for reporting purposes is to print an individual copy of the charts for each participant removing the names of all other participants from the charts. An alternative is to print all the charts with the participants' names replaced by a code, and then to tell each participant which code they are. The disadvantage of anonymized studies is that it limits the analysis that participants can do for themselves and makes contact between participants for clarification or advice more difficult: the facilitator usually having to act as go-between.

Another possibility is to provide only summary statistics such as averages, quartiles, best in class figures, etc. and, of course, the participants' own data.

Partially anonymized studies are those studies where some of the data are anonymized and other data are not. The most common split is that financial data are anonymized and other data are not. This facilitates communication and understanding on non-cost issues such as safety, environment, training, and failures but inhibits understanding on financial issues.

There are a number of methods of coping with the charting in this situation. The charts with participant names can be placed in a common section in the report, and the anonymized charts placed in the participant specific section. The anonymized charts can be produced as described above.

An issue with partly anonymized studies is that participants may be able to determine the codes on the anonymized charts. For example, if there is a manpower cost chart and a manpower hours chart, it can be reasonably straightforward to deduce that the participants with high manpower costs will be the ones with the high manpower hours. Once the code is cracked for one chart it may be possible to fill in the names for all anonymized charts.

One way to make identifying anonymized codes more difficult to discern is to change the coding for every chart or every group of charts (e.g. use one coding for marketing, one for sales, one for customer services, etc.).

15.3 WHEN TO START WRITING THE REPORT

Although report delivery is often the last major task of the facilitators, it should be begun very early in the planning process. Even at the stage of developing a proposal it is useful to draft out at least the likely contents of the report, along with sample charts, tables, and text. While this may seem too early in the process, it helps to focus on the end product (as far as the facilitators are concerned) and therefore identify what information will be needed to complete the report.

There are many benefits to starting the report early including:

- It can help potential participants understand one of the main deliverables of the study.
- It encourages a great deal of thought to be put into the content and structure of the report, and hence analysis, data collection, and other required information.
- It can act as a repository for information and ideas collected during the project. In particular, as the analysis begins and conclusions and recommendations emerge the report can be built up. As the report is written it may become apparent that certain charts, analyses, and information would help enhance the value of it, these can then be accommodated in the study.
- It helps avoid the daunting task of having to write the report in what often turns out to be a very limited timescale.

15.4 THE DELIVERY PACK

It has been tacitly assumed that the main deliverable for this phase of the study is a report. However, the delivery pack may consist of other items instead of or as well as a report. For example, in an open (non-anonymized) study, a deliverable could be a database, either issued after the study, or updated on an ongoing basis as participants' data are validated, and downloadable from a website. In addition, the facilitator may provide charting and analysis tools. Another option is to supply an electronic presentation.

It is important is to discern what will be of most use to participants and then provide it in a user-friendly manner, rather than being tied down to the conventions of how other studies deliver the results.

In addition to the physical deliverable, participants may require individual debrief sessions with the facilitators in which the findings are presented. This not only helps maintain the momentum of improvement, but also gives the facilitators an opportunity to maintain contact with the participant and find out more about their needs and aspirations. It also provides a good opportunity for facilitators to learn more about the participants' organizations: their challenges, practices and other operational situations that can help them add value to future studies.

15.5 THE USE OF PRE-REPORT SUMMARIES AND REVIEW MEETINGS

If the study is a one-off study or the first of a series it is worthwhile to issue a draft report for participants to review. Required changes, additions, extra analyses, mitigating factors, etc. can be discussed, possibly at a participants' review meeting, before issuing a final report.

Case Study: Benefits of a Pre-Report Review Meeting

One facilitator has been running a number of annual benchmarking studies for a number of years. Before issuing the report a review meeting is held to present the results to participants and elicit their comments. Participants are able to review their own performance levels, request changes in charts, text, analysis, etc. The key benefit for participants was that they received a useful preview of results several weeks before the report was issued, during which time they began to act on findings.

Where a review meeting is held this has a number of benefits for both participant and facilitator:

- ✓ It provides another point of contact for facilitators to learn more about participants' future needs.
- ✓ It builds on the relationships between facilitators and participants and between the participants themselves.
- ✓ It allows participants to get as much as possible from the report and to request extra analyses, charts, or other information.
- ✓ Often participants will explain why they think their performance is high or low and precipitate a learning discussion.

The disadvantage is that the meeting does take time to organize and, where the participant base is international, the travel time and costs may be significant.

SUMMARY

✓ The report should be designed to meet the needs of key users, usually management and staff.

✓ The best time to start planning the report is at the project planning stage. This not only helps the facilitators focus on a key deliverable, it can also be used to show potential participants during recruiting.

✓ Deliverables of this part of the study may include a report, databases, and/or presentations.

✓ A typical report will include:

  - An introduction.
  - A section tailored to each participant covering such issues as strengths, weaknesses, opportunities for improvement and recommendations.
  - Useful summary charts e.g.: spider diagrams, stacked bar charts and scatter diagrams.
  - Where some or all of the data cannot be shared openly the three common methods of anonymizing results are:

    * To tailor each set of charts for each participant naming only the participant's performance.
    * To replace each participant's name by a code, telling each participant only their own code. Coding may need to be changed on every chart or group of charts.
    * To show each participant's results compared to summary statistics, the most common being average, best in class, and top quartile.

✓ In some situations it is useful to invite participants to give feedback on a draft version of the report.

ϒ ACTIVITIES

? What are the key needs of the customers? What other needs might they have?

? How can these needs be met? (e.g. should be included in the deliverables? In the case of a report, how will it be structured, what sections, analyses, charts, etc. should be included?).

The Benchmarking Process – Improving performance

Chapter 16

From Report to Improvement: Preparing for Improvement

INTRODUCTION

In this book we assume that a key deliverable is a report comparing performance levels of participants. Whether a report is a deliverable or whether participants receive data and carry out their own analysis to produce findings, the result will be similar: i.e. findings on which participants will choose whether or not to take action. However this point is reached, and whatever the deliverable we refer to the result as a 'report'.

For participants, the first step after receiving the report and other deliverables will be to review them to ensure that they are both understood and correct.

Having reviewed the report and clarified any uncertainties, the next phase of the benchmarking project is for each participant to use the results to progress the project to meet their own specific objectives. These may include non-improvement activities such as planning and budgeting and/or improvement activities. Improvement activities may require further data/information sharing with participants or even further benchmarking.

For the facilitators this is also a key point at which they should gather feedback from participants with a view to improving both their general benchmarking process and future projects with this study.

In this chapter we consider how:

- To review and validate the report.
- To use the results for non-improvement activities such as budgeting and planning.
- To use the results for improvement activities with or without further information sharing with participants.
- Facilitators can gather information to improve the benchmarking process.

DELIVERABLES

- A verified and understood report, including any further analysis. Other deliverables will be study dependent.
- A list of improvement opportunities for facilitators.

16.1 REPORT REVIEW AND FURTHER ANALYSIS

Whatever is included in the report it is important that the participants ensure that they understand the findings, analysis and other details and are satisfied that they are correct.

The first questions that are likely to be on most people's mind when receiving the report is how well they performed compared to the other participants and in which areas they are achieving relative good and poor performances. As participants review the findings they will naturally begin to delve into the areas of interest, cross checking data and charts, confirming that the data they submitted was correct and seeking logical explanations, especially for unexpected conclusions. This reviewing not only serves to help the participant understand the report but also acts as a sense check that the report is correct.

Recommendations, findings, conclusions and/or data that are likely to be acted upon should and usually will receive extra attention as it is important to glean as much information as possible from the report before committing resources to acting on it.

The report may not include sufficient detail for the participant to start taking action. They may need to carry out further analysis, for example to investigate in more detail the reason for performance gaps.

16.2 THE NEXT STEPS

Once the participants feel they understand and agree with those aspects of the report that impact them, the next steps in the benchmarking process depend on the objectives of the study and the content of the report. Broadly speaking there are two likely possibilities:

1. Use the results for non-improvement activities.
2. Embark on improvement activities.

16.2.1 Non-Improvement Activities

While the traditional reason for benchmarking has always been to improve performance, there is an increasing use of benchmarking for other reasons including:

✓ To give confidence/guidance to planners that their plans and budgets are realistic and to provide a learning opportunity where the plan or budget differs significantly from benchmarked data.

✓ To help justify budgets to management.
✓ To help validate new approaches/methods/ideas by answering the question: 'I wonder if anyone else has tried ... and what happened?'
✓ Especially for project benchmarking, participants can search for projects similar to the one being planned and use benchmarked data to help determine likely costs, resource requirements, typical problems and solutions.

Example: Benchmarking for Planning

O *theory*

Using benchmarking as a realistic input for budgeting is certainly better than one common method which could be labelled 'O theory', where O stands for Opinion. This represents budgeting by all non-data analysis methods. Typical methods include:

● 'last time + 10%', or, if there is a budget squeeze, 'last time − 10%'
● '... plus some contingency'
● 'I think we can do it for ...'

'O theory' is related to 'T theory', where the T stands for tough, though others may call it 'G theory' where G stands for 'Gung ho'

● 'I'm sure we can do better than that .... What's your stretch target?'
● 'We're on a tight budget, can't you do it for...'
● 'If you can't do it for.... I'll find someone who can'.

Such budgeting is based on opinions, beliefs and historic performances (which may or may not be appropriate). It has little to do with what is achievable and appropriate in the current commercial environment.

O *theory: from Strategic Benchmarking Reloaded with Six Sigma,*
Gregory H. *Watson. p. xii*

'It's no use management saying "cut costs by 10%" if the workforce do not have the data (and time etc.) to improve the processes. You could, I suppose, beat the vendors over the head, but this can backfire...'

                                                     Anon

Case Study: Benchmarking Manning Levels

One of the more common non-improvement objectives for benchmarking is to identify whether manning levels are appropriate.

One organization was in the process of a strategic review of its organization. They believed that manning levels were too high and should be reduced to a certain level. When the new staffing levels were announced the staff were concerned that the cuts would result in two few people to carry out the required workload, and in particular were concerned about the safety implications.

It was suggested that a benchmarking study be carried out using the proposed manning levels as the source data to obtain an independent view.

The results indicated that the cuts would reduce manning levels to below that achieved by the other organizations in the study and management agreed to reduce the cuts.

16.2.2 Improvement Activities

Participants using benchmarking as an improvement tool will usually draw up a list of potential improvement projects. It is possible that there will be more projects than the participant is willing or able to undertake at one time, and so the projects will need to be ranked and then one or more selected for action, perhaps using a method similar to that explained in ⇨ Chapter 3.

Having selected one or more improvement projects, there are two approaches that a participant is likely to consider in order to progress them:

1. Internal improvement activities.
2. External improvement activities – i.e. those requiring help from other participants.

It can be argued that external improvement activities always lead to internal improvement activities and are in reality just a method of collating ideas on how to improve. We will consider external improvement activities first.

16.2.2.1 External Improvement Activities

External improvement activities are fundamentally sources of information from which participants can gather ideas on how to improve practices and therefore performance. Data and information shared is limited only by what organizations are prepared and legally allowed to share.

There are many methods of learning from others which can generally be divided into: Figure 16.1

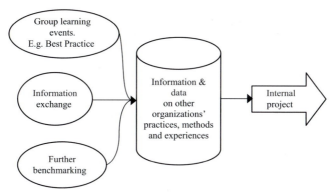

FIGURE 16.1   Group learning events, information exchanges and further benchmarking all lead to internal improvement projects.

- **Group learning events**, often called Best Practice Forums, at which participants present information on their practices to others in the study. ⇨ Chapter 17.
- **Information exchanges**, often reciprocal, of data, information, methods or anything else agreed between the participants. This type of event, usually between two participants, may itself take the form of a simplified one-to-one benchmarking study, or may be carried out informally perhaps by phone, email or meeting or a combination of methods. There are several examples of this type of benchmarking throughout the book, see for example ⇨ Part 4 case studies Dundee City Council and Best Practice Club.
- **Further benchmarking** studies with the same and/or different participants focusing in-depth at one or at most a few selected aspects.

These methods may or may not include site visits.

Further formal benchmarking studies are most likely to occur where the initial benchmarking study has a wide scope and therefore is likely to only include metrics summarizing performance in each area. In these situations it is likely that participants will realize that benchmarking at a high level will probably not give enough information to be able to improve performance without further benchmarking. (One of the objectives of the benchmarking study may have been to identify in which areas to initiate further detailed studies.)

Example: High level Benchmarking as a Method of Initiating Detailed Benchmarking Studies

A group of hospitals had benchmarked all their activities including patient admission, treatment, discharge, catering, safety, environment, patient satisfaction etc. Due to the wide scope of the study, in most areas there were only a few high level metrics. A participant who performed poorly on catering wanted to benchmark all aspects of catering in more detail and initiated a study which included both hospitals and other organizations.

Where further benchmarking is seen as the way forward, a good starting point for recruiting participants is the current benchmarking study. It is likely that a study focusing on one specific area will have a greater potential for attracting participants from other industries, as in the example above.

16.2.2.2 *Internal Improvement Activities*

After the initial benchmarking study participants may decide to carry out improvement activities without gathering any further external information. The case study below shows how one person responded to the results by driving through both immediate and long-term improvement activities.

Case Study: Benchmarking as a Driver for Improvement 1

One organization involved in a benchmarking study realized that their operating costs were high. One of the first steps that a senior manager took was to carry out an informal audit of the organization. There were many quick and easy improvements he was able to implement with little cost. For example, he discovered many items on long term hire that were not being used. In addition to returning these, he initiated a small cross-function improvement study to reduce the amount of time that hired equipment was not in use which resulted in further significant on-going savings.

These and similar activities were taken over the following few years and the participant could see their performance improving against others in the industry over the following studies.

It could be argued that it was not necessary to carry out a benchmarking study to make these savings, but the poor performance results from their first participation was the driver for taking improvement action.

Another organization followed a more extreme approach:

Case Study: Benchmarking as a Driver for Improvement 2

One organization had anecdotal evidence that their overall performance on staffing and cost levels was high. To verify the extent to which this was true they decided to benchmark against other similar organizations in the industry.

Realizing that they were indeed uncompetitive compared to other participants they embarked on a several year project of training the whole organization in improvement activities, putting together project teams to improve specific areas of the business. Three years after the original benchmarking project was complete they had made some improvements and were continuing to make significant organizational, cultural and physical changes to the workplace.

Probably the more usual internal activity is to initiate one or more improvement projects to investigate and improve specific aspects in which the participant has decided to improve.

'The charts show us where we are behind – where our performance is worse. We identified a £3m lag and we could do some simple things to improve – so we have hired a couple of engineers and are working on it.'

Best Practice Forum presentation

In all improvement situations – whether further data and information is gathered from participants after the initial study or not – sooner or later the participant will need to embark on one or more internal improvement projects. In ⇨ Chapters 1 and 18 we briefly discuss improvement projects but a detailed discussion of improvement methodologies, of which there are many, is outside the scope of this book, and the interested reader will find many relevant books and courses.

16.3 FACILITATORS' REVIEW

In most benchmarking studies the issuing of the report and/or other deliverables are a key and sometimes the final step for the facilitators. They too should be aiming to improve the benchmarking process.

Throughout the process they will have identified problems that have arisen, participants that declined to join the study or later withdrew, late data submissions, queries from participants, errors in data, analyses and reports. These should all be logged so they can search for ways to prevent their recurrence or mitigate the effects in future studies.

In addition to this internal data and information gathering, feedback from participants on how they viewed the study, how they would like to see it improved is critical not only to encourage participants to continue with the benchmarking group, but also to help improve the process to achieve a better result.

SUMMARY

Once the report and/or other deliverables have been received, participants are responsible for validating the report and then using the results to meet their benchmarking project objectives.

Report validation typically includes:

✓ Checking the data.
✓ Checking the analysis and conclusions.
✓ Checking understanding of the content.

The likely next activities could be one or both of:

✓ Non-improvement activities, such as budgeting and planning.
✓ Improvement activities, which may include further data or information sharing with participants or further benchmarking activities with the same and/or new participants.

Facilitators also need to use information gathered throughout the process and feedback from participants in order to improve the level of service for future studies.

⅄ ACTIVITIES

When planning the study:

? How will you use the outputs of the benchmarking study to meet your objectives?
? Will the outputs be sufficient to meet the objectives? If not, what else will be required?

After the report has been received:

? Are you able to meet your project objectives with the data and information provided? If not, what else do you need?

Chapter 17

Beyond the Numbers: Gathering Information From Participants

INTRODUCTION

Many benchmarking studies include information exchange activities scheduled to take place after data have been analysed and reported. The purpose of these activities is to allow the participants with superior performance to share with other participants how they achieve their performance levels.

In some benchmarking studies data are not gathered or analysed. In these studies the benchmarking activity is mainly confined to the sharing of, for example, practices, philosophies, lessons learned, tips and pitfalls or any other information that is agreed between participants. Any data that are shared are often incidental to the study.

In this chapter we discuss:

- Best Practice Forums.
- One-to-one information exchanges.
- How to organize and manage site visits.
- The benchmarking code of conduct.

17.1 BEST PRACTICE FORUMS

A common method of gathering useful information on how other participants achieve superior performance levels is through Best Practice Forums.

Best Practice Forum:

A Best Practice Forum is an event,
typically of one or two days,
at which the better performers (as identified in the data analysis)
in different benchmarked topics will
give presentations
explaining how they achieve their performance levels.

Many benchmarking studies include a provision for Best Practice Forums. This may be formalized by including a clause in the benchmarking agreement that if asked to do so, a participant will give a presentation on a topic chosen by the facilitator.

There may be one forum covering all topics, or there may be one forum for each topic or a group of topics. For example, an airline benchmarking study may have separate forums for aircraft maintenance, security, customer relations and delay management.

Successful forum events need to be well planned and managed. The starting point, as with many projects, is to consider the customers and their needs. The customers of the forums will include both the participants and presenters. The needs for each forum will need to be ascertained with the participants in the light of the report and are likely to include items such as:

✓ To identify practices that lead to superior performance.
✓ To understand why these practices work.
✓ To identify pitfalls and how to avoid them.
✓ To sketch out a route map for improvement.
✓ To network with other participants.

One of the benefits of these forums, and any cross-group meeting discussing performance, is that people talk to each other and it is often these informal discussions that lead to ideas and improvements.

Comment made at a Best Practice Forum

> ### Case Study: Offers of Help for Installing Software
>
> Meetings between participants do not need to be formalized, and may come as offers of help ...
>
> During a Best Practice Forum a facilitator was involved in a conversation in which a participant, who was beginning to introduce an industry standard software package into their organization, was asking others for their experiences of implementing the same software. He received comments on a number of implementation issues such as training, pitfalls and what preparation would be helpful. He also received invitations to visit and discuss the implementation with both users and software personnel.
>
> Throughout the implementation one of the participants continued to take an interest in the project and offer advice.

17.1.1 Organizing a Forum

If the forum is to be successful it must address issues that are of interest to the participants. The organizer of the forum, who will usually be the benchmarking facilitator, will have access to the benchmarking report and will be able to draw up a list of forum topics likely to be of interest to the participants. These will probably reflect the areas covered by the study and, in particular, areas where there is wide variation in performance levels.

In addition, there may have been issues identified by participants, those carrying out data validation, analysts or those writing the reports that would be appropriate topics for forums.

Having developed a list of potential forum topics, this may be circulated and participants asked to add any further topics they would like included. Once the final list has been developed the participants can be canvassed for their preferences, and a final list of presentations developed.

In addition to participant presentations, forums may include:

- Presentations from the facilitators summarizing the data and conclusions of the benchmarking study on the topic being discussed.
- Presentations from participants explaining how the use the deliverables of the study (particularly appropriate where the deliverables focus on data and not analyses and recommendations).
- Presentations from appropriate experts.
- Panel questions and answer sessions.
- Discussion on what changes, if any, the participants would like to see to the benchmarking study in light of the forum.
- Discussion on any follow-up activities the participants may like.
- Work groups to focus on subjects of interest to the members.

Recruiting participants to speak at forums can be difficult. Their reluctance may be due to lack of time, fear of speaking in public or a feeling that their performance and/or practices are not that different from everyone else's. In these situations the facilitators can help by discussing the presentation with them and even working with them to develop an appropriate presentation. Particularly in the situation where the participants feel they do not have a lot to offer, it is worth pointing out that one of the purposes of the presentation is to initiate discussion.

Where appropriate, the event could be held at the site of one of the presenters as this will facilitate demonstrations and allow company specialists to be brought in for part of the day, or be on hand to answer queries if necessary. In addition, during the forum delegates could visit key areas of the site to help gain a better appreciation as to how the organization functions.

Having finalized the topics and speakers, participants can be contacted to ask what specific issues and questions they would like the presenters to cover during their presentations.

Handouts and any other information should be available at the beginning of presentations. The forum should be minuted and minutes issued as soon as possible after the meeting. It may be appropriate for those not attending the forum to receive copies of the minutes and handouts and/or for them to be posted on the forum website (⇨ Chapter 12 for further comments on organizing and facilitating meetings).

17.2 ONE-TO-ONE INFORMATION EXCHANGES

One of the benefits of joining a benchmarking club is that participants have access to a network of people working on similar processes. This network can be used informally or formally to facilitate learning between organizations. Some participants are wary of giving away all the practices that help them to remain competitive and so will only share practices with an organization from which they can learn and will then 'exchange' information.

There are many possible formats of one-to-one information exchanges including, for example:

- As a formal presentation similar to a forum.
- In a similar manner to a review benchmarking study where one person or a team visit the sites to review processes, practices and other information producing a report with findings and recommendations.
- As a one-to-one benchmarking study.

The information exchange may be one-sided or reciprocal and may include questionnaires, further data gathering and analysis and may be subject to separate agreements. While these events usually involve two participants, it is possible to include three or more.

At their simplest, information exchanges can take place during other benchmarking meetings, via on-line forums, emails, or Best Practice Forums. Two or more participants will discuss topics of interest and may pass on information and ideas that others adopt or act on in their own organization. Identifying when this happens is difficult for facilitators, unless they happen to be involved in the exchange, as there is no requirement to tell the facilitator of the event.

At a more formal level participants may require the help of the facilitator to broker agreements. There are many methods of achieving this. A typical method that works well with anonymous studies is that after the report has been issued, participants contact the facilitator with a list of issues that they would like to discuss with the better performers. The facilitator will then pair or even group participants where exchanges of information between the groups can take place. Normally the exchanges will take place during a site visit, but information could be passed anonymously through the facilitators.

As the participants get to know one-another there is often a shift away from the formal organizing of information sharing to one of the less formal interactions. Participants may phone or email each other and arrange site visits over lunch. If there are areas of interest to several participants they may set up their own working groups.

Case Study: Working Groups

During a Best Practice Forum a facilitator overheard one participant describing how a group of three participants, who happened to be located close to each other, had carried out a series of visits to discuss and work together on improving inspection planning and procedures.

17.3 SITE VISITS

Case Study: Organizing a Visit

One organization phoned the facilitator of a benchmarking study of which they were a member and asked if the facilitator knew which organization was the best at a certain activity. On confirming that the facilitator did know, the caller asked if the facilitator could organize a meeting with this target organization. Meetings were arranged and held and the inquiring organization identified and implemented improvements.

There are a number of times during a benchmarking study that a site visit is likely to occur including:

- A marketing visit by the facilitator or organizer to explain the details of the study.
- Where participant meetings are held at participants' sites (this has the benefit of spreading the work of hosting the meeting, travel and to allow participants to familiarize themselves with each other's sites.)
- For data and information collecting prior to reporting.
- Where Best Practice Forum meetings are held at a participant site, usually at the site of a presenter.
- For information and data exchange visits, either as a follow-up visit or as a one-to-one benchmarking visit.

Each of these types of meeting has different purposes and goals. The character of individual meetings will be different depending on such factors as the industry, culture, as well as the number of and relationship between participants. The protocol for all site visits follows basic politeness and legal norms, but there are some steps we can take to help ensure that the needs of both visitor and host are met.

Rather than address each type of visit separately we will only consider the last, i.e. visits for the purpose of information and data exchange, which are usually carried out by participants of the study. For other types of visit the items discussed will either not be relevant or will be covered by contract or normal business protocols.

The process for arranging and completing a site visit follows the same process as the benchmarking process described in this book, albeit that some of the steps and tasks will be very much simplified or may not apply. There are three phases to covering a site visit (Figure 17.1)

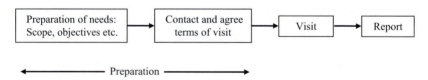

FIGURE 17.1   Process for carrying out a successful site visit.

1. Preparation.
2. Visit.
3. Post-visit report.

17.3.1 Preparation

As with any business meeting, careful preparation is a key to success. When visiting another organization preparation is important not only because time is likely to be limited and the visitor will want to maximize the benefit of the visit, but also so as not to waste the hosts' time.

The preparation phase includes identifying:

- The scope of the visit.
- The justification for the visit (though this may not be shared with the target participant).
- The objectives of the visit.
- What data and/or metrics and other information are wanted.
- What we would like to see is, e.g. documents, charts, organization charts, procedures, equipment, software, questionnaires.

As with all benchmarking, it is important to check with the legal department that the proposed sharing of information is legal (⇨ Chapter 21).

The next step is to contact the potential host to agree the terms of the visit. This also follows the process described earlier in the book.

If the potential host agrees to the visit there may be some negotiation regarding what data and information can be shared, the number of people that can visit, and the duration of visit. The host may or may not ask for a reciprocal visit or for data or information (usually only required for information exchanges). Once agreement has been reached it should be formally documented. It is likely, and will facilitate a successful visit, if:

- A list of data and information to be shared is sent to the host beforehand. This will give the host time to gather it and possibly even supply the appropriate information before the visit.
- A confidentiality agreement is drawn up, agreed and signed (⇨ Chapter 21).
- An agenda is agreed. This will help the host ensure that appropriate people and information are available when required.
- A list of people who will be visiting, along with their job titles is sent to the host.

The agreement should also clarify what post-visit documentation will be supplied.

It might be appropriate for the visitors to invite their hosts to a meal on the eve of the visit, which will help build up relationships before the meeting.

17.3.2 Conducting the Visit

It is worthwhile contacting the host a few days before the visit to confirm the arrangements and address any changes or concerns that have arisen.

The visit is likely to begin with a meeting covering such issues as introductions, confirmation of the agenda and timings of the day, and last minute issues that may have arisen, for example, unavailability of personnel or breakdown of equipment. This is also a good time to confirm whether it is allowable to take photographs, computer printouts, sample documents, notes, etc.

It is likely that there will be a wrap-up meeting at the end of the day. If there are any outstanding issues that were not covered during the day it may be possible to agree how they may be closed out. It is also the time to confirm what, if any, follow-up contact there will be.

17.3.3 After the Visit

Even if no follow-up activity has been agreed, it is important to write and thank the host for their time. It is also useful to supply a list of the information that was gathered and, if agreed, to supply a report of the visit.

Case Study: Unexpected Benefits from Site Visits

One of the many advantages of site visits is that it is sometimes noticing incidental things that the host is not even conscious of that leads to learning:

One organization arranged a follow-up visit to another participant. As they entered the operations building of the host there was a mirror on top of which was the caption 'Meet your safety representative'. The visiting organization copied the idea to highlight that whatever management may do, ultimately it is the individual who is responsible for ensuring they follow safety guidelines and consider the safety implications of their actions.

17.4 BENCHMARKING CODES OF CONDUCT

Many of the issues outlined above, and throughout the book are addressed in the Codes of Conduct that have been developed by organizations such as the American Productivity and Quality Centre (ⓘ available from www.orau.gov/pbm/pbmhandbook/apqc.pdf. The European benchmarking code of conduct, which has a similar scope, is available at www.efqm.org).

The table in Figure 17.2 discusses the eight principles of the European code of conduct and where each item is addressed in this book. Most of the code of conduct is based on good, common business sense. The other key areas cover confidentiality/legality and preparation.

Clause of Code of Conduct	Reference
1 Principle of Preparation Demonstrate commitment to the efficiency and effectiveness of benchmarking by being prepared prior to making an initial benchmarking contact. Make the most of your benchmarking partners' time by being fully prepared for each exchange. Help your benchmarking partners prepare by providing them with a questionnaire and agenda prior to benchmarking visits.	Covered in Phase 1 and Chapter 17 Preparation of scope, objectives, required data/metrics and other information. Share appropriate aspects with potential participants. Research who you want to benchmark with and why.
2 Principle of Contact Respect the corporate culture of partner organizations and work within mutually agreed procedures. Use benchmarking contacts designated by the partner organization if that is its preferred procedure. Agree with the designated benchmarking contact how communication or responsibility is to be designated in the course of the benchmarking exercise. Check mutual understanding. Obtain an individual's permission before providing their name in response to a contact request. Avoid communicating a contact's name in open forum without the contact's prior permission.	Chapter 17 Much of this is common good business practice. Include any specific requirements from either side in the agreement.
3 Principle of Exchange Be willing to provide the same type and level of information that you request from your benchmarking partners provided that the principle of legality is observed. Communicate fully and early in the relationship to clarify explanations, avoid mis-understanding and establish mutual interest in the benchmarking exchange. Be honest and complete.	Chapter 17 Clarify what is to be shared by both parties in the agreement. Ensure that you have the answers to questions that are likely to be asked and the corresponding information for your organization that you are asking from the other participant.
4 Principle of Confidentiality Treat benchmarking findings as confidential to the individuals and organizations involved. Such information must not be communicated to third parties without the prior consent of the benchmarking partner who shared the information. When seeking prior consent make sure that you specify clearly what information is to be shared, and with whom. An organization's participation in a study is confidential and should not be communicated externally without their prior permission.	Chapter 21 Adhering to confidentiality is not only common good business practice, but often a legal requirement. Confidentiality should also be confirmed in the agreement.

FIGURE 17.2   The European benchmarking code of conduct

5	**Principle of use**	Chapter 21
	Use information obtained through benchmarking only for purposes stated to and agreed with the benchmarking partner.	A common issue here is the sharing of information with joint venture/business partners who are not involved in the benchmarking study. Agree with other participants what, if any, information can be shared and under what circumstances.
	The use or communication of a benchmarking partner's name with the data obtained or the practices observed requires the prior permission of that partner.	
	Contact lists or other contact information provided by benchmarking networks in any form may not be used for purposes other than benchmarking.	
6	**Principle of Legality**	Chapter 21
	If there is any potential question on the legality of an activity, you should take legal advice.	Also refer to national and international laws such as the Treaty of Rome as appropriate in your country.
	Avoid discussions or actions that could lead to or imply an interest in restraint of trade, market and/or customer allocation schemes, price fixing, bid rigging, bribery, or any other anti-competitive practices. Don't discuss your pricing policy with competitors.	
	Refrain from the acquisition of information by any means that could be interpreted as improper including the breach, or inducement of a breach, of any duty to maintain confidentiality.	
	Do not disclose or use any confidential information that may have been obtained through improper means, or that was disclosed by another in violation of a duty of confidentiality.	
	Do not, as a consultant, client or otherwise pass on benchmarking findings to another organization without first getting the permission of your benchmarking partner and without first ensuring that the data is appropriately 'blinded' and anonymous so that the participants' identities are protected.	
7	**Principle of completion**	Chapter 17
	Follow through each commitment made to your benchmarking partner in a timely manner.	
	Endeavour to complete each benchmarking study to the satisfaction of all benchmarking partners as mutually agreed.	
8	**Principle of Understanding and Agreement**	Chapter 21 and legal agreements.
	Understand how your benchmarking partner would like to be treated, and treat them in that way.	
	Agree how your partner expects you to use the information provided, and do not use it in any way that would break that agreement.	

FIGURE 17.2 (*Continued*).

SUMMARY

✓ There are many methods of gathering information and learning from other participants. At one extreme a Best Practice Forum is characterized by the better performers giving presentations on how they achieve their performance levels. At the other extreme, individual participants meet informally to discuss issues of common interest either during or outside participant meetings.

✓ In this chapter we discussed how to plan and run Best Practice Forums and site visits.

✓ When visiting organizations normal business protocols should be followed. In addition it is important to agree what information will be shared by the host, and be prepared to respond to reasonable requests made by the host. Benchmarking Codes of Conduct for completing such visits have been developed and are readily available.

¶ ACTIVITIES

? How will you gather the information you require (or may require) from participants in the benchmarking study?

Chapter 18

Implementation Considerations

INTRODUCTION

For many organizations a key benchmarking objective will be to improve their performance levels by changing the way they work – their practices.

Each organization is different. Each has a different culture, structure, commercial environment, business drivers and practices. Therefore every organization will need to develop an appropriate methodology for changing the way it works.

In this chapter we focus on issues and methodologies that will help make effective changes. Throughout it is important to bear in mind that what works for one organization may not work for another. We cannot assume that there will be any benefit in copying another organization's successful strategy unless we have an understanding of their culture and processes. ⇨ Chapter 19, copying without understanding.

18.1 KNOW YOURSELF

Much of the process of benchmarking will require that an organization understands how it operates and how it performs. One of the benefits of planning a benchmarking study is that it does precisely this. Before the report is ever received participants will have:

✓ **Identified the purpose and objectives for initiating or joining a study**. Unless the participant has been coerced into joining the study it will have considered questions such as:

? What do we need to do to improve? Why?
? Why do we need to initiate/join a benchmarking study?
? What benefits can it bring us?
? What do we hope to learn?

To answer questions such as these requires some knowledge and evidence of actual or suspected weakness in the area selected for benchmarking.

✓ **Identified pertinent metrics, definitions and other information**. This will require knowledge of what information is available as well as some idea of which metrics are important for understanding how well the organization is performing.

✓ **Supplied data**. The process of data gathering encourages the organization to look afresh at the data they are using to manage key aspects of the topic being benchmarked.

✓ **Supplied information**. In addition to data, many benchmarking studies will collect other information such as process flow charts, facility/office/equipment layouts, operational philosophies, key concerns, operational/commercial restrictions. The fact that this information has to be supplied will at least encourage organizations to look afresh at their operations.

Case Study: Learning During the Benchmarking Process

After taking part in a benchmarking study participants were asked what had been the greatest benefit for their organization.

One participant replied that the greatest benefit was that it helped management identify data they needed in order to manage their organization.

The participant explained that 'At the beginning of the study all the participants came to a common understanding of the aim of our organizations and from this we derived metrics. We believed that these metrics were fundamental to the effective management of our organizations'.

'When we came to collect the data, we found it very difficult to extract it from our data management systems. This caused us a lot of concern – how was it, we wondered, that we were running the organization without using this key data'.

The major outcome from the study for that participant was to change the metrics they used for monitoring and managing organizational performance.

Exactly what an organization needs to know about itself will depend on why it is benchmarking. When benchmarking a specific technical problem, it may be that the self-knowledge required to resolve it is minimal. If the intention is to improve processes/practices, then it will be important to have detailed knowledge of what these process and practices are and why they function in the way that they do.

Probably as a minimum for improving, an organization will need to have an appreciation of:

✓ Detailed performance levels in the area being benchmarked.
✓ How its processes work (typically found in process flow charts and descriptions).
✓ Management style/culture/beliefs and systems.

✓ Restrictions on what can be changed, e.g. due to legislation, commercial environment.

For example, before moving to a self-directed cross-functional team-based approach to working, it is important to understand the culture of the organization. If the current culture fosters top down, individual-oriented management, expecting individuals to work as a team is unlikely to work until a culture of team responsibility is developed.

18.2 IMPLEMENTING IMPROVEMENTS

Many organizations will have adopted one of the many improvement methodologies currently available, or developed a version of their own. Though the details of these methodologies may be different, the basic principles are similar and one such methodology, Six Sigma, is summarized in ⇨ Chapter 1. In this section we focus on the latter stages of these improvement processes: implementation. The implementation steps for the improvement aspect of benchmarking projects is similar to those for process improvement projects and so are only summarized here (Figure 18.1). However, more careful planning is often required for implementing benchmarking project improvements because they are often imported from other organizations.

At this point in the project we have identified one, or hopefully more, potential improvements to practices, methods, processes, etc. which we want to consider for implementation. These may have been identified from a benchmarking report, discussions with other participants or our own observations during the benchmarking process. Our next step will be to select the preferred solution(s) and schedule them for implementation.

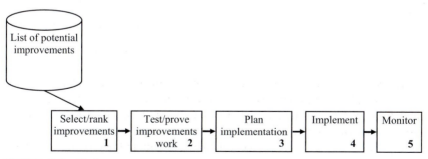

FIGURE 18.1   The improvement process.

18.2.1 Select/Rank Improvements

Selecting preferred solutions is itself a process which will consider issues such as the cost, expected benefits, capability of the organization to make the solution work, risks, and effects on other parts of the organization. The process is broadly similar to the project selection process as discussed in ⇨ Chapter 3, and will ultimately be determined by the organization's own processes.

18.2.2 Testing/Proving Improvements

Having ranked or chosen the preferred improvements the next step in the process is to demonstrate that selected improvement(s) work. This step is perhaps even more important with benchmarking than with non-benchmarking improvement projects. This is because we need to understand not only how other participants achieve their results but also why their methods work in their organizations and why they may or may not work in ours. This is such an important area of benchmarking that a whole chapter has been devoted to the problem of 'copying without understanding' ⇨ Chapter 19. However, what benchmarking does provide is objective evidence that a solution has worked elsewhere and its likely performance level.

There are a variety of methods of testing/proving that a proposed improvement is likely to work in practice including:

- **Trial the new practice**. Trialling is the preferred testing method as it gives the best evidence that a solution will work. Trialling can also give valuable information on likely problems, how to improve the practice, likely costs and benefits, implementation timescales, etc. Unfortunately trialling is not always possible.
- **Develop the process flow map of the current and new procedure**. Analyse the differences and check that, for example, removed and modified steps will not negatively impact the outputs. For each new or changed step consider what could go wrong, and how to prevent it from going wrong.
- **Brainstorm what could go wrong with the new process**; address these issues.
- **Conduct open sessions** with those involved in the process to ascertain from them why the proposed change might not work, their concerns about it and their suggestions.

18.3 PLANNING THE IMPLEMENTATION

Once the solution(s) has been proven, the next step is to plan the implementation. This step is necessary in all organizational changes. The advantage of benchmarking is that if the changes are based on practices in other organizations they may be able to advise on implementing the changes effectively.

Perhaps of all the difficulties of managing organizations, the one which causes the most angst is implementing new ways of working. The problem can often be traced to lack of preparation for changes. We may know what we want to change; we may have written new procedures, changed the process flow chart, and held a few meetings to explain the changes to the staff. Then we are surprised when the change does not bring about the hoped for benefit. As a colleague once put it, 'we make some changes, hope that magic happens, and then wait for the results' (Figure 18.2).

FIGURE 18.2   Sometimes organizations seem to be in the business of magic.

The problem is that often we do not plan for the change very well. It is important to document the new process and methods of working, keep people informed of what is happening, or going to happen (and this is often what organizations don't do enough of), provide training and answer questions. However, we often forget the more subtle aspects of change management.

Before implementing a change it is important that those impacted be **ready** to change and have the **capability** to change. It is usually helpful to explain to people why the current process needs to change – what is wrong with it – and then tell them what changes have been proposed, why they have been proposed, when they will happen and how their work will be affected. **Readiness** to change entails being willing to change, having the motivation to change and having a goal to aim towards. **Capability** to change is more related to organizational aspects and involves having the power to change, having the influence and authority to allocate resources. The following checklist addresses many of these issues:

- Who will have to perform the majority of the implementation tasks?

  ? Do they have the time?
  ? Do they have the skills?
  ? What are their priorities?
  ? Who might resist the changes? Why? What reasons for resistance have been identified? How will they be overcome?
  ? Who will lose out by the changes (e.g. lose status, staff, budget, influence)? What can be done to negate the losses?

- What operating standards have been established for the improved process?

  ? How were these standards derived?
  ? How have these standards been communicated?

- What new or revised procedures exist for the process?

  ? Are the differences obvious?
  ? Do the changes make sense as they are being performed?
  ? What have been the reactions of the operational staff and management to the procedures?

- What training is required?

  ? Will it be effective?
  ? How do you know?

- What could go wrong with the implementation?

  ? What fool proofing has been done?
  ? What contingency plans are there for things going wrong?
  ? What will be the impact of the change on suppliers and on operational support and managerial staff?
  ? What will be the impact on customers?

- What cross-departmental/divisional/customer issues might be thrown up by the implementation?

  ? Who has been involved in resolving these boundary issues?

- What management systems/cultural support mechanisms will discourage the change?

  ? What management systems/cultural support mechanisms are needed to support the change?

One common problem with making changes within organizations is the failure to address the cultural and managerial systems surrounding the planned change. For example:

- To encourage teamwork the reward mechanisms must support it. Rewarding individual performance, perhaps by employee of the month awards, does not support team-working.
- To encourage innovation, people who (responsibly) try new ways of doing things need to be rewarded, even when they fail. Penalizing people who try new ideas and methods which fail stifle innovation.
- If an organization expects its employees to treat customers well, the employees themselves must be treated well.

It is important therefore, to ensure that the management processes and culture support the new working practices.

Finally, one of the key issues that so often results in failure is that of not addressing the resistance to change within the organization. In most changes there will be winners and losers. People may resist change for a wide variety of reasons including: lost of status, not wanting or being able to learn new skills, fear of the unknown. It is important to identify and address this resistance to change to help ensure a smooth and successful implementation.

18.4 IMPLEMENTATION

If the planning and trials have been successful, implementation should simply be a case of implementing the plan. However, the plan should include provision for:

- Monitoring implementation to ensure it is going according to plan and taking any necessary remedial action.
- Checking for and dealing with any actual or potential problems that occur.
- Reviewing the implementation to ensure that the new procedures/practices are working as planned (e.g. by audit, reviews, focus groups, feedback sessions).
- Quantifying the performance change and comparing it with the expected change.

This is the same for all organizational changes – we want to check that the anticipated performance improvement has occurred. The improvement results will be a key element in evaluating the success of the whole benchmarking project as well as the implementation of the changes.

18.5 MONITORING

The final step of an improvement process is to install a monitoring process to ensure that the improved performance is maintained. This will typically include the use of control charts to monitor performance and possibly audits, reviews or other methods to ensure, at least in the short term, that the new methods of working are maintained and are continuing to provide the expected benefits.

SUMMARY

✓ Before implementing changes within an organization it is necessary to understand how and why the organization works and performs as it does.
✓ Much of this knowledge will become apparent as an organization takes an active part in planning the benchmarking study.
✓ Many of the implementation issues related to benchmarking are the same as those for other process improvement activities – so organizations that have an effective improvement methodology will be well placed to implement changes resulting from benchmarking.
✓ Many of the reasons for organizations failing to realize the benefits of improvement are focused on the planning and implementation of the changes. Specifically:

- Ensuring that people have the desire to change.
- Ensuring that people have the ability to change.
- Ensuring that management systems and cultural issues support the change and the new practices.

Chapter 19

Copying Without Understanding: a Risk Too Far

Gordon Hall
Deming Learning Network

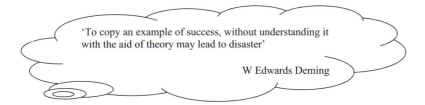

'To copy an example of success, without understanding it with the aid of theory may lead to disaster'

W Edwards Deming

INTRODUCTION

A common objective of benchmarking studies is to identify an organization that does something better than you do with the anticipation that copying their practices will lead to similar performance levels within your own organization.

Sometimes such copying is appropriate. However, sometimes it is not.

One of the risks in benchmarking is the temptation to copy what another organization does without first understanding why what they do works. In this chapter we discuss the importance of understanding the management theory that underpins an organization's practices.

19.1 THE RISK OF COPYING WITHOUT UNDERSTANDING

The risk of copying another organization's practices without first understanding why the practice works is illustrated in the following case study:

Case Study: Volvo Quality Circles and Suggestion Schemes

The classic example of failure to realize the implications of the relationship between management theory and results was the Quality Circle and Suggestion Scheme initiatives of the 1980/1990s. The manufacturing industries of 'The West' observed the success of Japanese companies in harvesting the thinking and ideas of the workforce, and thought that by simply copying the methods used in Japan that they too could reap the benefit.

The following is an extract by Bo Ekman, Head of Group Development at Volvo in the foreword to the book 'The Power of Learning' by Klas Mellander

The average Japanese worker turns in between 30 and 40 suggestions per year on ways of increasing efficiency and improving quality. The corresponding figure at an assembly plant in Detroit, the motor city, is one suggestion every seven years. After an intensive campaign, Volvo doubled their own number of implemented suggestions it received to one per worker per year.

Such examples were not unusual. The West, in the late 1980s and early 1990s, in copying Japanese methods, created a broad range of suggestion and quality circle initiatives. Yet the West got little benefit. Their basic thinking had not changed.

Case Study: Suggestion Schemes and Management Committees

In the late 1980s a large multinational oil company implemented a staff suggestion scheme. Proposals were made by filling in forms which were reviewed by various management committees. Forms included a financial analysis of the costs and expected benefits, etc. The typical time from suggestion to acceptance or rejection, by management, was between 3 and 6 months. If the suggestion were accepted the proposer would receive a bonus.

After a couple of years the scheme was withdrawn, due to lack of use, and very few accepted suggestions.

Unfortunately when quality circles and suggestion schemes failed to deliver the expected benefits, managers abandoned the concept rather than re-appraising its fundamental foundations. They were unaware of and therefore failed to see the difference in the management theories between the Japanese organizations they copied and their own organizations. Leaders in the West reverted to the status quo and continue to this day in having a poor utilization of the creativity and innovation of front line people.

All too often failure to reap the expected rewards was due to the fact that management's basic thinking had not changed. Senior management still saw themselves as the brains behind the operation and created complex approval structures before a suggestion could be implemented. Most suggestions got lost in the bureaucratic processes, and the workforce soon became disillusioned with 'management's' poor response times.

Konosuke Matsushita, founder of Matsushita,[1] encapsulated the problem in this quote in 1982:

> '*The reasons for your failure are within yourselves. Your firms are built on the X theory (hierarchical) model. Even worse, so are your heads. With the bosses doing the thinking while the workers wield the screwdrivers, you are convinced deep down that this is the right way to do business. For you the essence of management is getting the ideas out of the heads of bosses and into the hands of labour. We are beyond your mindset. Business, we know, is so complex and difficult, survival of firms so hazardous in an environment increasingly unpredictable and competitive and fraught with danger, that their continued existence depends on the day-to-day mobilization of every ounce of intelligence*'.

The issue that Matsushita raised is that the problem lies with the underpinning theories management hold in managing their people.

More recently team-working is being adopted by many organizations in the West. Groups of people who naturally work together are told they are now a team, some teams are encouraged to have a team hug, and in many cases this is all that happens. Once again many organizations are failing to reap the benefits experienced by the Japanese because they are ignoring the management theories underpinning it. While management still promote or in some other way reward individuals in the team, team members will continue to compete, not cooperate with each other. For teamwork to be successful other mechanisms need to support teamwork. For example, success and failure needs to be seen as due to the team, not individuals within the team. The whole team needs to be held responsible and rewarded, or otherwise, for their actions. For example, asking salesmen to work together as a team, and then paying a bonus to each individual based on their own sales will foster competition amongst the team, not cooperation. These and other similar actions are tantamount to tampering with management thinking, i.e. making isolated changes to management systems while largely maintaining command and control style management (called X Theory Management – see below).

The fact that sometimes copying without taking cognisance of the management theories does work has obscured the reasons for success or failure. When copying works it does so usually either because the management theory is immaterial, such as when copying mechanical processes, for example how to dismantle a piece of equipment, or because the theories are similar enough not to matter (Figure 19.1).

---

1. Matsushita's products are sold under the Panasonic, Technics and several other names.

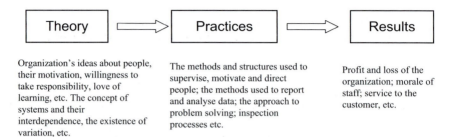

FIGURE 19.1   Relationship between Theory, Practices and Results in a work context.

19.2 THE RELATIONSHIP BETWEEN THEORY, PRACTICES, AND RESULTS

Everyone has beliefs and theories about all aspects of their lives. We have beliefs – theories – about speeding, politics, and work. We have theories as to why things are the way they are and theories about how things should be done. We live and act according to these theories and beliefs.

In business our theory might be that if we work hard and do a good job we will be promoted, in which case if we want to get promoted we will work hard. Alternatively if we think that in order to get promoted we need to get 'in' with the right person, we will attempt to do that. In general, we will act in the way that serves our interests.

If management of an organization believes that staff are mainly interested in getting their pay at the end of the week and are not bothered about the quality or quantity of their work, then they will act accordingly. They might, for example, introduce quotas, offer rewards such as employee of the week or month, or promote the 'best' worker. Unfortunately such actions may often appear to work in the short term when in reality what they do is distort the system as the following case study shows.

Case Study: Mis-Reporting Accidents

One organization was concerned about the level of accidents within the company. They investigated causes of accidents, supplied safety equipment, provided training and put up notices.

Eventually they decided to offer bonuses to shift workers who had 'accident free' months. The scheme was simple: if there are no accidents during the month on a shift, the whole crew would receive a bonus.

The accident rate declined, supposedly vindicating management's safety bonus policy. What was not clear for several years was that people still had accidents, but the accidents were covered up by the co-workers eager to earn their bonus. While the reported accident rate declined, the actual accident rate was unknown and opportunities to improve safety were missed.

One of the earliest modern documented links between theories, methods and results is Douglas McGregor's Theory X and Theory Y as explained in the following example.

Example: Theory X and Theory Y

Theory (X): The organization thinks that its people are inherently lazy and they need to be motivated, directed and controlled.

Practice: The organization will establish supervisory structures with vision statements, job descriptions, staff appraisals, targets and incentives. The data collected are invariably used to measure the performance of employees and encourage greater diligence.

Results: Outcomes as directed by top management, poor employee morale and commitment, High turnover of staff, employee compliance, poor take-up of innovative ideas from workface personnel, lack of entrepreneurial spirit.

In contrast:

Theory (Y): That people like work and welcome responsibility.

Practice: The focus of leaders is the design of systems to enable their employees to perform. Effort is put into the development of self-discipline and aligning the aims and values of the employees and that of the company. There are structures for listening to the workface personnel and implementing their innovative ideas.

Results: An innovative company that is at the forefront of developments in its field of operation; high morale and good staff retention; sound profit and loss figures.

(Adapted from Douglas McGregor's X & Y theories portrayed in his classic book – The Human Side of Industry 1960)

Organizations do not consciously select or develop the management theories which they adopt and are often not even aware of what their theories are. It is only when the organization makes the effort to consider specific questions aimed at exposing their theories that it begins to realize what they are. It may seem incorrect to consider that it is the organization that holds the theories by which it operates rather than the individuals within the organization. However, once we realize that the theories usually remain more or less static as people join and leave the organization, it becomes clear that theories are seldom dependent on or dictated by individuals; the more so the larger the organization.

More recently, Peter Senge, one of the foremost contemporary management thinkers of our time, reaches similar conclusions. He too recognizes the importance of the underpinning management theories, or, to use his term, 'Mental Models', of an organization. He uses the term Mental Models because he believes it is a more accurate term than theories. To explain the relationship between Mental Models and external results he uses the analogy of an iceberg (Figure 19.2).[2]

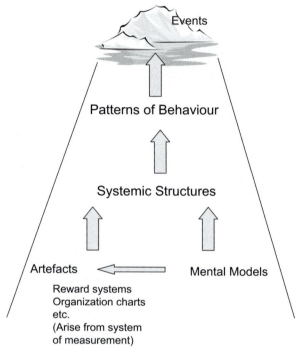

FIGURE 19.2   Senge: From Mental Models to Results.

In this model, Senge's Mental Models evolve into artefacts which under-pin the systemic structures of the organization. Systemic structures are the inter-relationships between variables that cause the organization to behave in certain ways. For example, an organization may wish to eradicate the negative effects of 'crisis management' only to find that the root causes are somewhere deep inside the organization in how they collect and analyse data and how they 'motivate' their managers. Usually systemic structures are hidden, but they result in certain behaviours. We are usually unaware of these structures but they influence the more subtle patterns of behaviour, which in turn lead to observable events or results. It portrays a similar message, in that if we

---

2. Presented at the "Society in Sync" conference, in Aberdeen, Scotland in 2005.

compare only the results we are only scratching the surface. To learn from a benchmarking exercise we require to appreciate the whole complex under-pinning assumptions of the organization.

'In a traditional authoritarian organization, the dogma was managing, organizing and controlling. In a learning organization the new dogma will be vision, values and mental models'.

Bill O'Brien
CEO Hanover Insurance

Example: From Mental Models to Results: Examples of Senge's Model

Example 1: From the John Lewis Partnership (a major high street retailer)

Mental Model: The primary focus of the company should be the well being and happiness of their staff.

Artefact: Fifty-one per cent of John Lewis is owned by the staff – or partners as they are called.

Structures: Management report to the front line staff. The staff have structures through which they can influence the company right the way up to the chairperson. They can technically dismiss the chairperson though it has never been done. The primary auditing function is in relation to adherence to the original vision and values of the company. A branch can make a good profit but if surveys reveal that the staff are not happy then an investigation is initiated.

Behaviours: Morale of staff is high, they feel valued.

Result: The interface between staff and customers is very positive. The company continues to grow and be profitable over the long term.

Example 2: Measurement

Mental Model: If you cannot measure it you cannot control it.

Artefact: Expensive management data are collected around those processes that can be described by numbers – especially in monetary terms. Little attention is paid to those aspects that cannot be measured such as the long-term effect of training, human resource policies, the detrimental effect of competitive tendering and bonus systems, the morale of staff.

Structures: Budgetary control; Performance management; accountability and blame, targets.

Behaviours: Poor respect for 'management', a '9–5' attitude, the distortion of data to fit performance criteria.

Result: A compliant and manipulative or political culture. Very poor use of staff's willing contribution to furthering the aims of the organization.

Contrast this with an Alternative Model

Mental Model: There is a full appreciation that variation naturally occurs in all processes.

Artefact: Statistical Process Control and in particular the control chart.

Structures: Management data presented in control chart form. Companies employ statisticians.

Behaviours: The company does not look for reasons for individual variances, instead they look to reduce the inherent variation in the system.

Result: The company make sound decisions based on accurate interpretation of data.

Example 3: Applying Mental Model to McGregor's Theory X

Mental Model: People do not like work and are naturally lazy.

Artefact: Extrinsic rewards such as bonus or qualifications so that we can motivate the individual. Procedures to give clear instructions.

Structures: Hierarchical supervisory and auditing structures to ensure due diligence and manage the extrinsic reward system.

Behaviours: The individual moves his thinking from how to do the job to the best of his abilities to how to secure the most from the reward system or how to be seen to comply with written procedures. The boss grows in feeling of importance as he uses the reward system and the procedures to control and manipulate the individual.

Result: A compliant culture that is poor at utilizing the thinking potential of work face staff.

If we wish to learn from a benchmarking exercise then comparison needs to include theory, the methods used as well as the results. Miss any one of these three headings and the comparison will be incomplete and is likely to be flawed.

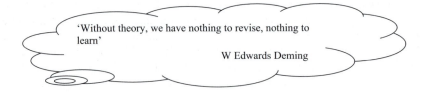

'Without theory, we have nothing to revise, nothing to learn'

W Edwards Deming

The implication is that managers, especially senior managers, need to be aware of their management theories before attempting to learn from another organization. It does not imply that one theory is better than another, just that we should be aware of the theories in use, and that learning starts by being willing to challenge these theories. As McGregor explained: *The purpose ... is not to entice management to choose (between theories). It is rather to encourage the realization that theory is important, to urge management to examine its assumptions and make them explicit. In doing so it will open a door to the future.*

19.3 HOW TO IDENTIFY THEORIES

The identification of the theories in use is not as difficult as it may appear. By examining the methods used and the general management language in an organization we can deduce the theories that underpin the management of that organization.

For example our motivation can be viewed as either intrinsic or extrinsic. Intrinsic motivation reflects our innermost drives; such as our pride, our desire to belong, our wish to contribute and our need to be valued. Extrinsic motivation on the other hand depends on external rewards such as a bonus system, an academic qualification or working for a promotion.

There are schools of thought that focus on extrinsic motivation – basic Behaviourism for example as developed by B.F. Skinner – which sees us as reactive individuals that respond to stimuli.

There are other schools of thought that feel that Behaviourism is too simplistic. They argue that Behaviourism has been based on research on caged, starving animals and the findings used to predict the behaviour of normal healthy humans. They argue that the human mind is much more complex than that of animals and they are much more than beings who simply respond to stimuli.

So if we look at the methods used by an organization we can deduce whether management think their staff need to be extrinsically motivated or whether they are aiming to develop their existing intrinsic motivation. If the organization thinks in terms of extrinsic motivation then they use such methods as incentive schemes, targets, budgets, appraisal of performance, the encouragement of competition for career development, audits, etc.

If on the other hand the company recognizes the value of intrinsic motivation then they will see that people come to work wanting to do their best and their focus will be on creating an organized and positive work atmosphere. The leaders will spend time designing robust systems that enable people to do a good job. Furthermore they will see front line staff as the customers of their systems design; and will have structures that listen to staff, and respond to their suggestions. They will also invest heavily in training and development. They will recognize the natural creativity of staff and have means of capturing continual improvement ideas. This focus on internal drivers can in fact establish a very challenging atmosphere.

See Figure 19.3 for more examples.

Theory	Characterized by	Practices and methods	Typical management language and tools
McGregor's X Theory:	Staff are inherently lazy and require to be motivated, directed and controlled.	• The aim or purpose is developed by "bosses" or central government and they issue directions as to the activities that should be carried out to meet the aim. • The use of written procedures drawn up by "management" and the use of inspection to ensure compliance with those procedures. Amendments to procedures required to be approved by the appropriate authority. • Measurement is taken to reflect the diligence of the individual. • Staff are motivated to perform through budgets, staff appraisal and concepts of accountability (extrinsic motivation). • Bosses are seen to be the decisions makers.	Hierarchical organization chart Supervision
Management of systems: mechanistic/reductionist thinking	We manage by breaking the whole into manageable parts. Our belief is that the whole is the sum of the parts.	• Identification with purpose is only at the highest level. • The organization controls the parts through a hierarchical management structure, job specifications, standards, inspection, budgets, targets etc. • Individual diligence is seen as the main variable in performance levels; hence a focus on qualifications, competence and concepts of accountability. • Formal communication is predominantly up and down the hierarchical chain.	Job specifications Budgets and targets
Measurement: Deterministic thinking	Unique figures have meaning. From tabulated data we can gain understanding by comparing a figure with budget and last period.	• The measurable outcomes are reported in tabulated form, with comparisons between this period's figures and that of the budget or of a previous period. • "Management" is predominantly through the review and monitoring of "accounting" measures. • There is belief that "if you cannot measure it you cannot control it". • The turning of a blind eye to the widespread manipulation of data to meet targets and incentives. Leaders using distorted data.	Accountability (and Blame) Staff appraisal
Knowledge: Experiential	We learn through experience.	• Managers learn through experience. • Training is in context of the development of personal skills. • New initiatives are usually the application of new methods or structures based on original organisational paradigms – often sceptically referred to as "Flavour of the month". • Management seen as a craft concerned with methods but not underlying principles.	Individual learning/training Standards & procedures
Organizational learning: individualistic	That the learning of the organization is developed by enhancing the learning of the individual.	• The intelligence of the organization lies with the individual employees, in particular the leaders. • To increase the capability of the organization you invest in training the individuals and leaders. • To measure the capability of the individual you focus on "Qualifications".	Auditing and compliance

FIGURE 19.3 Determining management theories.

Principle	Statement	Detail	Key concepts
McGregor's Y Theory:	Staff come to work wanting to do their best. Management's task is to recognize and develop self discipline and integrate staff with the aims of the organization.	• Extensive effort is undertaken to ensure that the aim of the organization or process is owned by the participants. • Those at the workface play a major part in the writing of the procedures. The upgrading of procedures is led by workface personnel. • It is recognized that the majority of the intelligence of the organizations lies with those at the work face, as they are nearest the task and have the greatest insight into the detail of the challenge. • Bosses are seen as systems designers to enable front line workers to be all they can be.	The system and its portrayal in flow maps Purpose of the system
Management of systems: Systems thinking	We manage the whole. The whole is the sum of the parts plus all the interactions between the parts. We look for cooperation between the parts	• The organization manages the whole by managing the parts and the interrelationships and interdependencies between the parts. • There is a positive focus on communication and co-operation between individuals, departments, suppliers and customers. • There is a strong focus on systems design. Bosses see their primary role as the design of robust systems that enable their workforce. • The major systems in the company will have been flow mapped. These flow maps will reflect the nesting of systems. • Measures are taken to reflect the effectiveness of the system.	The needs of the participants in the system - Internal customers Capability of the system System dynamics Interdependence and cooperation Communication
Measurement: Acceptance of variation	There is recognition that variation exists in all systems and that figures only have meaning in context. It is patterns that we seek to understand.	• The recognition of the existence of variation and that no single figure has meaning outside its context. • The portrayal of data in a control chart format. • The recognition of the difference between the variation occurring in the natural operation of the system (common cause) and that which reflects a special event (Special cause). • The use of measurement to identify the capabilities of the system, the average and the inherent variation in the system. • The use of the control chart to identify trends. • The recognition that the processes that provide numbers only account for 5% of the whole. Aspects concerning morale, training, use of data etc. are not measurable; feedback requires more subtle investigations such as listening and involvement.	SPC – variation – control charts
Knowledge theory, experimentation and experience	We gain knowledge through being aware of our underpinning theories (mental models) and having them open to be challenged by modern research.	• The recognition of the underlying theories in use in the organization. • The willingness to identify and challenge those theories and mental models. • The application of "Scientific Method." i.e. the postulation of a theory; experiments; confirming or disproving of the theory; and then application of the conclusions. • The development of "management" as a profession, fully aware of its theoretical underpinning.	Organizational learning
Organizational learning Organizational + individual	The recognition that the organization has an intelligence and learning ability that is separate from the individual	• The organization has an intelligence that is separate from the individual. That intelligence is characterized by the culture, structures, use of measurement, purpose, its relationship with its people and customers etc. • The organization knows which theories or mental models it is applying. It also recognizes that its theories may be different from those of individual employees. • The organization has a learning ability. It can identify and challenge the theoretical assumptions in use. • Its learning ability is reflected by the creativity employed to keep improving the systems and processes used in the company.	Motivation (intrinsic) Theories as a basis of knowledge

FIGURE 19.3 (*Continued*)

SUMMARY

✓ If we are to learn from a benchmarking study then we need to be aware of the underpinning theories (Mental Models) that characterize our comparison company as well as those used by our own company.

ᛉ ACTIVITIES

Identify the management theories in your organization. To help with this activity Figure 19.3 provides:

? A list of typical theories.
? A selection of typical characteristics of the theory.
? A selection of typical practices and methods used by an organization adhering to the theory.

The theories are divided into two groups. Typical management language and tools for each group are given on the right hand side of each table.

ⓘ For further reading see the bibliography.

Managerial and Organizational Aspects of Benchmarking

INTRODUCTION

Part 2 explained in detail a rigorous benchmarking process that would be appropriate for most types of benchmarking.

In Part 3 we discuss the supporting mechanisms for benchmarking activities. In Chapter 20 we discuss two managerial aspects of benchmarking. The first is managements tasks, roles and responsibilities for ensuring that benchmarking is used effectively within the organization. The second aspect is that of effective management of individual benchmarking projects.

Chapter 21 provides an introduction to typical legal issues. The key issue for most benchmarking studies is usually that of confidentiality. However, benchmarking studies that share data need to be aware of potential legal issues arising from anti-trust laws, the Treaty of Rome and other national and international legislation.

Finally, in Chapter 21 we discuss the use of an external consultant as a facilitator. In the situation where data needs to be shared anonymously the use of an external consultant is more or less a requirement. In other situations there are advantages and disadvantages in using consultants. The key is to ensure that the consultant is managed by, and answerable to, the participants.

Chapter 20

Management Roles and Responsibilities

INTRODUCTION

Up to this point we have considered the process of benchmarking. There are two key managerial aspects of benchmarking that are vital to the support and success of benchmarking that are considered in this chapter.

1. Management of benchmarking support activities within the organization.
2. Management of the individual benchmarking projects, discussed later in the chapter.

Most benchmarking studies undertaken by organizations require significant amounts of resources, mainly manpower, and for this reason they are usually only carried out after due consideration. Ad-hoc studies may be initiated because someone in the organization has a specific need and sees benchmarking as the best way of meeting that need. Nevertheless, because of the effort required, benchmarking is unlikely to become widespread and effective unless it is actively promoted and supported by management.

In this respect benchmarking is similar to any change or introduction of new methods of working: management has the responsibility for making it effective.

In this section we explore some of the activities and attitudes that will encourage and support benchmarking within the organization.

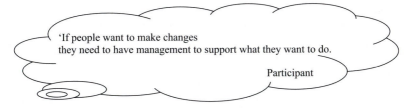

'If people want to make changes they need to have management to support what they want to do.

Participant

20.1 MANAGEMENT ROLES AND RESPONSIBILITIES IN PROMOTING BENCHMARKING

However much management may support benchmarking, this needs to be visibly shown through concrete actions that demonstrate ongoing commitment to other managers and the workforce. If management is perceived as not caring whether or how benchmarking is used in the organization, staff are unlikely to make the effort to benchmark.

One of the first actions that an organization, particularly a large organization, is likely to take is to appoint a Benchmarking Steering Committee or similar management team or person, with roles, objectives and responsibilities. This is often a good way to proceed, as it formalizes management's position regarding benchmarking. Regardless of whether such a team or person is appointed, it is still important that management demonstrates their ongoing commitment to benchmarking by taking an active interest and promoting benchmarking by what they say and do. Typical activities that demonstrate commitment include, but are not limited to:

✓ **Development of a policy** or other statement explaining the organization's intentions and view of benchmarking, along with plans and goals. In addition it would be appropriate to develop, publish and monitor benchmarking activities. One useful document to draw up is a benchmarking protocol and code of conduct setting out how the company views contacting, working with and sharing information with other organizations, some of which may be competitors. It should also include norms, protocol and processes for contacting and working with other organizations and responding to requests to benchmark received from other organizations.

✓ **Commission a Benchmarking Steering Committee** or person. This is typically a network of managers and others who are keen to support and be actively involved in benchmarking activities. This network would, for example, work together to pursue benchmarking activities and share learnings from benchmarking studies. In larger organizations this committee will be highly formalized with specified roles and responsibilities, discussed below. In smaller organizations this may be a part-time task assigned to one person. In all situations the roles are similar; the main difference will be in the formality with which they are carried out.

Case Study: Lunch-and-learn Sessions

One organization developed a very successful network that met monthly for lunch-and-learn sessions. Lunch was provided by the organization and the content of the event varied to cover topics of interest. For example, presentations were made at the conclusion of projects; speakers from other companies were invited to talk about their benchmarking experiences; and there were presentations on benchmarking topics such as data analysis, working with participants and legal issues.

✓ **Provide training**. Probably the first type of support that most organizations provide is training. Training is useful, if not necessary, and indicates management support. Training is likely to include a suite of events from awareness seminars and presentations, possibly for all staff, to training courses for those actively involved in benchmarking activities. A very powerful method of demonstrating management's support of benchmarking is for management to be amongst the first to take the training and be involved in later delivery of training events.

Case Study: Management Involvement in Training

One organization held various training events on benchmarking. A senior manager would always give a presentation on the importance of benchmarking to the organization. Delegates would often ask the manager probing questions about his involvement in, and experience of, benchmarking. The delegates really wanted to know whether the manager was "walking the talk" about his stated commitment to benchmarking.

✓ **Commission and support benchmarking teams**. Benchmarking projects may involve one or more key people from the organization. This person or team should be formally commissioned, like any project, with objectives, terms of references, etc. Those involved in benchmarking activities need to be released from other duties to the extent required to complete the benchmarking project. A key tool for facilitating this, the Charter, is discussed in ⇨ Chapter 3. In addition to commissioning teams, management is responsible for supporting them, releasing required resources and removing any roadblocks that may arise.

Case Study: Result of Lack of Management Support

One organization decided to adopt Quality Circles as a method of improving performance. The circle meetings survived for about 6 months before finally fizzling out. Management were disillusioned by Quality Circles and had 'proved' that they did not work. Staff were disillusioned with lack of management commitment and saw Quality Circles as just another management fad (⇨ Chapter 19).

Investigation revealed that the Circles were expected to meet in their own time, after work, for no extra pay, nor time off in lieu of overtime worked.

✓ **Take an active interest in benchmarking activities**. Management should take an active interest in benchmarking by, for example, monitoring benchmarking activities, attending presentations, asking for project updates and findings, recognizing and rewarding those involved in benchmarking, and trusting them to do a good job.

The actions listed above are typical benchmarking specific actions. There are other more subtle but important cultural issues that will support benchmarking such as:

✓ **Foster a questioning attitude**. The whole drive for benchmarking is facilitated by wondering how other organizations perform and how they achieve their levels of performance: those who want to ask and answer questions such as:

- ‘How do our competitors do this?’
- ‘Does anyone else have this problem?’
- ‘There must be a better way to ....’
- ‘If that organization can achieve what they do, why are we so poor at doing what we do?’

✓ **Foster a spirit of improvement**. Very much linked to the questioning attitude is the desire to improve. The attitude in the organization might be to expect that improvement and benchmarking are as much a part of the job as turning up for work in the morning.

✓ **Reward success, do not punish failures**. Rewarding those involved in benchmarking studies will increase their morale and demonstrates management's support. Rewards need not be financial.

The management of one organization gave team members a small gift such as a book and lunch at a local restaurant at the end of the project. The reward was a token of management's appreciation and was given to all teams completing a project, regardless of the impact of the project on the organization.

Case Study: The Result of Using Benchmarking as an Appraisal Tool

One organization benchmarked their performance on a regular basis. Staff believed that while their performance was above average nothing would happen, but when their reported performance was below the industry average management used the results to penalize staff. When the facilitators realized this, they understood why the data received from the organization was usually late, often inaccurate, and answers to their queries were often unclear.

In another organization those responsible for the area being benchmarked stated that they would never take part in benchmarking again because senior management had severely reprimanded their departments because of their poor performance relative to other participants.

All people tasked with benchmarking activities – managers, focal points, team leaders, etc. – need to be chosen with care. The first step is to consider what skills, level within the organization and personality traits (e.g. leadership, motivational, diplomatic, charisma, attention to detail) are required for the task. If benchmarking is going to require a cultural shift, finding appropriate people to lead benchmarking activities may be difficult because successful people in the organization will have been successful in the current culture: they may feel that they will not be successful in a different culture.

There will be many other methods and opportunities for management to show their support of benchmarking as the following case study shows, albeit that this was in support of safety:

Case Study: Demonstrating Priorities

At a monthly management meeting the senior manager opened by saying that he wanted to begin with the items related to safety. Once these items had been dealt with he excused himself, asking the other managers to complete the meeting without him. The lesson was not lost on his management team: he had demonstrated his interest in, and commitment to, safety.

20.2 ROLES, RESPONSIBILITIES AND TASKS OF THE BENCHMARKING STEERING COMMITTEE

While top management are responsible for ensuring that benchmarking is promoted and supported by the organization, the implementation of their wishes may be carried out by other people, especially in larger organizations, typically called a Steering Committee. The Steering Committee works with senior management and the division of tasks between them will vary from organization to organization. One general guideline is that the Steering Committee implements the wishes of the senior management and prepares key documents which are approved by senior management. The size of the committee will also vary, depending on the needs of the organization, from one part-time informal role for small organizations to several full and part-time members in larger organizations. Typical tasks of the committee, many of which are discussed above, include:

✓ Develop the benchmarking policy, which is signed off by senior management, and then implementing the policy.
✓ Set benchmarking goals and objectives.
✓ From the goals and objectives, develop, implement and monitor benchmarking plans to ensure that positive action is being taken to integrate benchmarking as the normal way of doing business.

✓ Plan and possibly deliver training.
✓ Develop a benchmarking protocol/code of conduct ⇨ Chapter 17.
✓ Commission, support and advise teams. Monitor individual project progress and provide practical support, particularly removing any roadblocks, such as resistance from individuals within the organization.
✓ Actively promote the use of benchmarking.
✓ Liaise with benchmarking groups outside the organization.
✓ Be a source of benchmarking information both internally and externally.

20.3 DEALING WITH RESISTANCE TO BENCHMARKING

Resistance to benchmarking can come at any stage of the benchmarking process from the first mention of the name, through to inviting participants to join a study, to accepting the results and acting on them.

Methods of dealing with resistance to a specific benchmarking project are outlined in the appropriate part of the book (⇨ Chapter 11 Gaining management approval, ⇨ Chapter 12 Recruiting participants, ⇨ Chapter 13 Acquisition of data. See also ⇨ Chapter 1) and many of them are appropriate for this section too. In this section we discuss the more frequent causes of resistance to the general concept of benchmarking within an organization and give some suggestions on how to deal with it. However, as with all situations the key is to understand the real cause of the resistance rather than the presenting argument. For example, a manager may resist benchmarking, declaring it a waste of time, taking too long, taking away resources that he urgently needs, not being relevant because his situation is unique. However, his real fear may be that his group will be shown up as a poor performer and so his managerial skills may be called into question. This example illustrates that much resistance to benchmarking, and indeed many changes in an organization, are emotional rather than logical.

✗   **We are unique – nobody else does what we do**

> **Case Study: We're Unique, Just Like Everybody Else**
>
> Some years ago I ran a series of courses on Quality Management. At some point in every course someone would mention that his organization was different. I would ask for a show of hands of everyone who thought their organization was unique – different to all other organizations. Every hand went up; always.

All organizations are different. They have different markets, economic, legal, environmental, etc. situations. Therefore, they could argue that benchmarking cannot work. A useful response is to give examples of organizations that have benchmarked successfully, preferably those that have written about their own situation. The experiences of Xerox described in the book Prophets in the Dark are particularly helpful, but also there are many examples in this and other books. Particularly useful are the case studies of cross-industry benchmarking such as the ⇨ Formula 1 case study in this book.

As one objection is satisfied there may always be another, and when there is, it is a sure sign that the objections are a smoke screen, often for a fear of being shown up as a poor performer compared to other organizations.

Case Study: Smoke Screen

One manager got very frustrated with other managers in his organization. He was asked to give an example of where benchmarking had worked within the industry. He quoted an example. This did not satisfy the protagonists, who then demanded an example of benchmarking not only within the industry but also within the same country in the same industry. After some research the manager found another example. The next request was to find an example in our industry, in our country of a company as big as ours . . ..

✗   **We don't have the resources**

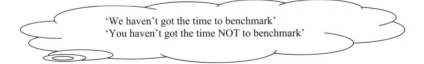

'We haven't got the time to benchmark'
'You haven't got the time NOT to benchmark'

Another common objection is lack of resources, usually time. Frequently it is genuine. Usually it is most valid where an organization is not in control of its business but reacting to situations as they arise. Processes are chaotic, things go wrong, there are emergencies, rush orders, late deliveries. If likened to a building, they are 'so busy putting out the fires, they never get round to improving the building'. These are usually the very organizations, or the parts of an organization, that would benefit most from process improvement of any sort, benchmarking being just one method of achieving improvement. To what extent is a lack of resources a valid objection? One management consultant related his experience . . .

> **Case Study: Short cut to Cutting Costs**
>
> ---
>
> A management consultant once confided: one of his clients was concerned at perceived over-staffing in the organization, yet he explained, everyone was working hard, often putting in overtime.
>
> He wanted to benchmark with other organizations to ascertain if and why their manning levels were high but people complained that they had no time or resources to benchmark.
>
> With growing concern over the financial status of the organization he finally decided to act. Rather than carrying out a detailed manning review or benchmarking study he employed a 'hatchet man' who, after experiencing the organization for a few months estimated the level of over-manning and cut staff accordingly.
>
> With staff reduced, effort focused only on the key areas of the organization, and to do these more efficiently, with the staff available, the organization finally benchmarked.

This case study seems to fit with the adage that the work will expand until everyone is apparently working hard and has no time for anything else. It often seems to be true both of individuals and organizations. Somehow organizations cope when people resign, retire or take sick leave: it is perhaps also possible to cope with a benchmarking project. So, while it is important to recognize the contribution people are making, it is also likely that there is slack somewhere that would free up people for a limited time to pursue a benchmarking project.

Management's responsibility is to ensure that resources are made available for benchmarking and they can demonstrate their commitment by changing priorities or even taking on additional resources for the duration of the study.

✗ Benchmarking is only for large organizations

It is probably true that proportionally large organizations have led the way when it comes to benchmarking. However, organizations of any size can benchmark – all it requires is a desire to improve and the resources to do so. In the book 'Benchmarking in Construction' page 182 by Steven McCabe, he states 'One of the most successful organizations to have used benchmarking was a small quantity surveying firm which employs only five people' In the book McCabe quotes the managing director as saying '... when you are as busy as we were, you don't have time to think about what you are doing. One of the things I would recommend to any firm is to get out and see a successful outfit carrying out their business'.

✗ We are doing well enough, we don't need to benchmark

This same argument can, and has, been used to resist most changes and improvements. However, past performance is not necessarily indicative of future success as many extinct organizations have discovered too late. The organization may have been performing well, but that does not mean that competitors are not in the process of taking business away from it. It is safer to benchmark when there are resources can be made available and avert a crisis rather than wait until a crisis is looming and funds are tight.

Case Study: We are Doing Well Enough (Prophets in the Dark, page xiii)

'To many in the outside world we still looked very much like the invincible copier giant and bedrock of the Fortune 500 that we had always been. It was evident that we were having a few problems. But still, we were making hundreds of millions of dollars a year. We were turning out fresh products. We were employing a hundred thousand people'.

Even seemingly highly successful organizations are at risk. Fortunately for Xerox the CEO, David Kearns, suspected Xerox was in deep trouble and acted, just in time, to save the company.

What is important in business today is not just where we are compared to our competitors, but how fast we are improving compared to them, as Figure 20.1 illustrates.

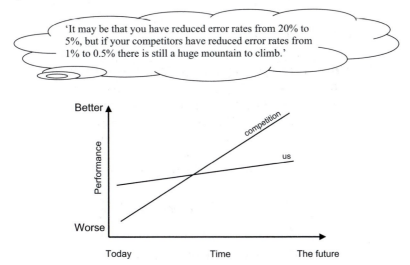

'It may be that you have reduced error rates from 20% to 5%, but if your competitors have reduced error rates from 1% to 0.5% there is still a huge mountain to climb.'

FIGURE 20.1   Being 'better' than the competition today is no guarantee of survival in the long term.

✗ Benchmarking is too difficult.

The answer to this is yes … and no. Most things seem difficult until you know how to do them. Fortunately there are books, organizations, consultants/facilitators and clubs to help you succeed. That does not necessarily mean it will be easy, but you will not learn until you try.

20.4 SUCCESSFUL PROJECT MANAGEMENT

Like all projects, benchmarking projects need to be managed.

Many of the project management aspects are discussed in the relevant sections on the book. In particular, for resource planning ⇨ Chapter 11, for managing meetings ⇨ Chapter 12.

In this chapter we consider the tasks of project management throughout the project and the need to identify the appropriate skills and therefore team mix for each phase.

The role of the project manager is to ensure that the project is a success and this usually includes the management of participant activities.

20.4.1 Team Tasks and Membership

During a benchmarking project there are many tasks that need to be carried out as have been detailed throughout the book. In smaller organizations or projects it may be appropriate that one person carry out all activities, albeit with help from others in the organization. For larger projects it will be more appropriate to commission a team of people to work on the project, partly because of the workload, partly to bring different skills and expertise to the project and partly to encourage support from all groups impacted by it.

Specific tasks and the associated skills required vary during the project are discussed below. In addition there are general tasks and skills that are required throughout the project, which are also discussed below. A useful way of viewing these skills is to group them into key skill areas:

- **Project management skills**: including budgeting, planning, monitoring, analysis, report writing and presentation.
- **Running successful meetings**: a key skill for benchmarking as much of the work will be done and managed through meetings, often with people from different organizations.
- **People skills**: a key skill to help maintain team cohesiveness, gain support from key people and groups, lead the project to a successful conclusion, liaise with participants.

20.4.2 Specific Project Tasks

From the point of view of skill requirements, a typical project that includes data benchmarking and follow-up activities is likely to consist of five phases as shown in Figure 20.2.

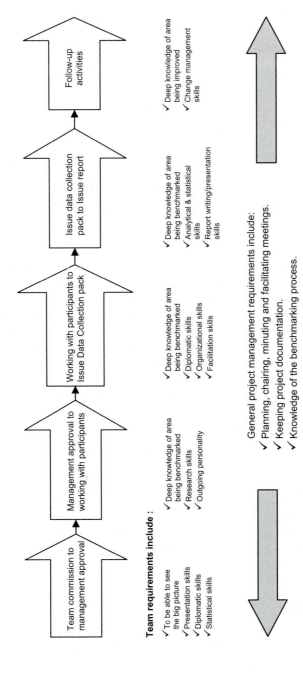

Team requirements include :

✓ To be able to see the big picture
✓ Presentation skills
✓ Diplomatic skills
✓ Statistical skills

✓ Deep knowledge of area being benchmarked
✓ Research skills
✓ Outgoing personality

✓ Deep knowledge of area being benchmarked
✓ Diplomatic skills
✓ Organizational skills
✓ Facilitation skills

✓ Deep knowledge of area being benchmarked
✓ Analytical & statistical skills
✓ Report writing/presentation skills

✓ Deep knowledge of area being improved
✓ Change management skills

Team commission to management approval

Management approval to working with participants

Working with participants to Issue Data Collection pack

Issue data collection pack to Issue report

Follow-up activities

General project management requirements include:

✓ Planning, chairing, minuting and facilitating meetings.
✓ Keeping project documentation.
✓ Knowledge of the benchmarking process.

FIGURE 20.2   Main team skills requirement phases for a benchmarking project.

Each phase includes different types of tasks and so has different skill requirements. Therefore, before beginning each phase the team membership and time commitment from each member should be reviewed.

For example, in the first phase, from team commission to management approval, one task is likely to be to assess the impact of the project on different departments and groups. One skill required to do this will be an overall appreciation of how departments and groups interact. It will probably also be necessary to enlist support from these groups, so diplomacy will be important.

Typical key skills for each phase are given in Figure 20.2. Not only is it important to identify people with the appropriate skills, it is also important to assign to the team people at an appropriate level. For example, in the first phase, a senior manager is more likely to secure cooperation for the project than a junior staff member.

20.4.3 General Project Tasks

In addition to the skill-specific project tasks, there are more general project tasks with their own skill requirements, such as steering and managing the team. Other general tasks include planning, organizing, chairing, minuting and facilitating meetings, keeping project documentation, and external liaison.

Meeting facilitation is a particularly important skill. It involves ensuring that all attendees are given the chance to air their thoughts and concerns and make a positive contribution so that the meeting can benefit from their knowledge and skills. At times the facilitator may need to diffuse tensions and seek agreement on contentions issues. He/she needs to be able to judge when to dwell on a particular point, and when to move the team on to the next discussion topic. The facilitator's work starts with the preparation required before the meeting to ensure that it will be as productive as possible.

20.4.4 Follow-up Activity Tasks

In many projects the team membership will not change a great deal until the final 'follow-up' activities phase. During the other phases it is likely that the workload for each individual will change, but the membership often remains quite static. For example, the analytical role will be heaviest at the analysis phase, while the people skills requirement will be at their least at this time.

'The 'follow-up' activities are those activities that the organization undertakes as a result of comparing data and/or other information between benchmarking participants. These activities are usually handed over to the relevant departments or groups who will set up their own projects with appointed team members. During this phase it is likely that either the benchmarking team will have been disbanded or it will remain in place with a monitoring and reference brief.

20.4.5 Cross-functional Representation

One of the common recurring issues within many organizations is the lack of buy-in from groups not involved in benchmarking (or other change) projects but expected to change their way of working.

A common and successful method of dealing with this issue is to include representatives from those areas likely to be impacted by the project in the benchmarking team. Not only do these cross-functional teams help reduce resistance to any recommended changes, but they bring knowledge and skills from the impacted areas. In addition these representatives are well positioned to lead, or at least be involved in, the follow-up activity teams.

20.4.6 Project Tools

In addition to the tools mentioned in the book there are tools specifically aimed at project and team management. These can be split into two types:

1. **Project planning, budgeting, monitoring.** The typical tools for this include Gantt charts and project management software which can be supplemented as necessary by other tools.
2. **Team management and facilitation**. There are a variety of activities and tools for dealing with the softer issues of helping a teamwork well together. They include ice breakers/team building activities; setting team and meeting ground rules; reviewing the agenda at the beginning of the meeting and giving (and receiving) feedback from team members. While many of these skills can be honed through training and experience, ideally a team facilitator who naturally has these skills should be appointed.

SUMMARY

In this chapter we discussed the key role that management plays in embedding an effective benchmarking approach within the organization.

Management should:

✓ Act as role models by taking part in, or at least showing an interest in, benchmarking activities.
✓ Develop a benchmarking policy, appoint key benchmarking roles such as benchmarking steering committee, focal point and provide resources (both manpower and financial) for benchmarking activities.
✓ Ensure that the organizational culture supports benchmarking.
✓ Identify and deal with resistance to benchmarking.

We also discussed key skills required for managing benchmarking projects, which are similar to managing other projects and include:

✓ Project management.
✓ Running successful meetings.
✓ People skills.

When selecting team members:

✓ Team members should be appointed on their ability to contribute to the tasks in hand.
✓ Team members should be drawn from key areas likely to be impacted by the project so as to help ensure buy-in to findings that may affect them.
✓ Team members may change as the required skill and knowledge requirements change throughout the project.
✓ Team facilitation skills are key to a successful team project.

Chapter 21

An Introduction to Legal Issues and Working with Consultants

INTRODUCTION

In this chapter we consider legal and contractual aspects of benchmarking and the use of external consultants as benchmarking facilitators.

Except for internal benchmarking studies where legal issues are unlikely to apply, it is important to be aware of the legal issues surrounding benchmarking. The two key issues are confidentiality of data and information shared between participants and national or international laws of data and information sharing.

Throughout the book we have referred to facilitators of the benchmarking study. The facilitator(s) may be members of participant organizations, or may be external consultants. Many benchmarking studies, and most of the larger studies, will use external consultants as facilitators and here we discuss some of the advantages and disadvantages of doing so.

21.1 LEGAL ISSUES

Legal issues can be very complex and will depend on the laws in the countries in which the benchmarking studies are taking place as well as international law. Because of the variability of national legal requirements, this chapter is an introduction to legal issues that are likely to arise. The reader should also refer to the benchmarking code of conduct, ⇨ Chapter 17, and the legal department in their own organization for specific advice.

The two most common legal issues related to benchmarking are:

1. **Sharing information in a manner not agreed by participants**. This usually relates to sharing information with joint venture partners, parent companies, or other parties who are not themselves involved in the study. It is controlled by confidentiality and employment agreements

2. **Sharing information illegally**. This usually relates to sharing of cost information amongst participants which may be considered anti-competitive. It is covered by national and international law.

Legal issues are likely to fall into the following three legal structures:

1. Confidentiality.
2. Intellectual property.
3. Competition.

21.1.1 Confidentiality

Confidentiality relates to benchmarking in three different ways (Figure 21.1):

1. **Employees**: employee confidentiality issues are governed chiefly by their contract of employment, and apply to everyone involved in the benchmarking study.
2. **Participant organization**: restrictions on what data and information can legally be shared between participants is governed by national and international law. Within these restrictions, the data and information that participants choose to share, and how it can be used is governed by the participant agreement.
3. **Facilitator and other consultants**: external facilitator and consultant confidentiality is governed by contract of engagement with participants. Internal facilitator confidentiality is governed by participant confidentiality agreement and contract of employment.

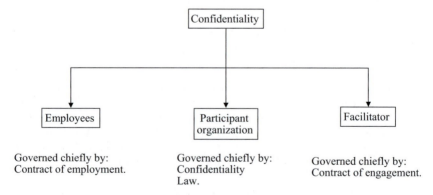

FIGURE 21.1   The three aspects of confidentiality.

Employees: The contract of employment will be organization dependent and will usually prohibit employees divulging certain information about the organization for which they work. Benchmarking activity will usually be covered by this agreement and is unlikely that any additional agreement will need to be drawn up. Internal consultants employed by a participant will also be subject to employee confidentiality agreement.

Participant confidentiality agreements are participant dependent and usually stipulate that information shared between participants may not be shared with third parties unless agreed in writing. The agreement to share data and information within the group of participants is implied by participants joining the study.

There are many ways of using and sharing benchmarking data within a study. For example:

- Data may be shared openly – i.e. each participant has full access to every other participant's data.
- Data may be shared anonymously – i.e. each participant has full access to every other participant's data, but the name of other participants masked.
- Summary data are shared. For example, the participant's own data and summary statistics such as the average, best in class, top quartile, etc. figures are shared.

In any study a combination of these sharing methods may be used, for example, one common scenario is to share non-cost data openly and cost data anonymously.

Confidentiality agreements typically require that:

(a) Data and information be kept confidential within the membership
It is usually agreed that participants will not share data with third parties. However, this can result in some difficult situations and so the following issues may need to be considered:

? Can data be shared with parent companies and/or co-venturers? If so, under what conditions? Sharing with parent companies or co-venturers may be difficult if they are themselves potential participants but choose not to join the study. This would result in an organization's data and information being shared with a potential participant without that organization reciprocating. One solution to this difficulty is that the parent companies or co-venturers may be shown, but not allowed to keep, anonymized results with only the participant's results highlighted and will not be allowed any other access to data, information, reports etc. pertaining to the study.

? Can data be shared with a company that takes over a participant, and if so, under what conditions?

? Some organizations may wish to publish results on their intranet and/or in their company magazines. The risk of this is that others, perhaps not aware of confidentiality/legal issues may misuse the data. Under what conditions, if any, will internal publication of results and data be allowed?

? How long does confidentiality last? Will there be some aspects of the study or some third parties (e.g. the media) that will never be allowed access to any aspect of the study? Will a participant's data be allowed to be included in future studies even if that participant is not involved in the study?

? Some organizations may wish to use external consultants for example to carry out further analysis. Under what conditions will the consultant be allowed access to data?

One possible solution to the sharing of data with non-participants is that it be anonymized. Not only is it unlikely that the non-participant will need to know the source of specific data, but it also protects the participant from any inadvertent (or otherwise) mis-use of the data by the non-participant.

(b) Best practices be shared

One of the issues that should be agreed before the study begins is the details of whether participants will be expected to share practices, and if so what they will share and under what conditions. Attempting to write such an agreement before the results are known is usually quite difficult, and often the requirement would simply be that a participant agrees to share practices if asked to do so. Most organizations will take this in good faith and share what is reasonable to do so, on the understanding and trust that others will reciprocate and keep shared data and information confidential. A separate agreement may be drawn up at a later date covering specific sharing events.

(c) All data and information to be shared be specified beforehand

This detailed agreement is usually only specified for site visits where there has been little, if any, data sharing beforehand, e.g. One-to-One benchmarking and information exchanges. The agreement will typically specify what data and information will be shared ($\Rightarrow$ Chapter 17).

21.1.1.1 *Code of conduct, confidentiality agreement and benchmarking/participant agreement – a summary*

There are three terms commonly used to in relation to guiding benchmarking activities. Some have been mentioned earlier in the book and we draw them together here for convenience.

The Code of Conduct was explained in some detail in $\Rightarrow$ Chapter 17. It sets out in general terms how organizations should relate to one another during benchmarking activities. Some organizations will simply adapt one of the standard codes while others may adapt them to their own specific requirements and expectations.

> **Benchmarking Code of Conduct**
>
> A Code of Conduct
> sets out the general principles of behaviour
> that guide the decisions of and interactions between
> those seeking to benchmark
> in such a manner as to respect the rights and expectations of all participants.

One of the principles in most codes of conduct concerns respecting confidentiality of data and information divulged by other participants. This is a key issue in many benchmarking studies and will vary from study-to-study. For this reason many studies will draw up a confidentiality agreement which states how shared data and information may be used and disseminated by participants.

> **Confidentiality Agreement**
>
> A Confidentiality Agreement is
> an agreement between participants of a benchmarking study
> that sets out how the data and information shared during the study
> may be used and disseminated by participants.

A participant agreement (also known as a benchmarking agreement) sets out the responsibilities and expectations of members of a benchmarking study. Agreements are often documented and may include or refer to the confidentiality agreement, a code of conduct or other related documents. They may include for example, financial arrangements, agreement to adhere to the benchmarking plan and/or an agreement that if asked to do so a participant will present at a best practice forum.

> **Benchmarking Agreement**
>
> A Benchmarking Agreement
> sets out the responsibilities and expectations
> of members of a specific benchmarking study.
> It may refer to related documents such as a confidentiality agreement
> or code of conduct and
> is intended to ensure the outcome of the study meets participant expectations.

21.1.1.2 *Facilitator and consultant contracts*

External facilitators and consultants are normally excluded from sharing any data or charts with any third party without permission from the participants. However, a general exception may be made for marketing the study to potential participants. Under this circumstance, the data are usually anonymized and/or historic and/or incomplete.

For a database study where the consultant holds the database, there will usually be an agreement regarding the use of data restricting both consultant and participant.

Facilitators holding data or information on participant organizations need to ensure that it is secure. To minimize the risk of illegal access to data and information held on computers or websites facilitators use a password system. In addition, they may monitor those accessing the data from outside the company. They will also need to prevent loss of data through regular back-ups.

21.1.2 Intellectual Property

Intellectual property refers to creations of the mind such as inventions, symbols, names, images, designs of literary and artistic works. It can be split into two types: copyright and industrial property.

Copyright, which is likely to be of lesser concern in benchmarking studies, covers artistic inventions. Those most likely to be encountered in business are books, published articles, presentations and courses. The rights of the creators of these artistic inventions are covered by copyright (called author's rights in some countries) and the wider body of law called intellectual property. Where the creator of an artistic work, for example a course, is an employee of an organization, the copyright may be held by the individual or the organization depending on local laws and individual contracts. Put briefly copyright law prohibits unauthorized copying of materials, as explained at the front of books and other publications.

Industrial property refers to inventions such as industrial designs, inventions and trademarks. In benchmarking, the one we are most likely to encounter is innovative process designs: novel methods of achieving an aim. These inventions are usually covered by patent. Utility models are similar to patents and apply in about 30 countries worldwide. Utility models usually apply to inventions that are less complex than those requiring patents and/or have a short commercial life. The purpose of the patent and utility model is to stop unauthorized use of ideas by third parties for commercial gain.

Normally, an organization will not share information that they do not want a benchmarking participant to know. However, to ensure there are no misunderstandings it is always wise to document what information was shared and how that information may be used.

ⓘ For further information see www.wipo.net

21.1.3 Competition: Legal Requirements

National or international law may prohibit the sharing of certain data. Some governments prohibit or severely limit the sharing of any data, especially with foreign organizations. In addition, some countries impose embargoes on other countries, and these embargoes may prohibit the sharing of data even when it is through an international study which includes participants from both embargoed and non-embargoed countries.

Two key pieces of law that apply to benchmarking are The Treaty of Rome, for European countries and Anti-Trust laws in the USA. The basic aim of laws such as these is to prohibit price fixing or the setting up of consortia to control pricing. The penalty for breaching these laws can be extreme, and many organizations are concerned about sharing any cost information out of fear of inadvertently breaching the law or even being accused of breaching the law.

The legal departments of many organizations are so concerned by laws such as these that their instinctive reaction to any sharing, especially of cost data are that it is not legal. However, a little persistence in explaining what it is proposed to share and a request for a detailed explanation as to why the data cannot be shared will often result in a reasonable agreement being reached.

The usual concern about sharing, particularly from the legal aspects, is the sharing of cost data. However, in benchmarking, the focus is often more on identifying who is achieving the best performance levels and how they are achieving them. The actual cost value of itself may be of lesser interest.

21.1.4 Legal Precautions

While confidentiality and legal issues may seem daunting, in reality they are not as onerous as they first appear. However, it is wise to take the time and trouble to ensure that the benchmarking activity is legal and agreed by a confidentiality agreement (Figure 21.2).

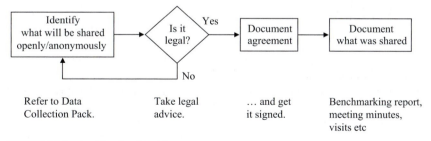

FIGURE 21.2   Avoiding legal problems.

The first step is for the participants to identify what data and information they would like to share, and then take legal advice to ensure that it is legal. If they are advised that it is not legal, seek an explanation as to why it is not legal, and if necessary change the proposal. Once agreement has been reached, document what is to be shared. Where site visits are included in the benchmarking study and data or other information is shared during these visits, it is wise to document what was shared.

The people who sign the confidentiality agreement are not always the same people that use the data and information from the benchmarking study. The users may be unaware of, or not understand, confidentiality issues. Therefore it is prudent to at least outline the key elements of the confidentiality prominently on documents, files, websites etc. that they may access.

21.2 WORKING WITH CONSULTANTS

Depending on the type of study, participants may have the choice of facilitating the study themselves or using a consultant as an external facilitator.

The choices are between:

- One or more participants facilitating the study.
- Representatives from each participant working together to facilitate the study.
- Consultants owning and facilitating the study (e.g. in a consultant owned database benchmarking study).
- Consultant facilitated, but participant owned study.

In this section we investigate some of the advantages and disadvantages of using an external facilitator, and the roles of both consultant and participants.

21.2.1 Self-Facilitation or External Consultant?

Sometimes the use of an external consultant will be unavoidable. Typical reasons for having to use an external consultant include:

- ✓ Where a consultant holds a database or other key information to which participants need access for benchmarking.
- ✓ Where data needs to be shared anonymously, it would be difficult to ensure anonymity without employing an external consultant.
- ✓ Where an organization wants to join an established benchmarking club that is run by a consultant.
- ✓ Where a consultant is carrying out a benchmarking study on behalf of other clients and invites you to take part.

Sometimes the use of an external consultant will be helpful. Benchmarking consultants can bring a lot of knowledge, experience and contacts to a benchmarking study and their use should be seriously considered. Typical situations where using a consultant may be beneficial include:

- ✓ Where independence is a particular concern

✓ When an organization is new to benchmarking and not sure how to progress.
✓ When specific skills are required and not available to the participant. Typical examples include analysis skills, data collection pack development, finding cross industry participants and legal issues.

Sometimes the use of an external consultant is not necessary. Organizations that have in-house benchmarking skills may not need external help, for example:

✓ For internal benchmarking studies.
✓ When independence is not an issue.
✓ One-to-One benchmarking, information trades or small group benchmarking when the participants can work together.

21.2.2 The Roles of the Consultant and Participant

The key task of the facilitator is to carry out the wishes of the participants. However, a progressive facilitator will always be trying to improve the quality of the study by, for example, increasing the number of participants and improving data validation, reporting, access to the data/information, encouraging learning through a website and best practice forums. This brings obvious advantages to the participants and helps the consultant firstly by making the participants want to continue using them and also by making entry of a competitor more difficult.

A difficulty can arise for the facilitators if participants just accept what the consultant offers. What starts as apathy from a participant may well result in them withdrawing from a study that they perceive does not meet their needs. Clearly, this is both detrimental to the participants and the consultant. A good facilitator will elicit comments and feedback from participants, will aim to discover their stated and unstated needs as well as their aspirations, and will aim to meet them.

The participants should hold the facilitator accountable for their actions, and should require that the consultant meets their needs. The key method of doing this is simply to take an active interest in the project, for example by taking part in meetings and making proposals that will meet their needs.

In addition to carrying out the wishes of the participants, there will be a specified or implied code of conduct that all consultants should follow, including for example, confidentiality, care of participant information and data.

21.3 A WORD ABOUT BENCHMARKING CLUBS

Each benchmarking study has its own attributes and requirements, but as a group perhaps benchmarking clubs vary more than any other.

In some clubs a consultant acts as facilitator and dictates, perhaps after consultation with participants, the terms of membership. In these studies the consultant has firm control and ownership of the study and the participants have little or no control over how the study is conducted. The disadvantage is that it is difficult for participants to modify the study to better meet their needs.

At the other extreme, the consultant acts purely as a facilitator and encourages the participants to discuss and agree to the details of the study. The facilitator responds to the wishes of the participants.

In reality, clubs usually operate somewhere between these two extremes. The consultant will facilitate meetings at which the participants agree aspects of the study, but will themselves make recommendations based on their experience of carrying out their role and working with participants. Where the consultant has specialist skills, such as IT or statistical skills, they are also well placed to use these to help develop the study.

The following table (Figure 21.3) summarizes typical facilitator tasks and comments on the use of using external consultants as facilitators. The list of issues broadly follows the order of activities in a typical club benchmarking study which includes data comparison and a report.

Issue	Comment
Finding new participants	Best done jointly. Where a current participant knows of an organization they would like to join the study, it is often better that the participant recruits the new participant. Often the consultant will identify and approach new participants, perhaps seeking advice or support from existing members.
Participant agreement	For one-off studies the agreement may be drawn up between the participants. A consultant will probably have a standard agreement which participants can modify to meet their requirements. For on-going studies (i.e. clubs) it is likely that the consultant will manage the participant agreement on behalf of the participants.
Organizing, facilitating and minuting meetings	Best done by a consultant. This is usually one of the key facilitator administration tasks, and being independent, the consultant will aim to facilitate the meeting in a neutral way. Also, any issues with the process can be levelled at an independent body rather than another participant.
Developing metrics	Depending on the study, the metrics may be numerous and complex. It is important to have appropriate skills available to help ensure that definitions are clear, metrics are appropriate, manageable, useful and complete. Statistical skills are very useful for this (and analysis and reporting) aspect of the study.

FIGURE 21.3 Typical facilitator and participant roles.

Issue	Comment
Preparing data collection documents	An administration task best done by a consultant as they will probably also be responsible for supporting data collection, analysis and reporting (see below).
Help-desk support	Best done by the consultant, and probably only possible to be done by a consultant if some or all of the data are confidential. In addition, the consultant will aim to ensure data are submitted impartially and as agreed.
Expediting data	Consultant can act as an external lever to expedite data, and depending on the agreement, may be able to apply financial or other incentives.
Analysis and report writing	This is probably the single most important reason why consultants are used. Confidentiality, if required, can be assured, data will be analysed and, where appropriate, recommendations and conclusions drawn independently.
Data management	For anonymity reasons it may be necessary to use a consultant as the data manager.
Continuation of study	While participants may leave and join the study, the consultant is likely to be the one continuous presence in the study. This provides continuity and secure controlled access to previous studies data, reports and other information.
Data ownership	Use of benchmarking data by the consultant should be as agreed and documented.
Study ownership	Ideally, the study is 'owned' by the participants and facilitated by the consultant. In practice, the consultant is likely to drive the study forward, and be at the forefront of recommending and implementing improvements based on the wishes and agreements with participants. However, it would be at least theoretically possible for the participants to take the data and facilitation away from one consultant and give it to another. In database benchmarking studies, the consultant has more control over the data as it often gets added to their database to be used for the next client.
Site visits	Organized principally by the participants, though an independent observer may be asked to be present as a monitor. The exception to this is in studies with little or no data comparison where a company seeks (usually one) other organization to visit.

FIGURE 21.3   *(Continued).*

SUMMARY

Legal issues

✓ Intellectual property rights belong to creators of the idea or organizations. It is chiefly governed by copyright and patent law.

✓ Confidentiality is often a key issue in a benchmarking study and is governed by confidentiality agreement.

✓ International competition laws and national law governs what information can be shared between organizations.

✓ When embarking on a benchmarking study it is important to confirm that data and information being shared does not contravene legal codes.

✓ The key document that governs the relationship between participants is the participant/confidentiality agreement. Drawing up a confidentiality agreement, vetting it for legality, getting it signed and abiding by it should be all that is necessary to ensure a litigious-free study.

Use of consultants

✓ External consultant facilitators do have a role to play in many studies. An experienced consultant can bring specific skills, experience and impartiality to a study.

✓ Consultants have a key role to play where independence and/or anonymity is a requirement of the study.

✓ Participants should ensure that their needs are being met by taking an active interest in driving consultants to meet their needs.

✓ In on-going benchmarking studies the participant agreement is likely to be managed by an independent facilitating consultant.

PART IV

Case Studies

INTRODUCTION

Throughout this book we have illustrated the theory and practice of benchmarking with numerous small case studies and examples. In this part of the book we present longer carefully selected case studies to illustrate a wide variety of benchmarking experiences and methods employed by different types of organization.

The case studies are:

Caleb Masland, Managing Director IT Benchmarking, Information Management Forum LLC (Chapter 22)

Information Technology is a notoriously difficult topic to benchmark because IT organizations vary widely in the services they offer, the work that they carry out and their financial structures. In these situations benchmarking against standardized data is usually of limited use. Citigroup, one of the largest international financial service organizations, wanted to benchmark their IT activities but were wary of using a standardized IT benchmarking model. They chose to hire a specialist IT consultancy, Information Management Forum (IMF), to work with them to benchmark and improve their IT service. IMF began by gaining an understanding of Citigroup's aim and situation. They helped identify key metrics and gather data which could be compared with selected data from other organizations.

Pat Stapenhurst, Director, Rushmore Reviews (Chapter 23)

The Drilling Performance Review (DPR) is an example of a large (over 200 participants) consultant facilitated benchmarking club, which broadly follows the methodology explained in the book. It is unusual in that it is a project benchmarking study where individual projects are benchmarked both within a participant's organization and between participants. Projects can be benchmarked across years, which raises intricate data management issues for the 25,000 projects held on file. Participants are allowed direct access to data through the internet which requires complex access restrictions, and the development of novel web site features. This case study also explores the challenges that have arisen over the last 15 years as drilling technology has become more complex and the number of participants and wells has increased.

Paul Carroll, Corporate Planning Manager, Dundee City Council, Scotland (Chapter 24)

Paul explains the background to benchmarking within councils in the UK including the use of league tables, external audits and transparency to the public and media. As a specific example, Paul then explains how the previously intractable problem of developing a One-Stop-Shop for customer enquiries was finally resolved by researching and then visiting other councils that had successfully developed similar services.

Ray Wilkinson, Director, The Best Practice Club (Chapter 25)

With the growth in benchmarking and the use of the web, on-line communities of organizations wanting to learn from each other have developed. One such community is the Best Practice Club (BPC). In this case study Ray explains the services that the club offers to its members and illustrates how the services have been used by participants through three case studies. In the first case study The Highways Agency sought and received offers to benchmark and share practices from several organizations on how to develop a single coherent operational management system. In the second case study, a retail member wanted to validate findings and recommendations of an internal strategic planning review. Finally, through its on-going contact with members, the BPC identified a growing interest in Health, Safety and Environmental issues and initiated a Special Interest Group which meets on a regular basis to benefit members with a continual trickle of improvements from which they all benefit.

Chapter 22

Benchmarking IT at Citigroup

Caleb Masland
Director, IT Benchmarking, Information Management Forum LLC

22.1 BENCHMARKING DRIVERS

Large corporations, particularly multinational organizations delivering a variety of products and services through multiple business lines, are regularly looking for opportunities to increase the efficiency and effectiveness of their IT operations. Depending on the internal culture and business climate, the Information Technology (IT) organization can initiate benchmarking comparison studies for a number of reasons.

- Uncovering opportunities for cost reduction: If the IT organization is in a position where optimization of spending is necessary as part of an IT or business driven cost reduction programme, benchmarking can be used to uncover cost reduction opportunities. Unit cost comparisons with world-class performers provide data for gap analyses and working target creation, while strategic delivery strategy analyses provide mechanisms for reaching those cost targets.
- Improving service quality: If the IT organization is in a position where service quality improvement is necessary to strengthen the IT–business relationship, benchmarking can be used to uncover service quality improvement opportunities. Comparisons with high-quality organizations provide data for gap analyses similar to cost comparisons, and strategic delivery strategy analyses provide mechanisms for changing service quality.
- Improving technology support productivity: As with cost and quality improvements, benchmarking provides operational targets and mechanisms for reaching those targets through peer group comparison and gap analysis.
- Building an IT service catalogue: If the IT organization is attempting to build a service catalogue, benchmarking can be used to set reasonable prices for IT products and services within the organizations. This analysis also allows the

IT organization to provide comparative evidence as to the 'real-world' price for similar services in peer organizations.
- Improving IT transparency: If the IT organization is working to improve the transparency of their operations (for any of the above mentioned reasons or as a separate initiative within the IT group), benchmarking can provide evidence that the IT products and services are comparable to industry peers and that the organization is striving for continuous improvement with the business in mind.

22.2 CASE STUDY BACKGROUND

The Citigroup Technology Infrastructure (CTI) manages one of the largest and most diverse technology footprints in the world. Technology provides critical business competencies that allow Citigroup to compete with other world-leading financial services conglomerates. Because of the nature of Citigroup's business, where merger and acquisition activities are a regular part of business, the size and diversity of the resultant IT operations created a large number of uncovered opportunities for improvement. Understanding the need for continuous improvement, transparency of IT operations, and alignment with business strategy, CTI chose to engage a strategic partner for comparative benchmarking analysis.

22.2.1 Engaging a Strategic Benchmarking Partner

The use of a strategic partner enabled CTI's improvement initiatives by providing external reference information for comparison and by adding a level of credibility to the operational cost, quality, and productivity goals that resulted from the external comparisons. Credibility of outcomes is critical in an organization that is working to align IT with business strategies and create transparency of spending and operational management. CTI engaged Information Management Forum, a group of senior IT professionals that used their experience from careers as IT executives to create opportunities for both quantitative and qualitative analysis for client organizations. IMF assisted CTI in reaching their goals by providing direction as to the critical information needed for comparison, facilitating data collection and analysis, and assisting with communication for key stakeholders within Citigroup beyond CTI.

22.2.2 Engagement Approach

Because no two technology organizations are exactly alike, IMF and CTI agreed to take a customized benchmarking approach. The customized approach requires that the benchmarking facilitator have a high level of understanding about the key business and IT drivers for the client organization, and creates additional need for detailed analysis of the client organization data by the benchmarking facilitator. However, this approach ensures that a true 'apples-to-apples' comparison is made through the comparative benchmarks, and the resultant outcome measures and improvement recommendations are truly actionable. From CTI's perspective, this was a necessary step to ensure that the resultant internal stakeholder communication could be easily translated into an operational action plan. The alternative to a customized approach is to follow a strict data collection protocol that would map CTI's IT environment to a proprietary benchmarking model. This alternative can lead to large portions of information being left 'out of scope' or the need for assumptions about comparability across CTI and external reference information. CTI determined that the risks of the alternative, stricter approach were too great to follow this methodology for benchmarking (Figure 22.1).

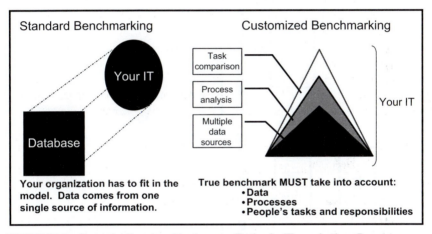

FIGURE 22.1   Customized benchmarking is more effective for IT organizations. Because companies can take myriad different approaches to IT service delivery, only a customized benchmark will yield comparable results.

22.3 DETERMINING KEY MEASURES, ENGAGEMENT PROCESS AND OUTCOME DELIVERABLES

By taking a customized approach to benchmarking, CTI was able to inject their strategic goals into the process of developing key measures, engagement processes, and outcome deliverables. Each of these was developed through strategic objectives discussions that served to kick off the benchmarking engagement. Through these discussions, CTI shared their key goals for benchmarking with IMF. The IMF facilitators then aligned these goals with the benchmarking approach, based on their knowledge of key IT drivers in CTI's industry and their previous experiences with IT benchmarking in general. For some IT operations areas, a number of industry standard benchmarking metrics were utilized, while others were developed specifically to fit CTI's stated goals (Figure 22.2).

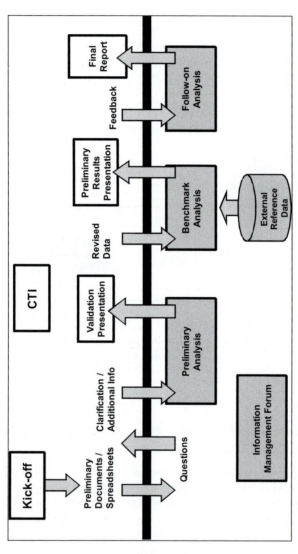

FIGURE 22.2   The customized benchmarking approach is predicated upon regular communication between the client and the benchmarking facilitator. In this case, IMF and CTI determined several checkpoints to ensure that alignment was maintained between client goals and the benchmarking methodology.

Information gathering was completed by a small team of individuals within CTI. The IT Finance organization provided IMF facilitators with financial and technical volume information as well as staffing reports. Because the customized approach was taken, the information was provided to IMF benchmarking facilitators in raw form, such that the facilitators could gain a deeper understanding of CTI's operations and extract the relevant information for comparison.

Regular review and realignment of CTI and IMF was critical for success. At several predetermined check points, and through ad-hoc discussions as necessary, the two parties discussed the state of data collection, analysis, and revision activities to ensure that study activities were still aimed at achieving CTI's goals. These regular discussions were also used to develop the metrics and reporting structure for study deliverables.

Through this process, the key deliverables chosen for reporting were:

- Detailed overall and unit cost comparisons (for a given benchmarked IT service area such as Servers, Desktop Computer Support, etc.) based on the total cost of ownership
- Productivity comparisons based on the total number of supported company employees or units of IT service (such as Servers, Desktop Computers, etc.)
- Quality comparisons based on industry standard metrics (such as IT service repair response time, avoidance of errors or problems, etc.)
- Outcome recommendations and performance improvement suggestions based on the industry knowledge held by IMF benchmarking facilitators and regular interviews with CTI management concerning key strategic IT objectives based on business drivers (Figure 22.3).

In order to facilitate continuity of operations within CTI, the bulk of the engagement effort was done by the IMF benchmarking facilitators. IT benchmarking, if not properly managed, can be particularly derailing of critical IT operations initiatives – due to the pace of the typical IT organization, benchmarking initiatives that require too much effort within the company can distract key managers and reduce overall productivity. In this case, CTI and IMF avoided this pitfall by allowing IMF to drive the project and complete the majority of data analysis based on their experience with previous benchmarking work and their expertise in the IT industry as a whole. As previously mentioned, a key to maintaining this approach was frequent and open communication about the status of the benchmarking project between the key project participants on both sides.

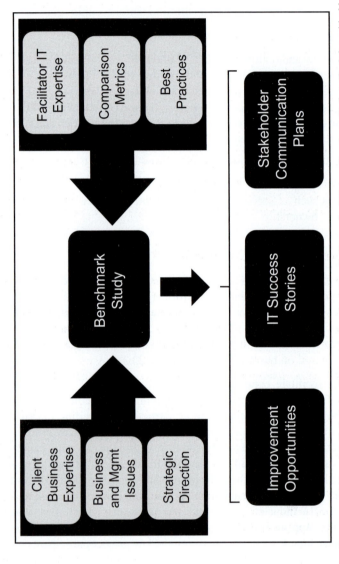

FIGURE 22.3 For a successful benchmark, both the client and the benchmarking facilitator should be equal partners. Both parties bring critical knowledge to the table, each piece of which is necessary to reach actionable outcomes.

22.3.1 Initial Data Collection

After meeting for a kick-off session (involving key data collectors and the project sponsors within CTI and the IMF benchmarking team), the initial data collection effort was undertaken. The ease and speed of this effort can be largely dependent upon the organizational structure of the benchmarked organization – for some organizations data owners are clear and information is readily at hand due to well-built and up to date systems. Within other organizations (particularly those that have no prior experience with a significant IT metrics programme or previous benchmarking work), information can be harder to discover and provide to the benchmarking facilitators for analysis. In the case of the CTI-IMF engagement, CTI had a history of internal measurement as well as the previous experience of building a catalogue of IT services for the users of IT services throughout Citigroup. This previous experience, in combination with the availability of information through intelligently designed financial analysis systems, led to a relatively quick data collection effort of only a few weeks. These data could then be analysed by the IMF benchmarking facilitators in preparation for the build-up of and eventual comparison to a group of peer organizations.

22.3.2 Reference Group Creation and Validation

Comparability is absolutely essential to the success of IT benchmarking initiatives. Depending on the technology deployment philosophy of a given organization in combination with the business climate in which that organization operates, dramatic differences can develop relating to the types of technologies used and the relationships between different technologies within the organization. A large amount of effort must be taken early on in the IT benchmarking initiative (ideally as soon as enough information has been gathered for the client organization to determine key factors of comparability) to match potential comparator organizations to the client. In this case, the IMF benchmarking facilitators analysed the information initially provided by CTI to draw out and understand these key comparator traits. Through their history of IT benchmarking, IMF found that the following are most critical in ensuring comparability (in order of relative importance):

- Volume of IT service delivered (the type and number of servers, desktop computers, telephones, etc.)
- Total volume of supported IT service users.
- Service quality.
- Business or industry of client organization (in this case, diverse multinational financial services).

After the reference group companies were selected for comparison, the IMF benchmarking facilitators went through a process of initial data comparison.

This involved ensuring that components of cost included in the CTI metrics were also present in the reference information. The benchmarking facilitators completed most of this work effort and CTI was responsible for responding to scope questions when needed.

The data for both CTI and the reference companies were then compared at a high level. As a part of the analysis, IMF benchmarking facilitators factored out the lower performing companies within the reference database, and compared CTI to the mean of the top quartile of comparable companies. The base metrics for comparison for the high level validation were the annual cost per unit of IT service (i.e. servers, desktop computers, help desk calls, etc.), the unit productivity per support employee (servers per support employee, desktop computers per support employee, etc.), and industry standard metrics of service quality. The high-level validation information was then presented to the project sponsors at CTI so they could validate the methodology, the comparison metrics, the reference group selection, and the CTI information presented. Detected anomalies were then set aside for re-analysis.

22.3.3 Detailed Analysis and Presentation of Outcomes/ Recommendations

After validation of the methodology and reference group selection with the CTI management personnel, the IMF facilitators completed additional analysis. This level of analysis focused on comparing CTI to the reference group used for validation, but at an additional level of detail. Comparisons were now made by breaking down the above-mentioned cost comparisons to the components of hardware, software, technology maintenance, personnel, external services, and other miscellaneous costs. In completing these detailed comparisons, the IMF benchmark facilitators were more able to uncover the major drivers of high or low performance in comparison to the reference group. These differences were then used to make recommendations concerning improvement opportunities.

The detailed comparisons and recommendations were then built into a final results presentation, which was presented to the CTI management team by the IMF facilitators. During this meeting, the CTI management team could ask questions about the implications of the comparisons, and gather information from the IMF facilitators to use in formulating improvement strategies.

Following the results presentation, the IMF facilitators remained available to the CTI management team. Follow-up activities included the formulation of a communications plan with key individuals at Citigroup within and beyond CTI, and additional discussions concerning follow-on studies required completing improvements recommended by the IMF facilitators. IMF remained engaged to assist with the completion of these follow-on activities.

22.3.4 Engagement Results

While the specific cost savings opportunities and strategic recommendations are confidential, the engagement approach as described above yielded a return many times the cost of the engagement for CTI. At a high level, the following directions were recommended and followed, to CTI's great benefit:

- Consolidation of technology infrastructure assets in areas of under- or no utilization to reduce immediate spend on basic 'keep the lights on' services
- Increased investment in key technology growth areas to take advantage of scalability and maximize long-term cost effectiveness
- Process re-evaluation and standardization to increase quality (short term) and reduce rework costs (long term)
- Regional re-alignment of IT operations services to take advantage of economies of scale and various regional economic factors
- Re-negotiation of service and/or purchase contracts with key IT vendors in areas where industry peers are purchasing similar services at lower cost.

In addition to these benefits, which yielded significant monetary advantage, the process of completing a customized benchmark built an internal culture driven more by metrics and quantitative comparison than was previously enjoyed. Not only did this arm CTI senior management with critical information for future decision-making, but it also increased the capability of the organization as a whole to make fact-based and informed decisions rather than relying on simply qualitative or subjective factors. This 'soft' benefit of IT benchmarking has positioned the organization for greater success and long-term health. Additionally, the organization is now well positioned for future comparative improvement studies, as the benchmarking process described has trained key individuals and their teams on the Critical Success Factors for this useful method of operational management.

All images © 2008 Information Management Forum.
www.theimf.com

Chapter 23

The Drilling Performance Review –
The Evolution of a Benchmarking Club

Pat Stapenhurst
Data Management Director, Rushmore Reviews Joint Venture

INTRODUCTION

The drilling of oil and gas wells is a highly specialized industry. This has resulted in drilling engineers and related personnel recognizing the benefits of learning from each other. In 1970 BP made the first oil discovery in the UK which led to the development of the Forties field. By the 1980s Aberdeen in Scotland had become a major oil city as oil companies discovered and developed oil fields in the North Sea. Even in these early days drilling engineers and others in the oil industry were informally sharing information and ideas.

In 1989 operators agreed to share data more formally in what was to become known as the annual Drilling Performance Review. At this time data collection and analysis was rudimentary and carried out by drilling engineers within the oil companies themselves. By 1993 these engineers were finding the administration of the study too time consuming and it was taken over by two small independent consultancies – Rushmore Associates Ltd and The Sigma Consultancy (Scotland) Ltd. Acting as a joint venture called The Rushmore Reviews these companies continue to facilitate the Drilling Performance Review.

The Drilling Performance Review, or DPR as it is called, enables its participants to carry out 'project' benchmarking. To drill a hydrocarbon well involves many different types of activities and expertise in many areas – geology, drilling fluids, cementing, wireline work, etc. It may take a couple of days or a few weeks, or many months to drill the well. A multitude of factors influence the time required: the length and design of the well bore, its location (e.g. on land; in ultra-deep water), geological factors and the experience of the drilling crew, to name but a few.

A large amount of information is collected about each well drilled, and the data are presented on the DPR website. The DPR database now contains searchable data on more than 25 000 wells drilled since the year 2000, plus archived data on thousands more. The data are available: the challenge for participants is to benchmark drilling performance on their wells, each of which has a unique set of properties and activities.

In describing the Drilling Performance Review we will consider all the steps normally required for running a benchmarking study:

1. Setting the scope and objectives;
2. Deciding the performance metrics and data requirements;
3. Selecting benchmarking partners (participants);
4. Collecting and validating data;
5. Analysis and reporting;
6. Follow-on action.

The DPR has run each year since 1989, and membership has grown consistently during this time. The following description highlights the challenges faced by the facilitators, Rushmore Reviews, partly as a result of this growth but also in response to advances in the Oil and Gas industry. We will show how the DPR has adapted and evolved to meet these challenges.

Terminology

The term 'Operator' has been used in this account to refer to an operating company that drills oil or gas wells in one or more countries around the world.

The term 'Participant' indicates a company that is operating in one country and participating in the DPR. For example, Lowon Gas is an Operator, while Lowon Gas UK, Lowon Gas Gabon, Lowon Gas Peru and Lowon Gas China are amongst the Participants of the DPR.

(Note: Lowon Gas, and other operator names used in this case study, are fictional and for illustrative purposes only)

23.1 STEP 1: SCOPE AND OBJECTIVES

23.1.1 Types of Wells

The subject matter for the Drilling Performance Review is the drilling of hydrocarbon wells, that is, wells drilled for the purposes of oil or gas production. Some of these will have been drilled to find viable reservoirs (exploration wells) or gather information about a reservoir (appraisal wells); others are drilled to bring hydrocarbons to surface (production wells) or to enhance production by raising reservoir pressures (injection wells – water, for example, is injected into the reservoir). All come within the scope of the DPR.

23.1.2 Drilling Activities Covered by the DPR

In 1989 the focus was entirely on the activities which commenced when the drill bit started cutting (this is called 'spudding' the well), and ended when the Total Depth (TD) of the well bore had been drilled. This period between spudding and reaching TD is called the 'dry hole period', the period before hydrocarbons start being produced. The basic activities which take place during this period are: drilling; pulling the drill bit out of the well bore and running in casing; and cementing this casing in place. (Figure 23.1 shows how casings may be used in a well bore.) Sometimes rock samples are extracted (this activity is called 'coring') or measurements are taken of down hole conditions (an activity called 'logging').

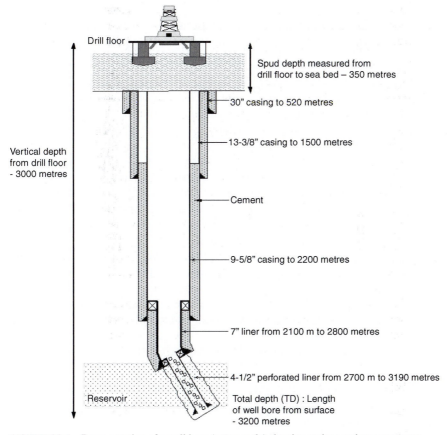

Drill floor

Spud depth measured from drill floor to sea bed – 350 metres

30" casing to 520 metres

13-3/8" casing to 1500 metres

Vertical depth from drill floor - 3000 metres

Cement

9-5/8" casing to 2200 metres

7" liner from 2100 m to 2800 metres

4-1/2" perforated liner from 2700 m to 3190 metres

Reservoir

Total depth (TD) : Length of well bore from surface - 3200 metres

FIGURE 23.1   Representation of a well bore (not to scale) showing casings and measurements.

In the following years participants became interested in knowing more about the pre-spud activities on the well, that is, the activities carried out before drilling starts. This was particularly of interest for wells not drilled from the surface but 'kicked off' in a new direction from part way along an existing well bore, as more pre-spud preparation is required in these wells. (This type of well is often called a sidetrack.)

Participants also wanted to know how much time was spent coring or logging, and they wanted to see data on what happened when the well bore reached its target: was the well bore plugged and abandoned or was equipment installed to make it a production well; how long did all that take? All this information helped to give a more complete picture of the drilling of the well.

The time window thus widened to encompass most of the activities between the arrival of the drilling rig at the location where drilling is to take place, and the rig departure time, after completing the task of drilling the well.

23.1.3 Geographical Scope of the DPR

In 1989 the only area covered by the DPR was the North Sea and data were only collected for wells drilled offshore.

In 1994 a participant asked the facilitators to initiate a similar study in Australia. Asia followed soon after. In 1997 they were asked for an Algerian study, which the following year was extended to encompass the whole of Africa. It was a natural step to start a South American study.

The facilitators were less enthusiastic about extending the scope to cover the Gulf of Mexico, because benchmarking already took place between operators there. However, a participant offered to sponsor this move. Data were also collected from the Former Soviet Union, so that by 2000 the DPR was described as a global study. It still did not cover US and Canadian land well drilling, but this was mainly due to the sheer quantity of wells drilled in those two areas. It is only in 2008 that operators in these areas have been invited to participate in the DPR.

The expansion into new regions has been accompanied by a need to take into consideration the different languages, different working practices, legal systems and currencies around the world, as discussed below.

The expansion of geographical scope has contributed to the increase in the numbers of wells submitted each year, from 200 in the early days to 3500.

23.1.4 The Objectives of the DPR Participants

In 1989 the main objective of the participating operators was to benchmark drilling speeds and costs with their peers who were drilling similar wells in nearby locations. 'Feet drilled per day' and 'cost per foot drilled' were, and still are, key performance indicators.

Nowadays participants have a variety of uses for the drilling data. These include:

- Identification of the 'best in class' operators with regard to speed and cost.
- Analysis of the performance gap between the 'best in class' well and 'my' well (gap analysis).

- Well planning, particularly for wells to be drilled in areas new to the participant.
- Analysis of non-productive time during drilling.
- Analysis of certain elements of well design (e.g. lengths of sections drilled before the well bore needs to be cased).
- Information about the use of certain techniques.

23.1.4.1 The objectives of the facilitators

The facilitators aim to:

- Provide a high volume of complete and accurate drilling data.
- Provide data from a wide spread of operators, both large multi-national operators and small independent operators.
- Give good data coverage, from as many countries as possible, of both on-shore and off-shore wells.
- Maintain transparency and identification of data, i.e. all calculated fields can be verified; no data are anonymous.
- Uphold the principle of sharing: 'participants must put data in to get data out'.
- Respond to the changing requirements of participants and anticipate their future needs.
- Provide a friendly and personal service to participants.
- Invest in research and development to improve the usefulness of the data.

The facilitators believed that if they were successful in achieving their objectives more participants would join and remain with the study, making it a profitable business. High participant retention rates have shown this to be the case.

23.2 STEP 2: PERFORMANCE METRICS AND DATA REQUIREMENTS

In the early days of the DPR only six pieces of information were collected for each well drilled. These were:

- Who drilled the well (operator).
- For what purpose (exploration, appraisal or production).
- At what depth did drilling start (spud depth, measured from the floor of the drilling rig).
- At what depth did drilling finish (measured along the well bore).
- How long did drilling take.
- How much did it cost.

With this data it is possible to calculate:

- The length of well bore drilled (in the DPR this is called the 'drilled interval').
- The drilling speed (feet drilled per day).
- The cost 'efficiency' (cost per foot drilled).

Drilling speed and cost efficiency remain important metrics for benchmarking drilling performance. However, the desire for more in-depth analysis, and the additional uses to which the data are now put, has resulted in a considerable increase in the number of data items collected for each well drilled. Up to 120 data items may need to be provided for one well. In addition a chart showing the planned and actual daily drilling progress (a 'time versus depth' chart) is required. (Figure 23.2)

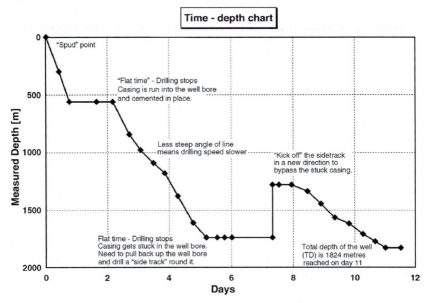

FIGURE 23.2　A time–depth chart, illustrating some of the information that can be gleaned from it.

The increase in the amount of data collected for a well enables the DPR to be used in more ways than originally envisaged. For example, it is now possible to analyse the proportion of 'non-productive' time during the drilling of different types of wells. Another example is that it is now possible to make approximate comparisons of the rate of drill bit penetration (speed of drilling) when drilling different diameter holes.

23.2.1 Choosing the Data Fields

The data requirements have always been decided by the participating operators themselves on a yearly basis, at the Annual Participants' Meeting. This meeting considers requests for new data items to be added to the list of requirements. Changes to data definitions, or clarifications, are also discussed and agreed by the participants who attend the meeting. Occasionally decisions are made to remove data items which are difficult to define, difficult to collect or obsolete.

Challenge: To hold an annual meeting was relatively easy when the few participants were all located in one country. Now there are over 200 participating operators, spread around the world.

Response: Since 2003 all proposals for changes to data fields are posted on the Rushmore Reviews website forum in advance of the meeting. Any participant can make a proposal, and all participants have the opportunity to comment. All proposals are then discussed by those participants who are able and willing to attend the annual meeting, and it is here that the final decision is made. The study is thus driven by the participants themselves, not by the facilitators.

23.2.2 Defining the Data

As described above, the data definitions are discussed by the participants at the annual meeting, after which they are firmed up by the facilitators.

Challenge: When selecting subsets of wells for benchmarking purposes it is often useful to include wells drilled in previous years. In order for the current data to be comparable with historic data it is necessary that the data definitions should be the same, yet the definitions are revised each year.

Response: Because comparisons with historical data play an important part in drilling benchmarking every effort is made to keep the definitions of key variables stable. However advances in drilling techniques may require clarifications to the data definitions so that the basic variables can remain unchanged.

For example, the definition of a key data field, the 'dry hole period' has become increasingly complex over the years. It used to be defined as the period between 'spudding' the well and reaching the total depth. Changes to drilling techniques have blurred the start and end points of the drilling process and it has been necessary to clarify the definition to take this into account. In some wells time is spent installing production equipment before the drilling has reached its target. This time has to be excluded for the purposes of recording the dry hole period – again the definition has been amended. 'The dry hole period' took less than 40 words to define in 1993 but now requires over 500 words for a full clarification of what is included and excluded. Yet the data remains basically the same and is comparable across the years.

Challenge: The data requirements must be possible for all participants to fulfil, and data definitions need to be unambiguous and comprehensible by all.

Response: It has been important that the requested data are readily available to operators, i.e. that operators already record the data as part of their daily reporting systems. The fact that the participants themselves agree the data requirements and discuss the definitions helps to ensure this.

Sometimes this has meant that desirable data items are omitted because they cannot be consistently provided. For example, a key determinant of drilling performance is rock hardness, but there is no accepted standard method of calculating a rock hardness value for each well bore. Even if there were, it is unlikely that most participants would have the source data to enable them to make the calculation. Because of this the DPR does not include any rock hardness data.

Another example is the horizontal distance between the end of the well bore and the start, which was collected up until 2003. If a well bore follows a direct path the calculation of the 'horizontal displacement' (HD) is straightforward (Figure 23.3). For well bores which twist, spiral or bend back under themselves (e.g. 'S'-shaped bores) the horizontal displacement of the end of the well may not be as great as the horizontal displacement of some other point in the well bore. Some participants offered different ways to capture the horizontal displacement, using 'wrapped' or 'unwrapped' measurements. Other participants were only able to supply one type of measurement, and it was not always obvious which measurement had been given. Because of this the field was eventually dropped from the data requirements.

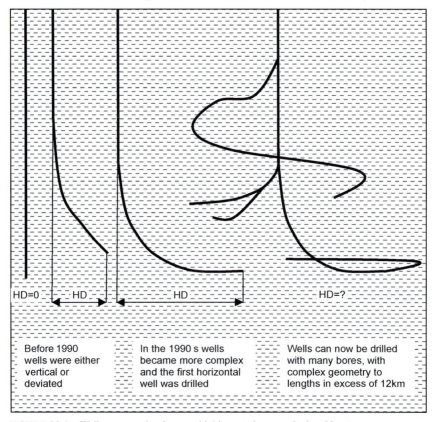

FIGURE 23.3   Well geometry has become highly complex over the last 20 years.

The global participant base means that not all the people who collect and submit the data are native English-speakers, yet the definitions are only supplied in English. English is widely spoken in the drilling industry, and is the language used for reporting on the DPR website. While language may sometimes cause a misunderstanding of data definitions it is expected that the rigorous checking process will identify problems of this nature.

23.2.3 Issues Concerning Cost Data

Challenge: Collecting data from many different countries and operators means that drilling costs are calculated using many accounting practices and submitted in a variety of currencies. For benchmarking purposes drilling costs need to be comparable across a range of wells.

Response: Only two costs are required to be given for each well: the cost specific to the drilling itself (the 'dry hole' cost corresponding to the 'dry hole period'); and the total cost, that is, the overall cost of constructing the well. The facilitators offer guidelines as to which types of costs might be included in each of these values but these are not prescriptive. There is also a means by which participants can indicate broadly what types of overhead costs have been included in their dry hole costs.

Cost data are often not available until many months after the well has been drilled. Because of this participants can indicate that they have submitted 'preliminary' costs which are later updated with the final costs.

With regard to reporting in different currencies the facilitators apply a consistent method for obtaining an exchange rate to convert all costs to US dollars. This is linked to the month in which the drilling finished.

Challenge: A more difficult challenge is to provide a means of comparing costs across years – converting 'money of the day' into today's values. This is a problem for which no solution has yet been agreed.

The facilitators are quick to point out that the published cost data are more variable than the time data for reasons given above. Because of this participants are more likely to use time data for benchmarking purposes while drawing on cost data for a broad 'ball park' figure.

Challenge: The anti-trust laws, or competition laws, are designed to prevent restrictions to free trading. It has been argued that the sharing of any cost data between oil companies might be seen as anti-competitive.

Response: This is a challenge that has surfaced a number of times in the history of the DPR, as lawyers seek to protect their organizations from the accusation of anti-competitiveness. The response by the facilitators has always been that the cost values collected are too high level, too 'broad brush', to be used in an anti-competitive way. Most lawyers accept that this is true. However, this has not prevented participants, particularly in America, from asking to be exempted from submitting cost data. For a while there was a provision for not submitting cost data from certain countries, and participants who did not submit their own cost data were restricted from seeing any other participant's cost data.

In 2008 it is mandatory for all participants to provide cost data as there is now general agreement that this is not anti-competitive.

23.2.4 Issues Concerning the Volume of Data

Challenge: The objective of the DPR is to collect full data on every well drilled by the participants. Some participants drill a few complex offshore wells in a year, each well being individually designed and planned. Other participants drill hundreds of straightforward land wells to a few basic designs. Submitting 120 items of data for each well is an onerous task for them and the data arguably adds little value because the wells are all similar.

Response: The challenge has been to satisfy both types of participants, i.e. to collect enough data to adequately describe the complex drilling projects without making data collection so time-consuming that the high quantity drillers decide not to participate.

The solution, suggested and agreed by participants, was to collect a subset of data from participants who drill many similar wells in a year. If a participant drills many wells but only a few basic types or designs, they only need to submit full data for 1 well of each type. A reduced set of data is collected for the rest of the wells.

Challenge: Operators 'exploring' for hydrocarbons in new areas are often highly secretive and their data are confidential. Some of the required data, particularly that relating to the reservoir formation or the depth and location of the well, can be highly sensitive yet it is mandatory.

Response: Due to its sensitive nature the provision of the geological data is optional. The well's location (latitude and longitude) can be approximate if the information is confidential.

Another concession, agreed by participants, is that only a small subset of data need be submitted for wells which are considered highly confidential. The amount of data collected is enough to allow basic performance indicators to be calculated. After a few years, when the confidential status of the well has been relaxed, the full data submission is requested and published.

23.3 STEP 3: SELECTING PARTICIPANTS

From the participant's point of view selecting other operators with which to benchmark requires the identification of operators who drill similar wells. In 1989 the DPR participants were a group of operators drilling in the North Sea. Many aspects of the wells they drilled – the location, supply lines, geological conditions, regulations – were similar.

By the year 2000 DPR participants were drawn from around the world. Each participant would like to find in the DPR database enough wells 'similar' to their own to enable them to make meaningful comparisons. This puts the onus on the facilitators to recruit as wide a spread of participants as possible.

The marketing carried out by the facilitators has always focused on discovering the benchmarking needs of the operators. This involves visiting potential participants, learning what they already do with regard to performance evaluation and demonstrating how access to a large database of drilling data might add value.

Some of the reasons that operators give for *not* participating are:

- They do not want to share their data with competitors.
- They do not need to benchmark as they are already the best (but how can they be sure if they do not know what their competitors can achieve?).
- They believe no other operator drills comparable wells.
- They cannot provide the required data as they do not record it in their own systems.
- They do not have the time or resources to extract and submit the data.
- They do not have the time or resources to carry out benchmarking analysis.

Many of these objections can be worked through with the operators, in ways indicated at various points in this case study. However, the process of recruiting requires patience and persistence, as it sometimes takes place over several years.

There are a few conditions which determine who can be a DPR participant:

- The DPR is only open to Oil and Gas operators – drilling contractors or service companies cannot be participants unless they are also operators in their own right.
- The operator must have drilled at least one well.
- The operator must agree to submit 100% of the required data on 100% of the wells they drill during the years in which they participate (and in the country in which they participate.) It is not acceptable to submit just the 'best' wells.

One of the basic principles of the DPR is that of sharing: in a sense participants submit data as part payment for their right to see other participants' data. Operators may not drill every year but providing they participate when they do drill they can access data as 'sleeping' participants in the intervening years.

Only participants can access the DPR data – the facilitators do not sell data to non-participants.

23.3.1 Reducing the Recruitment Burden

Up until 2000 each operator participated in a regional study and participation was for one year at a time. Data for each year and each region were kept in a separate regional file, and participants only had access to the data in the appropriate file. This was a limitation, especially if a particular region was poorly subscribed to in one year.

Since 2000 the study has been 'global'. All the data are now stored in a single database and each participant can potentially access data from any other participant, regardless of where in the world they are drilling. They can also access data from any year in which they participated in the DPR. (The pre-2000 data remain outside the database, but is now freely available to any participant regardless of the years in which they were participants.)

An obvious next step has been to recruit operators on a global basis. Rather than negotiating separate contracts with each operator in each country where they are drilling, it is more efficient to set up a global contract with the operator's head office. This covers the submission of data on all the wells drilled by the operator throughout the world during the year. An incentive for setting up this type of contract is that a 'global' participant gains the right to freely share DPR data with its units in countries that are not currently drilling.

Another drain on the sales team was the need to re-negotiate contracts with participants every year. It is the nature of a benchmarking club that every participant is keen to know which other operators have 'signed up'. Being able to show that major oil companies have not only 'signed up' but have contracted to participate for the next 3 or 5 years has been a great help in the recruitment process. The more operators that submit data, the more desirable access to the database becomes.

Some long-term participants have now agreed to 10-year contracts. For other operators, who have not participated before, a multiyear contract which extends backwards in time is the quickest way to gain access to a large quantity of data. For example, an operator can agree to submit data on all wells drilled in the past 5 years and thus, as soon as that data are validated and published, gain access to 5 years of DPR data.

23.3.2 Maintaining the Sharing Principle

Challenge: Suppose a multinational operator agrees that just one of its operating units (i.e. in one country) will participate in the DPR. When that unit gains access to the DPR database what is to stop them passing the data to all the other units in the operator group that have not submitted well data?

Response: To preserve the basic principle of sharing, a complex set of rules has been devised which govern each participant's access to data. These access rules ensure that if an operator has drilled in a particular country but has not submitted data, then they may not access any other participant's data from that country. Furthermore no participant gains access to any other participant's data until their own has been accepted as valid and published on the website.

These data access rules are better understood by using an example. Let us suppose that an operator, Lowon Gas, is drilling this year in the UK, Gabon, Peru, China and Russia. Suppose that only Lowon Gas UK and Lowon Gas

Gabon have contracted to participate in the DPR, and only Lowon Gas UK has actually submitted its drilling data. When the UK data have been validated and published Lowon Gas UK may access this year's DPR data for wells anywhere in the world, with the following restriction: no access to any Peru, China, or Russia wells (as those countries are not participating) and no access to any Gabon wells until the Lowon Gas Gabon data are published.

Challenge: The access rules determine which subset of wells an operator can view, stretching back over the years of that operator's participation. However, it is not uncommon for operators to merge or acquire other operators. Over the last 10 years there have been a number of mergers and acquisitions, where one or both parties have been global or partial participants in the DPR.

Response: This has necessitated another layer of complexity being added to the access rules, to ensure that no-one loses access as a result of a merger. The situation can be further complicated when one operator merges with another, and then in a later year takes over a third operator: this has happened more than once.

Defining the access rules continues to be one of the most 'difficult' areas of the DPR. They have added to the complexity of the website, which has to determine which well data each user is allowed to see. They are also difficult to explain to participants and prospective participants. However, maintaining the 'sharing' principle is considered very important, as it ensures that each participant does not have an unfair advantage over its competitor.

23.3.3 Agreeing Terms and Conditions of Participation

The terms and conditions of participation are reviewed at the annual Participants' meetings. Sometimes they need to be adjusted in response to a particular political situation which has arisen in one or more countries.

Two examples are:

1. Operators in US embargoed countries are not allowed under US law to have access to US drilling data;
2. Governments in some countries will not agree to the release of drilling data from sensitive areas.

In each case the terms and conditions are amended, with the agreement of participants, to allow for exceptions to be made in the affected countries.

Each participant has to agree to the same set of terms and conditions as every other participant. There are many companies who would like to vary those terms and conditions – the response to such a request is 'only if all other participants agree to adding this variation to the terms and conditions'. Because the participants themselves determine the terms and conditions, and in a sense 'own' the DPR, the facilitators are not at liberty to make exceptions. This has been one of the strengths of the DPR: that all participants are treated in the same way and have to abide by the same set of rules.

23.4 STEP 4: DATA COLLECTION AND VALIDATION

Although operators may contract for 3 or 5-year participation, the data collection process works on an annual cycle, January to December. The data definitions discussed at the annual participant's meeting apply for 1 year.

In 1989 and for the next 12 years, the annual data were submitted at the end of the year, and reporting could not start until all, or the majority, of data had been collected and validated. This was up to 6 months after the end of the year.

23.4.1 Quarterly Submission

The data are now collected on a quarterly basis. This has the advantage of spreading the workload, both for the operator providing the data, and for the facilitators validating it. Data can be collected soon after the well has been drilled which means engineers are more easily able to answer queries because the work is fresh in their minds. As soon as a participant's quarterly submission is complete it can be published. This in turn increases the participant's access to data on the website, as they can now see the data which other participants have submitted for that quarter.

There are quarterly submission deadlines. At the request of the participants there are financial penalties attached to the 3rd and 4th quarter deadlines. The participants prefer timely submission of data by other participants, so that they can begin their benchmarking analysis more quickly. For this reason they wanted an incentive to motivate people to submit data on time. In addition, having a penalty gives the people collecting the data more leverage for requesting resources from managers to help them meet the deadline.

At the start of each year participants are asked to indicate how many wells they plan to drill in each quarter of the year. This information enables the facilitators to know which operators to approach for data in each quarter. Participants also provide 'contact' details at the start of the year, that is, the names of people within the company who can supply the well data and answer questions about it. This information is confirmed with each quarterly data submission.

The DPR came into being before the internet and emails were the norm. In the early years participants saved data onto floppy discs which were posted to the facilitators. Occasionally a print-out of the data was sent by fax for manual data entry.

In 2008 data are entered in a 'workbook'. This is a formatted spreadsheet which contains the detailed data definitions. The completed data workbook is returned to the facilitators by email.

Challenge: Many operators find that they do not have the resources for collecting and submitting such a large quantity of data. Sometimes the people who are available to collect data do not have sufficient technical knowledge to understand the data requirements.

Response: The facilitators suggest three approaches to solving a resource problem:

1. Most operators use one of three or four popular database systems for capturing drilling data. The companies who have written these systems have now also written extraction software specific to the DPR. Much of the DPR data can be automatically extracted and the workbook populated. As the workbook usually changes slightly each year, these extraction programmes are upgraded on an annual basis.
2. The facilitators offer training to help the people who are involved with data gathering to understand the data they are working with. The facilitators are also available to discuss different types of drilling scenarios and to explain how to enter the data into the workbook.
3. The facilitators offer manpower to extract data from the operator's daily drilling reports, directional surveys and other documentation and compile the workbook on their behalf.

Challenge: the drilling data captured in the operator's own databases may not be in the format required by the DPR, or definitions of key terms may not correspond with the DPR data definitions.

Response: The definitions developed by the Drilling Performance Review are becoming an industry standard, and some participants have decided to bring their own definitions in line with those of the DPR. Because the DPR definitions are decided jointly by its many participants it is unlikely that the DPR definitions can be changed to suit one particular participant.

23.4.2 Data Validation

The quality and completeness of the data is one of the reasons for the success of this study.

When the facilitators took over the running of the study in 1993 there were less than 20 data items collected for each well drilled. Data validation consisted of a visual inspection to see if the values 'looked' correct. Any apparent errors or inconsistencies were cleared up by phoning the operator.

Because of the small number of data fields and the relatively few wells (just over 200 in 1993) this method of validation was reasonably effective.

However, as more data fields were added to the requirements the potential for errors increased. The validation clearly had to be automated. An automated checker was created which read each row of well data on the spreadsheet and checked for errors. This checker is still in use more than 10 years later, updated on an annual basis to reflect changing data requirements.

The main types of errors checked are:
- Missing data, i.e. mandatory fields are blank.
- Incorrect data format, e.g. text given in numeric fields.
- Incorrect code given for coded fields.
- One data value is inconsistent with other values, e.g. the given water depth is too deep for the type of drilling rig specified.
- Extreme data values, i.e. values look too large (or small) compared to the usual values given for that data item. (This is a warning rather than an error.)

For each field and each different possible type of error there is a corresponding error message. (There are more than 400 different error messages). When errors are identified in a data submission the appropriate error messages are automatically compiled into a query report.

Before the report can be sent out each error needs to be manually cross-checked and the report edited by a skilled data analyst. The automated checker may flag an inconsistency, but it cannot always identify the exact nature or source of the error – which may be obvious to the human eye.

In 1996 a participant suggested that the well data should be supplemented by a 'time versus depth' chart (Figure 23.2). Operators routinely produce these charts when drilling, as they show the depth reached along the well bore at the end of each day.

To the skilled eye a time–depth chart can provide information about
- The length of well bore drilled.
- The time taken to drill the well.
- The rate of penetration of the drill bit (drilling speed).
- The amount of 'non-productive' time.
- The amount of 'flat' time.
- The number of 'sidetracks' from the main well bore.

The graph is flat when no advance is made in lengthening the well bore. This occurs when casing is being run into the bore hole and cemented into position, an operation which is usually necessary to prevent hole collapse. 'Flat time' may also occur when things go wrong – for example when equipment breaks down or gets stuck in the well bore.

At the annual meeting other participants agreed that collecting a time–depth chart would provide much useful information without adding too much to the data collection burden (because such charts were generally already being produced as part of the well report). Time–depth charts became a mandatory requirement of the DPR.

Collecting time–depth charts has added to the manual validation task, as they need to be cross-checked against some of the numerical well data.

'Common sense' checks are made by comparing the data for one well with that of other wells drilled by the same operator in the same area. For example we would expect water depths for all the wells drilled from the same platform to be similar.

23.4.3 Expediting

The expediting and validation of data is very time intensive. The facilitators have made a big investment in personnel to carry out this checking and querying process to a high standard. The process of sending out queries, keeping track of them and 'chasing' participants to send back the replies is both time-consuming and challenging.

A database has been set up to record

- how many wells are expected each quarter from each participant,
- how many wells are actually received,
- the date when queries are sent out,
- the date when chase-up emails are sent,
- details of phone calls made,
- changes to contact details received.

Automated reminders can be set up in the database, to ensure submissions are not overlooked. When a quarterly submission is finally validated, that too is noted in the database, and that triggers the publication of the data on the website.

It is only when all the data for a quarter have been received from a participant, and satisfactorily validated, that it is accepted for publication. This can be a lengthy process, but data accuracy is a key goal of the DPR. However, the ultimate responsibility for the accuracy of the data lies with the operators. The source of the drilling data is on the rig-site where activities and timings are captured in the operator's system. It is important that, at the point of entry, engineers record accurate values and allocate appropriate activity codes.

A by-product of participation in the DPR is that some participants, or prospective participants, have had first to improve their own reporting systems (and hence their own management information) before they can guarantee that the required data are available or of an acceptable quality.

23.5 STEP 5: DATA ANALYSIS AND REPORTING

The DPR was conceived as a yearly study for operators in the North Sea. In later years studies in other regions were set up. The process for publishing regional results was this: when all the data had been submitted for a region a standard set of charts was plotted, using pre-defined categories to group the wells. The charts and data were printed in a report which was sent to all participants in that region; they also received a copy of the data on a floppy disc (in later years this was a CD).

The main performance metrics plotted were 'feet drilled per day' and 'cost per foot drilled'. Other charts showed average amounts of non-productive time; time lost due to adverse weather conditions ('waiting on weather' time); time spent taking core samples; and time spent gathering down-hole information (logging) for each operator charted. All these activities have a bearing on the key performance indicators. There were further charts in which regression lines were drawn to allow the estimation, for each operator and type of well, of the average time, or cost, to drill 10 000 ft.

Figure 23.4 shows an example of one of the types of performance charts produced. For each operator having wells of the type being plotted the average values are calculated. In order to give an idea of the spread of values the maximum and minimum values are also shown.

There were a number of short-comings in this reporting process, which continued for more than 10 years:

1. The report could not be compiled until all, or most, data had been collected and validated. This might not be until 4 or 5 months after the end of the year. Any operator who was too late submitting data, or too slow replying to queries on the data, missed the publication and had to wait several more months for a revision to take place. Furthermore the charts were static, and were out-of-date as soon as more data were received after the last revision was made.

2. The charts were pre-defined, and so was the way in which wells were categorized. However, when categories had too few wells for meaningful charts to be drawn, they were combined with other categories. Participants had no input into how the wells were categorized or the decision as to which wells were most appropriate for comparison with their own wells. Sometimes participants found that the dataset did not contain any wells comparable to the types of wells they were drilling.

   Broadly speaking the wells were grouped according to:

   a. the type of well drilled (e.g. exploration or production; new hole or a sidetrack/branch from an existing well bore);
   b. the type of rig which drilled the well (e.g. fixed or floating);
   c. certain geological characteristics (e.g. high pressure, high temperature);
   d. different countries or parts of the regions were sometimes grouped.

3. Producing the report was labour-intensive – a typical regional report might contain over 200 charts. For each chart a decision was made on where and how to combine the categories. Then each chart, once plotted, was manually adjusted so that the axis scales were not distorted by extreme values. Other adjustments were made to ensure the plotted points were not so crowded as to make it impossible to distinguish them.

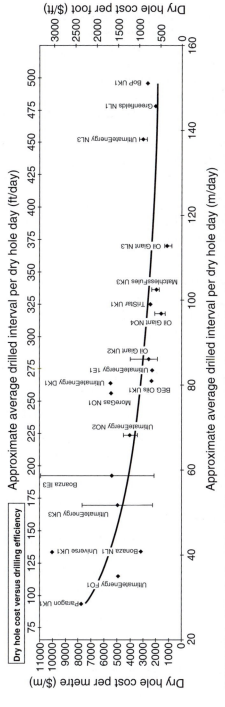

FIGURE 23.4 Example of a performance chart. Each operator's average values are indicated by a diamond, while the maximum and minimum "cost per metre" values for each operator are at the ends of the vertical lines. The numbers of wells drilled by each operator are shown after the operator names.

4. Each operator participated in one regional study and only received the report for that region. The charts for a particular region were plotted using the units (metres or feet) and the currency most preferred by that set of operators. So the Europe charts were plotted in feet and GBP; the Africa charts in metres and USD. While this suited most participants, some were left converting the results to their own preferred units or currencies.

5. Towards the end of the 1990s access rules were developed (as described above), restricting a participant's access to data to reflect the operator's global participation status. This greatly increased the already arduous task of producing charts. Each operator now needed their own report which only showed the data values from those countries to which they were entitled access.

6. In addition to the performance charts, each report contained a printout of the data and a full page time–depth chart for each well in the dataset. Thus the reports were lengthy, resulting in a considerable printing, packaging and delivery task.

Challenge: It was clear that the issues listed above need to be addressed.

- Participants wanted to be able to analyse 'this year's data' before the end of the year, to help them plan for the following year.
- They wanted control over the selection of appropriate 'offset' wells (wells drilled in similar conditions) for comparison purposes.
- Sometimes participants wanted to see data from regions or areas where they had not yet drilled any wells.
- The increased quantity of data collected, and the increased number of participants, made the burden of manually producing charts too great.

Response: The solution was to turn to a web-based system, which could allow each participant to simultaneously access data from any area of the world to which they had access rights, and for any year in which they had participated. This has greatly increased the pool of data available to a participant for analysis: a long-term global participant can today (in 2008) access more than 22 000 wells on the website.

Along with providing viewable and downloadable data the website also allows participants to select, sort and chart data in either metres or feet. Basically all the charts which were produced in the original annual reports, and some additional ones, can now be produced dynamically on the website. Data are received on a quarterly basis, and published on the web each week as the validation is completed. This means that well before the end of the year there is a good supply of recent data available for analysis.

The participant can search for the set of wells which most closely match his, in terms of the criteria which are most important for his benchmarking exercise. The data for these wells can provide the input into a set of performance charts.

These might display operator average values, or values by well, and quartile lines or average lines can be superimposed. An additional feature allows the well locations to be plotted on a map.

Charts can be quickly and easily drawn, and revised at a later time if further relevant data are published. Figure 23.5 is an example of a performance chart. On the website this chart is interactive, allowing the user to highlight points in colour and display additional information about the operators.

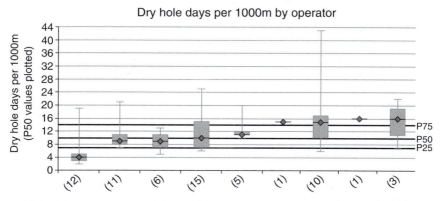

FIGURE 23.5   Example from the website of a dynamically produced 'box and whisker' chart. Each diamond represents the median of the operator's values, while the boxes indicate quartile values, and the lines maximum and minimum values. The numbers of wells are shown in brackets below the points. Note: on the website both operator names and numbers of wells are visible below the chart. Here operator names have been removed to preserve confidentiality.

Access to the website is controlled by usernames and passwords. The username determines the access rights of the user, and ensures that only the correct set of wells can be viewed or downloaded. Additional security is provided through regular monitoring of the site by the facilitators. This same monitoring system also enables the facilitators to learn how participants use the website and to make improvements where navigation issues are identified.

23.6 STEP 6: FOLLOW-ON ACTION

The DPR does not cover all the stages of a benchmarking club to the extent that the facilitators do not make any recommendations to Participants on how they should change their drilling practices to optimize their performance. The facilitators do not carry out in-depth analyses of this nature.

What the DPR provides is a large database of well data, complete and with a high level of accuracy. The rules of participation require that each participant must submit data on all wells drilled in a country, both the successful, fast ones and those which were beset with problems and failures. The data thus cover a wide spectrum of drilling performance.

How each participant uses the data is dependent on their own objectives and analysis skills. The facilitators provide the tools on the website to make the task easier. Participants can however download required data from the website to carry out more in-depth analyses using other software.

While the main deliverable for the DPR is data, and the ability to analyse it, an important secondary product is the set of contact details. From the beginning it has always been intended that a participant should be able to contact other participants to get further information about particular wells. The annual reports used to contain a list of participant contacts, and this has been continued on the website: for each well on the website the name of a contact person within the operating company can be retrieved along with their phone number and email address.

Challenge: Keeping contact details up-to-date is challenging in an industry where personnel are highly mobile.

Response: It has been described above how participant details are collected and confirmed on a quarterly basis. Unfortunately this is not always possible if an operator ceases to participate in the study, or indeed, if the operator ceases to exist! Every effort is made to record changes of contact names, and to correct email addresses or phone numbers, but inevitably there are some gaps.

23.6.1 Best Practice Meetings

It became clear to the facilitators that participants wanted a more formal way of sharing experiences and networking with their counterparts in other operating companies. As a result 'Best Practice' meetings were introduced, and these are now held annually.

In addition to talks by outside speakers, participants give presentations to show how they have used the data to benefit their organizations. Workshop sessions allow various aspects of drilling performance to be discussed. For example in one session participants shared ideas on how to categorize the causes of non-productive time, something for which there is currently no industry standard.

The meeting enables participants to get to know each other informally. All staff members from Rushmore Reviews attend one or more meetings, so that they can meet some of the people whose data they validate. Talking to the 'customers' means they can learn more about the issues associated with submitting of the data or navigating the website. They can also pick up ideas for improvements. One of the highlights of the annual meetings is a formal evening meal at a Scottish castle or ancestral home.

23.6.2 Normalization

A recurring question has been how to normalize the data. The three simple types of normalization used in the DPR are:

1. **Categorization**: The time and cost to drill a well is known to be dependent on a number of data items that are categorical. For example, wells drilled offshore are generally more expensive than wells drilled onshore. There are also different types of rigs which can drill wells including drill ships, production platforms, barges, land rigs, etc. Each type of rig will have constraints and advantages, and the costs and drilling speed of each will vary. Some of the ways in which data were categorized for producing charts has been described above.
2. **Unitization**: The two key metrics used in drilling are the cost and time to drill a well. However, the single most important factor determining both cost and time is the length drilled. Therefore, the two metrics most commonly used to measure cost and time are cost per metre drilled and metres drilled per day.
3. **Percentage**: Reporting a percentage is a special case of unitization. As with many industries, drilling equipment can fail or for other reasons be unable to work. In the drilling industry there are various similar measures variously called 'downtime', 'trouble time' or 'non-productive time', all of which are measured as percentages.
4. **Selection**: Still the question remains – how is it possible to make a meaningful comparison of the drilling performance on two very different wells? In order to make such a comparison there needs to be a way of taking into account the relative difficulties or complexities of the wells.

There has been previous work to develop indices, some specific to a certain aspect of drilling (e.g. the tortuosity of the wellbore) and some more general. None of these is particularly suitable for use as a normalization tool, either because they depend on subjectively determined factors or because they only consider one aspect of drilling.

For this reason participants of the DPR asked the facilitators to develop an index based purely on a statistical analysis of real drilling data – one that could be used to normalize well data. To give a simple example: if we can show that this well is twice as 'difficult' as that well, we would expect the drilling to be half as fast. If the drilling speed is in fact more than 'half as fast', then we have shown that the drilling performance is 'better' on this well.

By applying regression analysis to the available DPR data, the facilitators have modelled drilling speed, which can be said to reflect the difficulty of drilling the well. This model can be used to derive an index based on predicted drilling speeds. When the work is complete it is hoped that the Rushmore Drilling Index will provide a good, globally applicable index. Clearly the model is influenced by the type of data available, and the fact that the DPR data do not include measures of rock hardness or tortuosity is a limitation. However, it is hoped that the breadth and quantity of available data will enable the index to be a useable indicator of drilling difficulty.

23.6.2.1 *How the data are used*

Normalization is a useful technique for comparing otherwise disparate wells. However, most participants analyse their performance by selecting a subset of wells similar to their own. These may be drawn from several years, and from more than one country, providing they are similar with respect to the features that matter. Some participants have used the data to check whether their drilling speed is in the top quartile compared to their peers. If they find they are not they look for reasons why other operators are performing better. Others take the analysis further by finding the 'best in class' well and analysing the gap between their own drilling performance and that of 'the best'.

The data have also been used for an in-depth analysis of the drilling speeds in different sections of the well bore. A typical well bore starts with large diameter hole at surface. When a certain depth is reached this will need to have casing cemented in place, and a smaller diameter bit is therefore required for drilling the next section. This is repeated for each section. So a well bore which is 30 inches diameter at surface may be only 6 inches diameter when it reaches its target, and along the way there will have been several changes of drill bit size (see Figure 23.1).

Another use of the data is in planning future wells. When planning a well it is necessary not only to know the likely time to drill the well, but also to have an idea of the limits, i.e. the fastest possible time, and the slowest likely. By looking at the best drilling speeds for each diameter of hole it is possible to put together a theoretical 'composite' well, which is the 'best of the best' and has a composite speed faster than any actual well in the selected dataset. This has been used by some participants, along with the average and 'worst-in-class' speeds, to look at the distribution of possible drilling times.

This use of the data is not benchmarking, but once the participants have access to the database, providing they keep within the confidentiality agreements, they can use it for their own objectives.

Other participants may not use the data when drawing up their plans and budgets, but instead they use the data to justify their plans to management. If the planned times or costs are too far out of line compared to the benchmark data, they will not gain approval.

A number of participants use the data for internal benchmarking – for comparing one well against others they have drilled in the same field; or for

benchmarking one business unit against another. This should not be surprising –
once a participant has gone to the effort of collecting and submitting the data,
why would they duplicate the work in order to set up an internal benchmarking
study?

23.7 CONCLUSION

The Drilling Performance Review is an example of a long-running, successful
'benchmarking club'. This case study describes the way in which the study is run
and shows how it has evolved to meet the challenges which accompanied its
growth. Along the way, and at the request of the DPR participants, a second study
was set up – the Completions Performance Review (CPR). This covers the next
stage in the construction of a production well: the installation of the equipment
required to bring hydrocarbons to the surface. The CPR has its own story of
growth and development, but that will not be told here.

The main reasons for the success of the DPR are:

- A marketing strategy which has successfully recruited a large number of
  international operators with long-term global contracts.
- The strict application of the rules of participation – all participants agree to
  the same set of terms and conditions.
- The completeness and transparency of the data.
- The investment in in-depth checking and validation of the data.
- The maintenance of a large database of historic data.
- The investment in adding functionality to the website.
- A reputation for friendly personal customer service.

Finally, and key to its success, the DPR is driven by its participants.

For further information about the Drilling Performance Review see:
www.RushmoreReviews.com.

Chapter 24

Benchmarking Local Government Services

Paul Carroll
Corporate Planning Manager, Dundee City Council, Scotland

INTRODUCTION

Dundee City Council is a unitary local authority. It provides all local government services for the city from education to street cleaning (Figure 24.1). The population of the city is around 143 000 and the council employs just over 8000 people. Dundee City Council is one of 32 similar local authorities throughout Scotland. There are 442 local authorities throughout the UK, although in England they are not all unitary.

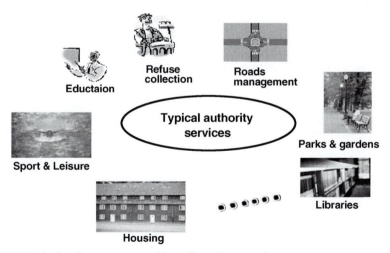

Eductaion

Refuse collection

Roads management

Typical authority services

Parks & gardens

Sport & Leisure

Housing

Libraries

FIGURE 24.1   Local government provide a wide variety or services.

The Benchmarking Book: A How-to-Guide to Best Practice for Managers and Practitioners

The city council can report that over 75% of its performance indicators improve year on year. In 2007 Dundee was in the top six local authorities in Scotland for its rate of improvement. It also compares favourably with the other cities and urban authorities with respect to performance statistics collected by Audit Scotland from all local authorities.

As a public sector organization the council is not threatened by takeover or with extinction if it makes a financial or performance failure. Equally significant is that a major improvement in performance will not be rewarded by growing the council's market share or income, as the market is fixed to a geographic boundary and its statutory duties. Council managers are motivated by career progression, peer comparison and generally enter their professions from a commitment to that type of work, e.g. teaching, social work, environmental management.

24.1 BENCHMARKING – THE GENERAL SITUATION

To a large extent the services provided by a local authority are prescribed by statute, heavily regulated and regularly independently audited. The data collected and reported by the local authorities, as well as audit reports, are in the public domain and therefore finding comparative data should be straightforward (see www.audit-scotland.gov.uk). In some cases the data have been collated into league tables, or highlighted in what the media will refer to as a 'postcode lottery' of different standards applied by different local authorities (e.g. for school inspection reports see www.hmie.gov.uk).

In the council we view benchmarking as a performance improvement tool, not simply as a performance comparison tool. While the data exist and can be compared it cannot be assumed that this will necessarily lead to appropriate analysis and improvement.

While such 'league table' data can be used destructively (i.e. to apportion blame and criticize) it can also be an effective way of quickly spreading information throughout the public sector about where best practice may lie.

Case Study: Recycling

In the current environmentally conscious culture, authorities are keen to monitor and increase domestic recycling rates. When one authority shows a significantly higher proportion of re-cycled waste than other authorities we would investigate to discover how they were achieving this performance and, where appropriate, seek to adapt and adopt their practices.

Identifying good performance is not difficult, firstly because the authority achieving good performance levels will advertise its success, and secondly because interest groups, politicians and newspapers will be keen to feature good (and more especially poor) performance levels.

Authorities typically respond in one of two ways to such information. They either:

1. Question the validity of the data, the report, the conclusions or the applicability to their own situation and will resist external pressures to change; or
2. Follow-up, by aiming to discover if there is anything they can learn and implement in their own authority.

The type of response is generally driven by the leadership within the authority.

In some situations the Audit Commission in England give scores as to how well a council performs. While this has obvious potential advantages it can lead to effort being spent appealing against the score to save its reputation. For example, if a school were to be labelled as a failing school its reputation would be tarnished. It is likely that many teachers would prefer not to work there and that parents would prefer not to send their children to that school. Therefore, a school would normally want to appeal such negative labelling.

Achieving continuous improvement is a statutory duty in the 2003 Local Government Scotland Act (www.opsi.gov.uk/legislation/scotland/acts2003/asp_20030001_en_1) and is referred to as part of achieving Best Value. This brings with it an expectation that performance will improve over time. Through a process of a three-yearly Best Value external audit, a local authority has to show that it is improving and that it has put management practices in place to underpin continuous improvement. Benchmarking is a prime method by which it can comply.

24.2 TYPICAL BENCHMARKING ACTIVITIES

Benchmarking is one of the methods that external government auditors ask about. The main issue is whether there is a proactive search for data or process comparisons from other organizations for the explicit purpose of identifying whether there is scope to improve. This should lead to learning how to improve from those performing better.

There are a number of ways in which councils in Scotland have addressed this requirement to benchmarking including:

Authority Benchmarking Club

A group of around 10 councils in Scotland have formed a benchmarking club called ABC. The club is dedicated to working together to improve services by sharing and comparing data, processes and innovative solutions to common problems. They believe that sharing experiences, and identifying and adopting examples of best practice, is fundamental to their delivery of Best Value principles.

'Best Value commits us all to continuous improvement. We cannot achieve this without a clear understanding of where we stand now, or a vision of possibilities which are open to us. The most effective way of developing both is to compare ourselves with our peers and to share ideas. I congratulate the ABC Benchmarking Partnership on its continuing success in helping members do just that.'

Andy Kerr MSP, Minister for Finance & Public Services

(For further information see www.eastrenfrewshire.gov.uk/benchmarking).

Business Models

Other authorities, including Dundee, used models such as the European Foundation of Quality Management (EFQM) to identify areas where they can improve. Organizations such as Quality Scotland provide support and networking to help find out who is good at what.

Best Value Reviews

Many councils, for example Dundee, run a programme of Best Value Reviews of their services (details available from the council's websites, e.g. www. dundeecity.gov.uk). A review typically consists of three stages:

1. Review of performance against previous levels, other councils, industry averages etc. as appropriate for the particular performance indicator.
2. Based on a performance gap analysis, areas of potential improvement are identified.
3. Possible improvement activities are identified, which may include detailed benchmarking with other authorities or organizations.

In general, comparison with, and learning from, non-authority organizations is more difficult than cross-authority comparisons. For this reason people are reluctant to benchmark with non-authority organizations.

Case Study: Servicing of Aircraft and Hospitals

Despite the reluctance to benchmark outside authorities, one study compared the servicing of commercial aircraft at airports with the servicing of hospital operating theatres and vehicle cleaning.

Therefore, benchmarking in Dundee City Council is mainly concerned with learning from new or good practices from similar professional functions within other local authorities.

As mentioned above, information about new and good practice is readily available through professional or regulatory documents (e.g. audits, press) and it is generally a case of following up on these heralded successes. Annual awards and supplier sales strategies can also be a source of information to enhance learning. In addition, Quality Scotland (www.qualityscotland.co.uk) organize Open Days for members to meet with perceived high performance organizations.

Clearly, the focus for learning will relate to those areas that can be replicated within the council. We will decide whether or not to visit another organization if their performance level is higher than ours (based on data comparison) depending on whether we have similar objectives. The following issues are examples of the most common areas of interest to council managers in a benchmarking study visit:

- strategic issues, e.g. objectives and performance measures set for our services,
- people issues, e.g. staff structure, grading and training issues,
- technical issues, e.g. software, in-house or outsourced contracts,
- environment issues, e.g. building layout (open plan/office, office design, location, etc.),
- communication issues, e.g. publicity materials used and methods of engaging stakeholders.

24.3 THE ONE-STOP-SHOP BENCHMARKING PROJECT

24.3.1 Background

A specific example relates to how Dundee City Council came to decide upon and then develop its plans for a One-Stop-Shop in its new city centre office. In local government a one–stop-shop relates to a strategy where there is one main office in a geographic area to which the public can visit or phone for all council services.

Historically each council service has been managed and run by independent groups within the council, e.g. housing, education, refuse etc. Any queries relating to these services had to be directed to that specific group, usually in different buildings. Where the enquiry related to several services, several visits to different locations might be required to resolve an issue and the enquirer might be passed between the services (Figure 24.2).

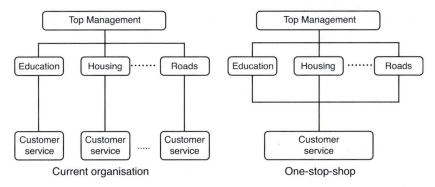

FIGURE 24.2   Moving from segregated customer services to One-Stop-Shop.

The benefits of a One-Stop-Shop include, for example:

- Customer only needs to go to one office to resolve all enquiries.
- Where more than one service is involved, they can work together with the enquirer in one place at one time.
- Economies of scale.

Although the concept of a One-Stop-Shop sounds straightforward there is no example of it being wholly and perfectly implemented. Those referenced as best practice are in reality just more advanced forms of 'work in progress'. The One-Stop-Shop idea is not new and there is around twenty years of practice in trying to achieve a One-Stop-Shop approach. Experience has shown that there are a number of significant difficulties including.

- Lack of a suitable building to house the One-Stop-Shop.
- Independently developed databases that cannot communicate.
- Historic 'silo' hierarchical organizational structures and mentality, developed over years of changing service, with its associated lack of communication.

Even if these are surmounted, there are practical issues such as whether to train all enquiry staff in all services, itself a huge task, or whether each person should only respond to queries relating to their own area (which could lead to some staff being idle while others have long queues).

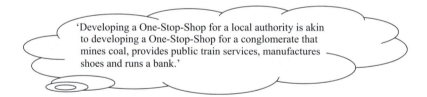

'Developing a One-Stop-Shop for a local authority is akin to developing a One-Stop-Shop for a conglomerate that mines coal, provides public train services, manufactures shoes and runs a bank.'

The internet and new information and communications technology (ICT) gave renewed interest in the concept as it promised to remove one of the main barriers: making data available from different department's systems to one group of front line staff.

Dundee City Council had enthusiastically adopted a new ICT Modernization Strategy in 2002, which included online transactions, a smartcard and a customer contact centre. The impetus was a government target to get all services online by 2005, and an additional grant that could be bid for from central government funds called the Modernization Government Fund. The idea that all councils develop a generic first point of contact with customers was firmly put back on the agenda during networking with practitioners working on modernizing government funded projects. These practitioners were asked to speak at conferences organized to help promote and facilitate the implementation of One-Stop-Shop. Another significant development around this time was the availability of software such as Customer Relationship Management.

Serendipity often plays a part in making something become more feasible. In Dundee's case discussion was taking place on the need to demolish its current city centre office block and replace it with a more efficient building. The One-Stop-Shop could fit into the plans for the new office building adding further justification to the decision to build it. So research into the business case for a One-Stop-Shop could turn its attention from the why to the how.

24.3.2 Benchmarking Visits

With all these developments and drivers a One-Stop-Shop project board was created. The first task for the board was to develop a business case following the authority's standard procedures. The business case included such drivers as:

- The consensus of opinion both from government and other councils was that One-Stop-Shops were the direction that needed to be followed.
- Other councils had, or were developing, One-Stop-Shops and anecdotal information concluded that their experiences were positive.
- The council's own research concluded that it would benefit from efficiencies of pooling resources.

The board organized study visits to some of these councils that were being most successful. The objectives of the visits included:

- To convince those still sceptical of the One-Stop-Shop approach.
- To learn from the experiences of others how to implement a One-Stop-Shop approach: in particular what would help effective implementation and what pitfalls and difficulties they had encountered.
- As team building events for those involved in the project.

Staff working for the Modernizing Government Fund (now part of the Improvement Service) at a national level recommended the best councils to visit. In addition, the UK government had funded early adopters of new approaches in a programme of 'Beacon Councils'. Newcastle was one of the 'Beacon Councils' for the new One-Stop-Shop approach. As well as Newcastle, members of the group visited South Lanarkshire, East Renfrewshire, West Lothian and Renfrewshire Councils.

Visits were well planned beforehand. A typical study involved around a dozen staff. Specific areas of interest, for example building design, were highlighted so that appropriate people from the council being visited could be made available. A typical agenda was:

- Introductions.
- Presentation of the One-Stop-Shop project.
- Meetings/tours with specific people and areas.

The study visits enabled the project board to visualize the type of facility, the use of new technology, the training programmes, the use of branding and the lessons learned from those involved in managing in an existing set-up. At this stage in a project the learning is not startling or new, but it can be deepened through being able to see and hear from people who have experience. Another advantage is giving the implementation team time to focus on the key issues. The conversations to and from the study visits are all part of forming a clear blueprint in the mind for planning a change in an organization.

Case Study: Learning from One-Stop-Shop Visits

Many things were learned during these visits and acted upon. For example:

- We had seriously underestimated the length of time it would take to train front-line staff. We had thought a few days would be sufficient, but other councils told us that up to six months would be necessary.
- We also learned from one council the importance of correct branding of the One-Stop-Shop. They had chosen a name for the service and advertised it heavily. However, some time after opening the shop they carried out a customer survey to ascertain awareness of the service and discovered that the public were confused as to what the service was: they had not realized that it was a One-Stop-Shop for council services.

The One-Stop-Shop in the new Dundee House is due to open in 2010. However, the current customer service teams are already implementing many of the lessons from the study visits.

SUMMARY

Councils are involved in a variety of activities that can be termed benchmarking.

- ✓ Government audits implicitly compare performances and comment on current performance levels. Some include a numeric score and recommendations for improvement.
- ✓ Data from audits are sometimes used to develop 'league tables' or similar comparisons.

✓ Councils are 'benchmarked' by the media who use data from the public domain to compare performances.
✓ Some councils have developed benchmarking clubs with the aim of comparing performances and learning from each other.
✓ Some councils use business models such as EFQM to identify potential areas for improvement.
✓ Information from a variety of sources often highlights councils that are performing particularly well in one area. This information can be used to select benchmarking studies and partner visits.

Chapter 25

The Best Practice Club

Ray Wilkinson Director

INTRODUCTION

Benchmarking and the sharing of Best Practices by nature require communication between organizations. Almost as soon as the internet became widely available, people realized that it would be an ideal medium for building virtual communities of organizations that wanted to work and benchmark together. Consequently many such organizations have sprung up.

These organizations, often called clubs or communities, act as a central resource for their members to ask for and/or offer help. Every club offers its own suite of products and services, but they all have similar aims: to facilitate the spread of good practices amongst member organizations.

In the UK one such well-known community of organizations is the Best Practice Club (BPC).

This chapter explains how the BPC operates, what services it offers and the typical benefits that its members have received. Ray concludes with three case studies illustrating not only different benchmarking and learning activities within the club but also different ways in which the club operates.

For further information see http://www.bpclub.com.

25.1 WHAT IS THE BEST PRACTICE CLUB?

The BPC is a web and face-to-face based community of member organizations seeking to improve their practices in order to become more effective and efficient for the benefit of all their stakeholders.

Do you remember the words of that eponymous advertisement: '... but I know a man who does ....'? How often have you gained satisfaction from helping others? Most people accept the importance of 'who' you know rather than 'what' you know. The club has taken the two fundamental truths that people like helping others and success is about who you know, and built a highly successful corporate networking model on them.

The BPC was founded in 1993 in response to the growing demand from industry and the public sector to form learning networks. Its aim was, and still is, to help organizations realize their Best Practice aspirations.

Today the BPC is:

- A large membership of complex private and public sector organizations.
- Mainly from the UK, but dealings with organizations as far afield as the Middle East, South Africa, the Far East and America.
- Drawn from over 20 sectors (the vast majority of the membership being from manufacturing, public sector, utilities, transport, construction or financial services).
- Half the member organizations have over 1000 employees and a quarter have 250–999 employees.

It has become a successful example of how organizations can work together formally and informally for mutual benefit. Every year the BPC helps thousands of employees from member companies with advice and information on a wide range of subjects from HR to Health & Safety, Customer Service to Business Strategy and Improvement. Benchmarking activities have covered many areas of business and recent projects include: maintenance services, call centres, corporate services, absence management, catering, product quality, Excellence Models, resourcing and complaint handling.

Over 90% of the club's membership has carried out benchmarking. Of those, 72% categorize their capabilities as either intermediate or expert. As might be expected the vast majority have had positive experiences of their benchmarking projects and have derived positive benefits from them.

25.2 HOW DOES THE BPC OPERATE?

The club categorizes benchmarking into two types: performance and process. It sees the former as the comparison of a subject organization's key performance indicators against a reference group of peer organizations. The latter involves a subject organization learning from the key processes of a small number of its peers.

Performance benchmarking is an excellent provider of reasons for change and will often act as a guide to 'what' is required, whereas process benchmarking provides the 'how' necessary to make the improvements identified. The resource investment required for a process benchmarking exercise will usually be significantly larger than that required for a performance benchmark.

A major issue with performance benchmarking in general is that cynics will often criticize the comparisons made for their lack of complete relevance, e.g. the reference data does not come from a source that is exactly the same as the benchmarking organization.

Because of this cynicism the BPC has shied away from the simplistic use of online questionnaires providing overly generic comparisons. Usually the reference group for an on-line performance benchmarking exercise is considered en mass and the specific performance of its individual organizations are not made known to anyone else (apart from the exercise facilitator). Such online exercises can be useful as quick healthchecks but they lack the rigour to provide meaningful drivers to improve performance. BPC members have commented in the past that: '... anonymous comparison does not lead to improved performance ...' and that they prefer to discuss the performance with the organizations concerned in order to gain maximum benefit. Thus the Best Practice Club has focused on providing practical process benchmarking opportunities both online and face-to-face.

The BPC facilitates communication and learning through:

- ✓ **The Forum**. The Forum is a facility on the BPC website on which members can post requests for information or help. Other members reading the notice can respond.
- ✓ Access to the **Business Performance Improvement Resource** (BPIR) information base.

Members of the BPC receive free access to BPIR, a portal to an extensive information base related to business excellence, benchmarking, best practices, performance measurement and improvement. The information base includes access to journals, articles, briefing notes and business excellence models (such as Baldridge) amongst others.

- ✓ **Programmed events**. The BPC organizes two types of programmed events: Workshops and Active Learning Days.

  Workshops, (called Best Practice Forums elsewhere in this book) allow up to three member organizations to showcase their experiences and knowledge on a particular topic. They are usually hosted at the site of one of the presenters to enable live demonstrations, and ensure that any questions can be answered by the appropriate people. Recent workshops have covered topics such as knowledge management, employee suggestion schemes and HSE strategic management. Workshops are organized and facilitated by the BPC and there are about 40 events (of all types) a year.

  Active Learning Days are courses, usually of one day, organized by the BPC and presented by vetted and approved associate trainers. Recent topics for Learning Days include: Balanced Scorecard, Lean Process Management, Internal Communications, Organizational Development and Internal Consultancy Skills.

✓ **Special Interest Groups**. Special Interest Groups consist of several members who choose to work together to deepen their understanding of a particular issue (⇨ Case Study 3). Interest groups typically meet every other month and are facilitated by the BPC.

✓ **Reactive events**. Reactive events are any event that takes place in response to a member request. They can be of any form but are usually one-to-one benchmarking events. Such events often begin with a member posting a notice on the website and other members responding. This may result in meetings between the two members, or between groups of members or a request that the BPC organize an event for the wider membership.

✓ **Facilitation**. The BPC often hosts and facilitates events. It will also facilitate benchmarking studies, when asked to do so.

25.3 ASCERTAINING MEMBERS' NEEDS

The BPC believe that a vital part of their role is to actively seek out and meet members' needs. They do this in a variety of ways including:

✓ Monitoring the members' Forum.
✓ Monitoring attendance and feedback from events, both informally and through formal events review feedback questionnaires.
✓ Building relationships with members through informal discussions as well as at events.
✓ Contacting members about activities in which that they think the member will be interested.

Information gleaned from these sources is used to identify events that the BPC believe members will be interested in, as well as to proactively put members with similar interests in touch with each other.

25.4 WHY ORGANIZATIONS JOIN FACILITATED BENCHMARKING COMMUNITIES

In today's fast moving and increasingly competitive world thoroughly networked individuals and organizations have a major advantage over their more insular peers. A manager within a large organization can often be lulled into a false sense of security and think that the world starts and ends with that organization; very rarely will they need to come into contact with anyone outside and thus their view of the world can become very parochial. It is to combat that inwardly focused and limited perspective, and to build change capability and an increased capacity to compete, that organizations engage in the BPC's benchmarking activities.

In addition to proactively seeking members' needs, the BPC receives over 130 benchmarking requests each year. The response to these requests range from simply providing advice and guidance, to facilitating and promoting learning from organizations that are more experienced, through to informal and ultimately formal process benchmarking exercises.

The fact that these requests result in nearly 300 responses from members is a tribute to the members' willingness to support and help each other by sharing their Best Practices. We also believe that the willingness of members to support each other validates the Club's networking model.

To show how communities like the BPC work we include three case studies:

Case Study 1: The Highways Agency

The Highways Agency was aware that there was a lack of coherent operational systems within the organization. People would carry out similar tasks in different ways. One result of this was that people joining a department would find out how to do things by asking those around them and often received conflicting advice. This lack of coherence led to significant waste in terms of effort (to find accurate advice) and quality of output.

The organization decided to develop and implement a formal common Quality Management System throughout the organization with the aim of streamlining operations and creating one source of information on how to carry out tasks.

They wanted to find out if any other organization had done something similar and posted a request on the BPC notice board. They received several replies and visited organizations, benchmarking with them, to learn from their experiences.

They received good advice on how to set about developing and implementing such a management system in addition to a number of 'lessons learned' to help them avoid the many potential problems.

Their development and implementation was successful and they avoided many pitfalls that they believed they would have fallen into. In addition they estimated they saved tens of thousands of pounds on external consultant fees and believe that because they carried out all the work themselves, rather than use consultants, their employees and management took more ownership of the system thus helping to make it a success.

After the project was complete, the BPC were delighted to receive this testimonial:

"In 2007 we took up the opportunity of membership with the Best Practice Club. We needed to find organizations of a similar make-up with which to benchmark. We put our requirements on the BPC site and within days we had contact details of three companies offering – no strings – help and advice. Meetings followed including a demonstration of a live Quality Management System, and clear advice on 'pitfalls' and 'practical solutions to problems' was given from real experience. This provided real momentum for moving our own internal system forward"

Mike Harlow, *The Way We Work Programme Manager, The Highways Agency.*

Case Study 2: High Street Retailer

A retail member organization had carried out an internal review of its strategic planning activities and had identified strengths and potential areas for improvement, particularly in the areas of scenario planning and strategic intelligence. In order to validate the findings of their internal review, and if possible learn how they could improve by studying what other organizations do, they were keen to benchmark their strategic planning processes and infrastructure with external organizations.

As a first step they approached the BPC for help in pulling together a small benchmarking group. As a result, three organizations facing similar issues, from both the private and the public sectors, offered to get involved. The BPC then facilitated an initial benchmarking session involving these interested organizations who were able to share key information on intelligence gathering, the interpretation and management of that intelligence, and the processes for implementing projects and actions resulting from the analysis of intelligence.

Following on from that initial session, other BPC members indicated an interest in the topic and so a Workshop was organized at which around a dozen organizations were present. A wide spread of sectors was represented including: Retail, Financial Services, Utilities, Construction, Energy, Health and Transport.

As a consequence of the value of the organizational learning that had taken place at this workshop several delegate organizations agreed to take the subject further and build on the original benchmarking group. They are now in the process of setting up a Special Interest Group which will meet every quarter to be facilitated by the BPC.

Case Study 3: Health, Safety and Environmental interest group

As a result of the growing awareness of the impact of Health, Safety and Environmental (HSE) issues worldwide and across a wide variety of industries, including those that would not normally be associated with HSE challenges, there was a growing demand within the BPC membership for a focus on this key topic.

In response to that demand, the BPC set up, and now manages, the ongoing facilitation of a Special Interest Group. This allows members of the community to review and benchmark their HSE performance, with a view to finding any areas of weakness and taking action to improve them.

This group meets six times a year at a variety of locations in the North East. There is a stable core of around 8 organizations in the group (who take turns in hosting meetings), with occasional temporary members for one or more of those meetings. The aim of the group is "To improve what we do in HSE".

Each meeting is on a different topic, agreed in advance, and will take the form of one or two members making presentations about their perspective on the topic in question. Key learning points are then discussed and experiences and practices shared. On a regular basis academics and professional specialists, e.g. lawyers, are brought in to take the lead during meetings when it is appropriate to do so.

Recent meetings have covered topics such as driver fatigue, implementing and maintaining a safety culture, new HSE legislation, contingency planning, contractor management and environmental issues. Thus, group members have the opportunity to share and learn from each other as well as being able to keep up to date with the latest HSE regulations and issues. Often learning takes place that is outside the normally perceived boundaries of an HSE professional's remit, e.g. people engagement and internal Public Relations.

The benefits are difficult to quantify and members feel that doing so is of little practical use: they are confident that the steady trickle of improvements has a positive impact on HSE and significantly outweighs the investment of taking part in the meetings and implementing changes.

Bibliography

Websites

On-line Benchmarking/Best Practices communities and services:

Best Practice Club (The) www.bpclub.com is an on-line community of organizations interested in learning from and benchmarking with each other. See their case study in Part 4 for more details.

APQC International Benchmarking Clearinghouse www.allianceonline.org is a benchmarking group aimed at not-for-profit organizations (nonprofits)

Benchmarking Exchange (The), www.benchnet.com provide a variety of services to those interested in benchmarking including an on-line benchmarking/survey facility.

Benchmarking Network (The) www.benchmarkingnetwork.com provides a variety of benchmarking services

Best Practices LLC can be found at www.best-in-class.com it is a research, consulting and publishing organization focusing on the health care industry.

British Quality Foundation (The) www.quality-foundation.co.uk promotes business excellence through numerous activities, based on a Business Excellence Model

Organizations quoted in the book offering Benchmarking Services include:

Information Management Forum (See Part 4 case study), are highly experienced IT consultants who offer a range of services. They can be found at www.globalinformationpartners.com/

Rushmore Reviews (see Part 4 case study) specialise in Drilling and Completions benchmarking and can be found at www.rushmorereviews.com

Juran Institute: http://www.juran.com/
Juran Institute is a management consultancy founded by Dr Joseph Juran. They provide training and consultancy in a range of quality management issues including Process Performance Improvement, Six Sigma, Change Management

and Benchmarking. They also run regular and bespoke benchmarking studies. They have offices in the US, Canada and Europe.

Solomon Associates are renowned for refinery benchmarking and consultancy services. www.solomononline.com

Excellence Awards:

There are numerous quality awards and associated models. We include three of the better known ones here:

Details of the Baldridge Award and model details can be found at: www.quality.nist.gov

Details of the Dubai Quality Award details can be found at www.dqa.ae

European Foundation for Quality Management (EFQM) Excellence Award and model can be found at: www.efqm.org

Codes of conduct:

The American code of conduct can be downloaded from www.apqc.org/PDF/code_of_conduct.pdf as a PDF file.

The European Code of Conduct can be found at www.efqm.org

Other websites:

Deming Learning Network (DLN) focuses is the official site for those wanting to know about the management guru Dr Edwards Deming. The sites includes a number of useful links, recommend appropriate books and other resources:

In the UK: http://www.dln.org.uk and http://www.deming.org.uk/

In the US: http://deming.ces.clemson.edu/pub/den/

Currency exchange rates current and historic, can be found at www.oanda.com

World Intellectual Property Organization, www.wipo.net provides information on intellectual property rights.

An example of an on-line benchmarking study can be found at www.croner.co.uk which benchmarks salaries.

Finally, Tim Stapenhurst can be reached at tim@sigma-c.co.uk

Books
Benchmarking books:

Camp, R. C. (1994). *Business Process Benchmarking Finding and implementing best practices*. USA: Brown (William C.) Co.

Robert Camp has been involved in benchmarking since his days at Xerox and has published many books on benchmarking, of which is this is a major contribution to the literature.

Kearns, D. T., & Nadlar, D. A. (1993). *Prophets in the dark – How Xerox reinvented itself and beat back the Japanese*. Harper Business.
The first half of this fascinating book details how the photocopier was invented and brought to market. The second larger part describes how Xerox, one of the leading American manufacturing companies took the market by storm as it mass produced photocopiers. However, while sales, profits and staffing were taking the company to new height after new height the Japanese were planning its downfall as they saw the waste in Xerox's processes and planned to take the market away from them. Fortunately David Kearns became aware of Xerox's vulnerability and acted by "inventing" benchmarking as a tool for identifying specific weaknesses with Xerox and learning new practices from world-class performing companies. This now legendary story ended with the survival of Xerox, and how it survived is a must-read for anyone interested in benchmarking.

Watson, G. H. (2007). *Strategic benchmarking reloaded with six sigma: Improving your company's performance using global best practice*. John Wiley & Sons. 'O Theory' is explained in the footnote on p xii.

McCabe, S. (2001). *Benchmarking in construction.* WileyBlackwell. The case study of a five-person organization successfully benchmarking is given on p. 182.

Other books:
Kaplan, R. S., & Norton D. P. (1996). *The balanced scorecard: Translating strategy into action*. HBS Press.
The Balanced Scorecard (BSC) is a management system that can help an organization translate its mission and strategy into operational objectives and metrics which can be implemented at all levels in the organization. This highly respected book quickly became a key for helping organizations determine, communicate and measure against key factors vital to success. If implemented correctly, the mission determines the metrics, and conversely, it should be possible to determine the mission by looking at the metrics. The authors demonstrate how senior executives in industries such as banking, oil and retailing are using the technique to evaluate current performance and target future performance based on financial and non-financial criteria such as customer satisfaction, internal processes and employee learning, and growth. If you are concerned that you may not be measuring the appropriate metrics, this book is worth reading.
Senge, P. M. (1990). *The fifth discipline*. Doubleday

This classic book on the Learning Organization is a milestone in business literature. It deals with the art, science and practice of organizations learning and improving. A key contribution relates to the soft issues such as teamwork and resistance to change. However, he also deals with systems thinking and management. The connection with benchmarking is that benchmarking is an improvement (learning) tool and so many of the ideas in the book are applicable to benchmarking, even though Senge does not discuss benchmarking in the book (however, as early as page 4 he quotes 'The ability to learn faster than your competitors may be the only sustainable competitive advantage.'). In addition to the Fifth Discipline, Senge later published The associated *Fieldbook: Strategies for Building a Learning Organization*, a large apparently daunting book that is surprisingly easy to read and which gives examples and explains further the ideas in the Fifth Discipline. These books are key management reading whether or not you are involved in benchmarking.

Deming, W. E. (2000). *Out of the crisis*. The MIT Press

This highly respected book was a landmark in management thinking that is still relevant today. Do not be fooled by the easy style of the book, it provides food for deep thought about the way we manage our organizations. It is an excellent starting point for thinking about what we need to do to help ensure long-term success in our organizations. Deming gives several simple examples of control charts but focuses much more on the implication of variation within organizations. He also explains the Red Beads experiment, an excellent tool for teaching many of the concepts and managerial implications of variation. For comments relating to the importance of soft issues, see pp. 121–126.

Stapenhurst, T. (2005). *Mastering statistical process control*. Elsevier, Butterworth-Heinemann. Includes examples of Control Charts used in benchmarking applications. Three detailed case studies are based on comparing performances between groups. See, Chapters 9, 10 and 13.

References for Chapter 19 Copying without understandings

1. The Fifth Discipline – Peter Senge – Century Business

   The most successful organizations of the future will be learning organizations. The organizations that excel will be those that discover how to tap into their people's commitment and capacity to learn at every level in the company.

2. The Fifth Discipline Field Book – Peter Senge & others – Currency Doubleday

   An extension of the above book concentrating on how to make the above ideas work.

3. Out of the Crisis – Edwards Deming – Cambridge University Press

    This book provides a full account of Deming's thinking on the primacy of management's role in improving quality and productivity. He demonstrates what managers do wrong and how costs, dependability and quality must be improved. This is not just another manual of techniques; Deming provides a theory of management that gets to the roots of the problems of industrial competitiveness, which face management today.

4. The New Economics – Edwards Deming – Massachusetts Institute of Technology

    The aim of this book is to provide guidance for people in management to successfully respond to the myriad changes that shake the world. Transformation into a new style of management is required. The route to take is what Deming calls profound knowledge – knowledge for leadership of transformation. Transformation is not automatic. It must be learned; it must be led.

5. The Human Side of Enterprise – Douglas McGregor – Penguin Books

    The landmark book that encapsulated the X & Y theory concepts of management. The book portrays two contrasting theories based on managements perception of its workforce. X theory where employees are seen as inherently lazy, and Y theory where it is recognized that work is a need, we are trustworthy and like responsibility. The structure evolving from these two contrasting perceptions is very different. X theory spawns a very hierarchical command and control culture while Y theory leads to an enabling and trusting atmosphere. McGregor's argument is that we capture so very much more of the talent and creativity of staff from Y theory perceptions.

6. Unleashing Intellectual Capital – Charles Ehin – Butterworth-Heinemann

    The organization that has become a world leader in applying the type of thinking McGregor described as Y theory (see above) is W.L. Gore. They are one of the most innovative companies in the world as well as regularly being recognized as one of the best companies to work for. Professor Charles Ehin was the Dean of the Gore School of Business at Westminster College. And in taking McGregor's thoughts further it reveals breakthrough principles for structuring Knowledge Age organizations. It offers a comprehensive framework to generate sustained levels of involvement and commitment.

7. The Seven Habits of Highly Effective People – Stephen Covey – Simon & Schuster

With penetrating insights and pointed anecdotes Covey revels a step by step pathway for living with fairness, integrity, honesty and human dignity – principles that give us security to adapt to change and the wisdom and power to take advantage of the opportunities that change creates.

8. Mastering Statistical Process Control – Tim Stapenhurst – Elsevier Butterworth-Heinemann

A book full of case studies demonstrating the importance of a basic knowledge of SPC when interpretting data. It provides the basis from which we can secure knowledge from data so that we can predict future outcomes. A practical book aimed predominantly at the service sectors.

9. Beyond Negotiation – John Carlisle & Robert Parker – John Wiley & Sons

We all work in a system – the supplier, the organization and the customer – our future lies in maximizing the system for everybody's benefit. If we compete within that system all we are doing is creating the opportunity for waste and conflict. John Carlisle has been awarded his doctorate for his work in the field of co-operation

10. Punished by Rewards – Alfie Khon – Houghton Miffin & Co

Our present-day culture wishes to control our staff and our children through rewards and punishment – unfortunately this has a detrimental effect on the motivation of our people.

11. Maverick – Ricardo Semler – Arrow

This is the inspiring story of a young man who took over his father's ailing company based on hierarchical thinking and transformed it into a company based on trust. A company where true democracy is approached.

12. Deming's Profound Changes – Kenneth Delavigne & Daniel Robertson – PTR Prentice Hall

This book helps us appreciate the origins of our management thinking which is still very much based on the teachings of Fredrick Taylor (1900s) – or worse still the corruption of Taylor's concepts. It expands on the transformation of our thinking that we should address if we are to compete in the modern world.

13. Future Edge – Joel Baker – William Morrow

In order to solve major problems facing us today, it will be necessary to break out of our existing paradigms or mindsets. This book aids that process.

14. Fourth Generation Management – Brian Joiner – McGraw-Hill

The first generation of management is simply doing it oneself. The second is instructing in detail others to do the work. The third is setting targets and allowing the employees to develop their own methods – it seeks to make employees accountable , this method is susceptible to distortion of the figures. The fourth generation of management is based on leadership understanding, through quality as defined by the customer, scientific method that includes the analysis of variation and team spirit both within and beyond organizations.

15. Leadership and the New Sciences – Margaret Wheatley – Berrett-Koehler

The book explores how new discoveries in quantum physics, chaos theory and biology contribute to our thinking of how we organize work, people and life.

16. A Simpler Way – Margaret Wheatley – Berrett-Koehler

The recognition that as humans we have the ability to self organize. We therefore need far less supervision and direction than is commonly assumed. It is a book full of hope.

17. Seeing Systems – Barry Oshray – Berrett-Koehler

Oshray weaves a remarkable explanation for the subtle, and largely unseen, ways in which our structures influence our behaviour.

18. Good to Great – James Collins – Century

Following the success of "Built to Last" James Collins collected a team to research,over a five year period, 1,435 fortune 500 companies on the underlying mechanisms that made some of these companies emerge from being good to being great. This insightful book is the outcome of that research.

19. The Living Company – Arie de Geus – Nicholas Brealey

At the heart of this book is a simple question with sweeping implications: What if we thought about a company as a living being? From the basis of this question and his lifelong commitment to Shell, de Geus develops his theme of living companies.

20. Guns, Germs and Steel – Jared Diamond – Vintage

The book is nothing less than an enquiry into the reasons why Europe and the Near East became the cradles of modern societies. Diamond shows definitively that the origins of this inequality in human fortunes cannot be laid at the door of race or inherent features of the people themselves.

He argues that inequality stems instead from the differing natural resources available to the people of each continent.

21. The Scottish Enlightenment – Arthur Herman – Fourth Estate

   Herman, an American with no particular connection to Scotland explores the enlightenment years of eighteenth-century Scotland with such names as David Hume, Adam Smith and James Watt. He traces how we developed a love of learning, and how that learning combined a rigorous understanding of theory and the nuances of the application of that theory. From this foundation, and through the fact that we Scots have travelled the world, we have produced an idea of modernity that has shaped much of civilization as we know it.

22. The Power of Learning – Klas Mellander – American Society for Learning

   A book on how we learn. It has major implications, therefore, on how we teach.

23. Profit beyond Measure – Tom Johnson & Anders Broms – Nicholas Breley

   Starting from a background in accountancy the authors are highly critical of the present practice of using figures to 'manage by results.' They recognize that we can only measure 5% of the whole. Using car/vehicle manufacturers Toyota and Scania they develop their argument to manage the whole system by seeing the organization as a living system. Their concept is "Managing by Means" guided by precepts that guide all living systems: self-organization, interdependence and diversity.

24. Birth of the Chaordic Age – Dee Hock – Berrett-Koehler

   Dee Hock was the founder and CEO of VISA. He makes a compelling case that all organizations are fundamentally based on flawed seventeenth century concepts that are no longer relevant to the vast systemic social and environmental problems we experience daily. He delineates a path to organizations he believes can harmoniously blend chaos and order, competition and cooperation.

Data Analysis and Presentation Tools

INTRODUCTION

There are a large number of tools for analysing benchmarking data and presenting the results of these analyses. Broadly speaking benchmarking uses these tools:

1. As an investigative tool, used for example to identify relationships between variables and investigate differences in performance levels. Statistical analyses and charts are the key tools.
2. To summarize data. Simple statistics such as the mean, quartiles and regression line are common tools, and these are often superimposed on charts.
3. To communicate conclusions to others. Charts are the main tool.

In this appendix we describe many of the more commonly used charts and analyses that are likely to be used by the non-statistician. In particular we explain the common pitfalls of regression and correlation analysis and how to interpret shapes of histograms. Many other statistical analyses have been omitted from this text because they belong in the realm of a trained statistician.

Before using a particular tool we should always be aware of what we are trying to achieve. This is particularly important for reporting purposes where the type of chart should be selected and then designed to highlight the message that the analyst wants to impart to the reader. Clear labelling and uncluttered charts, using simple standardized colour schemes (or black and white) are best. A good test of a well-presented chart is one that is self-explanatory with an obvious message.

In this appendix we consider the following tools and charts:

1. Tables of data
2. Histograms
3. Run charts
4. Scatter diagrams
   - correlation analysis
   - regression analysis
   - bubble charts
5. Control charts
6. Bar charts
7. Ranked bar charts and Pareto charts
8. Radar charts
9. Pie charts
10. Force Field Analysis

For each tool we describe what it is, its uses, limitations, and interpretation. Most of the charts can be readily drawn using standard spreadsheet or graphics packages.

A1.1 TABLES OF DATA

Uses: Tables of data are of most use for looking up and comparing key numbers or attributes between participants. Figure A1.1 is an example of the type of information that may be presented in a table for a railway benchmarking study.

Participant	A	B	C.....
Organizational relationship between railway operator and infrastructure provider	Same company separate accounts		
Cooperation between operator and infrastructure provider	The operator has a member on the board of the infrastructure provider		
Place of traffic control	Independent organization		
Policy objectives			
Incentives			
Source of funding for operations / investment			
Operator's key responsibilities			
Number of employees			
Number of km track			
Number of stations			
etc.			

FIGURE A1.1   Example of a table for an international railway benchmarking study.

Figure A1.2 shows an extract from a table of variances of plan versus actual values for two participants of a benchmarking study, Alpha and Beta. Organization Alpha reports data for 50 projects, while organization Beta 60. In addition to the plan and actual figures, the percentage variance has been included and was calculated using the formula

$$\text{Variance \%} = \frac{100(\text{plan} - \text{actual})}{\text{plan}}$$

Project Plan versus Actual Cost ($k)						
	Organization Alpha			Organization Beta		
Project no.	Plan	Actual	Variance %	Plan	Actual	Variance %
1	95	93	-2.2	172	172	-0.5
2	113	108	-4.6	154	140	-9.6
3	87	86	-1.4	161	158	-1.5
etc.						
.						
.						
48	120	129	6.4	168	191	11.9
49	107	108	0.5	165	161	-2.9
50	121	119	-2.0	167	159	-5.1
51				168	180	7.0
.				.	.	.
.				.	.	.
60				175	178	1.4
Mean			-0.62			+0.01

FIGURE A1.2   Project plan versus actual cost for organizations Alpha and Beta.

Using the table alone it is very difficult to interpret the data. With a little time we could find the maximum and minimum plan, actual and percentage difference. However, identifying outliers and trends, for example, is difficult. Tables of data like Figure A1.2 should only be used if the reader needs to know actual values and in these cases the tables would normally be included as an appendix. Tables can be useful for analysts who are spending significant amounts of time working with and referring to the data.

One statistic that is often included in tables such as Figure A1.2 is the mean, which for the Alpha percentage variance is −0.62%. This tells us that on average the Actual figure is 0.62% above the Plan. The mean variance for Beta is different. However, without knowing the variability of the data and carrying out the appropriate statistical tests, it is not possible to determine whether there is a *statistically significant* difference between the two sets of variances. The simple fact that one mean value is higher than another is does *not* imply that one is a significantly better performer than the other.

A1.2 HISTOGRAMS

A histogram is a graphic summary of the distribution of a set of data. In benchmarking, histograms are mostly used for analysis rather than presentation purposes.

The rationale behind the histogram is:

- Almost any set of data will show variation.
- This variation will exhibit some pattern.
- The pattern can give us clues as to what is happening in the process from which the data were obtained.
- It is difficult to see this pattern in a table of data.
- The histogram presents the variation in a way that helps us understand what may be happening in the process.

Figure A1.3 is a histogram of the percentage variance for the 50 values from Figure A1.2.

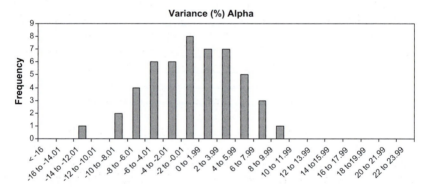

FIGURE A1.3   Percentage variance for Alpha.

The variances have been grouped into bands as shown on the horizontal axis, and the number of variances in that band, the frequency, plotted on the vertical axis. The group corresponding to the highest frequency (-2 to -0.01) is called the mode. Where the bars are more or less evenly distributed around the mode, as in Figure 2.1, the mode will be very close to the mean which, as discussed in the previous section, is -0.62%. The bands were chosen in such a way that a band boundary coincides with zero variance. This makes it easy to see how many projects were over budget (all the negative bands) and how many were under budget (all the positive bands). Since we hope or expect that the actual values will, on average, equal the plan, we could have chosen the bands so that zero variance was at the centre of a band, e.g. −1 to +1. Our ultimate choice of bands would depend on the point we want to make to the reader or the analysis we want to carry out.

We can also easily determine the approximate maximum and minimum percentage differences by looking at the bars furthest to the right and left of the histogram. In Figure A1.3 the minimum variance is about −13 and the maximum +9. The shape of the chart may also give us information about the process behind the data as explained below.

The histogram is also helpful in identifying outliers. In Figure A1.3, the only potential outlier is the lowest bar at about −13%. However, there was only one value at this extreme and it is not very far from the bulk of the data. It does raise the question: how do we decide if we have an outlier? The question is answered either by the control chart (discussed below) or by further statistical analysis. Should we decide to investigate the data value further, the first task is to check whether the data are correct.

However, as with the table of data, with a histogram we are unable to determine whether there are trends or other non-random patterns in the data.

A1.2.1 Interpreting Histogram Shapes

The shape of the histogram often contains important information about the running of the underlying process. Figure A1.4 gives examples of typical histogram shapes. Figure A1.4a is a histogram with a 'bell' shaped or, as it is usually called a 'normal' distribution. This is a very common shape of a well behaved 'in control' process: the data are evenly scattered around the average, with fewer points the further away from the average we go. While other shapes (especially skewed) are common with some types of data, variations from the normal distribution should usually be investigated.

Figure A1.4b has a double peak. A very common explanation for this distribution is that two processes have been combined upstream of the measurement point. For example, results of two shifts, machines, or suppliers may have been mixed before charting. If there are many peaks, or if the distribution is flat (Figure A1.4c and A1.4d), it may indicate that the outputs of many processes have been combined upstream of the measurement point.

The 'toothed' distribution, Figure A1.4e, may be due to measurement error or an unfortunate choice of grouping when constructing the histogram, or a rounding error. Sometimes it is also a due to a special case of the flat distribution. Data collection and grouping should be reviewed before investigating process related causes.

Figure A1.4f is an example of a separated distribution. This is frequently an extreme example of the double-peaked distribution where the distribution averages are very different. If one of the two distributions has fewer values, it is likely to come from a little used process, for example, overtime, a spare machine, a little used supplier.

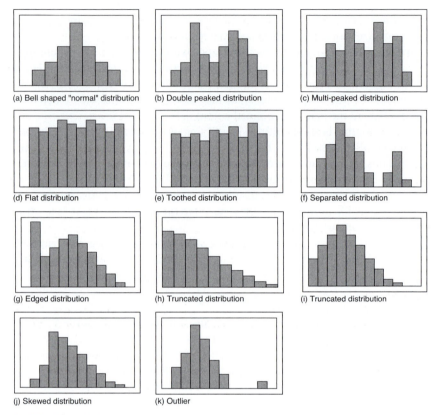

FIGURE A1.4   Typical histogram shapes.

The edged distribution, Figure A1.4g, is frequently seen where there is a specification limit or a target. People do not want to report values outside a specification limit and so may falsely report values just above, resulting in an unexpectedly high number of observations at this value. If the specification limit is at higher levels, then the 'edge' will be at the right of the chart. An extreme example of the edged distribution could occur, for example, when reporting equipment downtime. Some people not wanting to admit to having downtime report zero, while others report honestly. In this case there would be a single peak at zero, with a skewed distribution starting somewhere above zero. This distribution may also occur when, if there is downtime, the downtime is always significant, perhaps because of the time required to put the item back into service. This would occur for example, if a vehicle develops a fault, there is a significant minimum time to take it to a garage and for the mechanics to investigate, repair and return it to service. Even if the repair only takes five minutes, the vehicle may be out of service for hours.

Figures A1.4h and i are examples of truncated distributions where the data to the left have been omitted. The explanations could be similar to those for the edged distribution, Figure A1.4g, but in these cases the data have been discarded rather than added in at the edge of the distribution.

Figure A1.4j is an example of a skewed distribution, where the mode (most frequently occurring value) is not central, and there is one long 'tail' and one short 'tail'. This is a very common distribution in counts data where the average is low. For example, the number of accidents, failures or rejects typically follow this type of distribution. This is because although there is a practical minimum (of zero) there is no maximum. The skew can be in either direction, but that shown is by far the most common. Skewed data with the 'hump' on the right should be investigated.

Figure A1.4k shows how an outlier would appear on a histogram. An outlying value, shown on the far right, can occur with any of the above shapes and has the additional feature of a single, or, if there are a lot of data a small number, of data separated from the other data. The explanation could be as for that of a separated distribution where the little used process appears only once, but more likely the value is either a recording or measurement error, or more seriously a process aberration which needs to be investigated. This issue of process aberration, known as an out-of-control condition, is discussed below under control charts.

A1.2.1.1 *Grouped histogram*

Returning to the variance data, we could also draw a histogram for Beta's data and it would then be possible to compare the two charts. Alternatively we could draw the histograms on the same chart as in Figure A1.5a.

The highest frequency for Beta indicates that the mode is about +5%, i.e. the Actual value is about 5% less than the plan.

It is easy to contrast Alpha and Beta on the chart. As noted above, we can see that Beta has a 5% positive bias while Alpha has a small negative bias. We can also see that Beta has a much larger variability than Alpha. Beta's variances range from about −15 to +19, while Alpha range from −13 to +9. We can also see that while Beta has larger variability it also has a higher peak with a value of 10 compared to Alpha's 9. This has occurred because while Alpha has 50 projects, Beta has 60, and the difference in the number of projects makes detailed comparison difficult.

To overcome this difficulty we can change the vertical axis from frequency to percentage of projects, thus enabling a direct comparison regardless of the number of values, see Figure A1.5b.

The disadvantage of using percentage rather than frequency as the vertical axis is that we no longer have a simple reference to the number of values being charted. This is important as we would like a minimum frequency of 5 in all the cells, excepting the extreme tails, to ensure a reasonable representation of the underlying distribution.

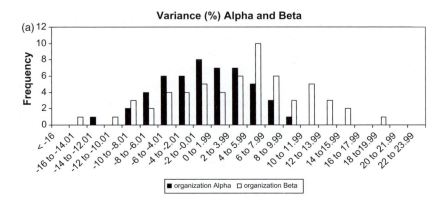

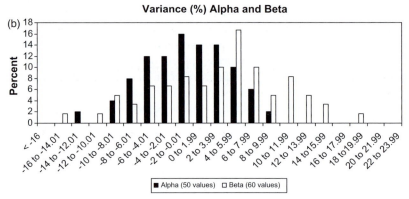

FIGURE A1.5 Histogram of variance for Alpha and Beta.

We can also see from Figure A1.5a and b that while it is possible to compare two or three distributions in this way, comparisons become more difficult the more distributions are included.

It is possible, though not often done, to group and stack histograms in a similar manner to the way in which we stack and group bar charts, which are discussed below.

A1.3 RUN CHARTS

Helpful though the histogram is, it does not tell us anything about the behaviour of the process over time. For example, it could be that all the low values were recorded first and the high values last, or that after every low value there was a high value. The simplest tool for revealing trends or patterns in the data is the run chart.

A run chart is a plot of a process characteristic, usually over time. Like the histogram, it is able to help us identify the maximum and minimum values and the mean. Unlike the histogram, it can also help us to identify non-random patterns including trends, seasonality and process jumps.

Figure A1.6a is a run chart of the plan data for Alpha. The horizontal axis shows the project number (in start date order) and the vertical the corresponding planned value. The chart includes a horizontal line representing the average. Just as with the histogram it is possible to spot possible outliers, though with the run chart we would be able to see if there was any obvious project relationship, e.g. if outliers occurred in consecutive projects.

A closer look at the chart shows that the vast majority of the first half of the data is below the average, while the latter half of the data are above the average (the plan could be measured, for example, in terms of costs, duration, etc.). This suggests that there may be a trend in the data. To investigate this further, we can re-draw the chart and include a trend (regression) line to highlight the relationship, as shown in Figure A1.6b. We could also calculate the correlation coefficient and carry out a regression analysis, both of which are briefly discussed below, but first we need to discuss scatter diagrams.

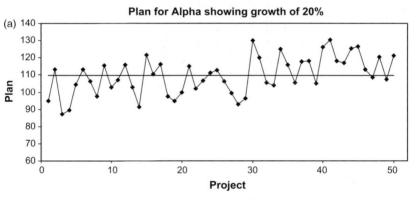

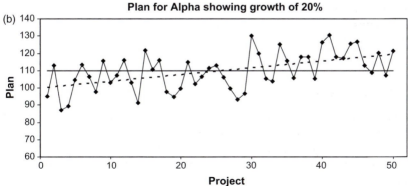

FIGURE A1.6   Run chart of Alpha's planned values. (a) Without trend line (b) with trend line showing a growth of 20%.

A1.4 SCATTER DIAGRAMS

A scatter diagram is a graphic representation of the relationship between two variables.

The rationale behind the scatter diagram is:

- Almost any set of data will show variation.
- This variation may be related to another variable.
- Relationships are easier to see in a scatter diagram than in a table.

There are four common patterns that are likely to occur as shown in Figure A1.7.

1. Figure A1.7a shows a positive relationship where both variables increase or decrease together.

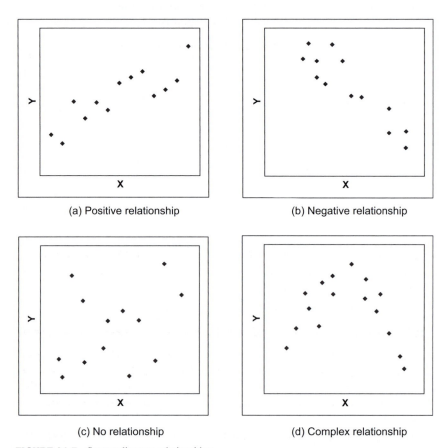

(a) Positive relationship

(b) Negative relationship

(c) No relationship

(d) Complex relationship

FIGURE A1.7   Scatter diagram relationships.

2. Figure A1.7b is a typical negative relationship where as X, charted on the horizontal axis, increases Y, charted on the vertical axis, decreases.
3. In Figure A1.7c there is no relationship between the two variables.
4. Figure A1.7d is one example of many possible complex relationships.

The run chart of plan data for Alpha showed that the planned values are gradually increasing over time. Where we are comparing planned with actual figures, it is possible that the variance, i.e. the difference between planned and actual figures will increase as the planned figures increases. To investigate this theory we can draw a scatter diagram, Figure A1.8, of the percentage variance against the planned figure. In this situation we are interested in the size of the variance, not whether the variation is positive or negative; therefore the value of the variance ignoring the sign (i.e. the absolute vale) has been plotted.

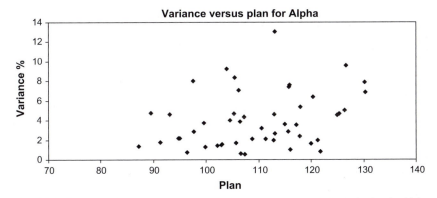

FIGURE A1.8   Scatter diagram of the relationship between variance and planned value for Alpha.

The diagram suggests that there might be a relationship between variance and planned value for Alpha. To investigate this we can calculate the correlation coefficient and draw a regression line.

A1.4.1 The correlation coefficient (r), r^2 and regression analysis

There are various methods of determining the strength of relationship between two variables. The most common is the correlation coefficient. The correlation coefficient, r, is the square root of the often quoted r^2. r measures the strength of the *linear* relationship between two variables, i.e. strength of a straight line relationship.

Figure A1.9 gives examples of various values of r and r^2. The charts plot two variables X and Y.

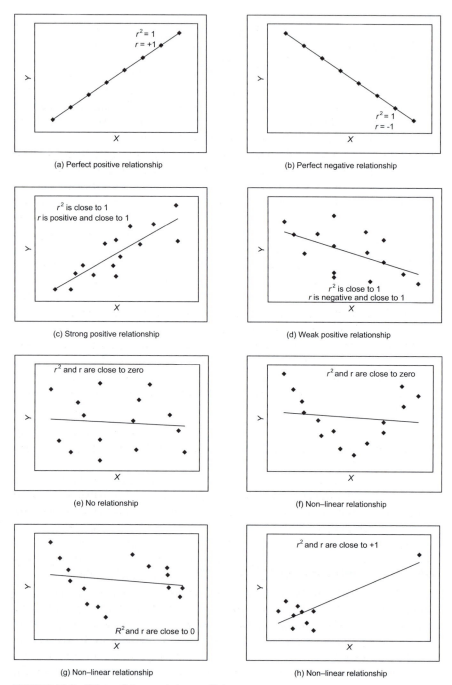

FIGURE A1.9 Values of the correlation coefficient, *r*.

- Figure A1.9a shows a perfect positive relationship. X and Y increase or decrease together and for any value of X or Y it is possible to predict exactly the other value. In this case $r = +1$ and r^2 will also equal $+1$.
- Figure A1.9b shows a perfect negative relationship. As X increases, Y decreases. As in Figure A1.9a for any value of X or Y it is possible to predict exactly the other value. In this case $r = -1$ and $r^2 = 1$. Comparing the r and r^2 in Figures A1.9a and b we can deduce that r indicates whether there is a positive or negative relationship, whereas r^2 does not.
- When charting relationships between two variables, values of r equal to $+1$ or -1 are extremely rare. Where they do exist there is usually some logical binding obvious connection and drawing a scatter diagram would be unnecessary. The chart in Figure A1.9c and Figure A1.9d are much more common results when comparing variables that are strongly related. In Figure A1.9c both r and r^2 are positive and $<$ (less than) 1. In Figure A1.9d r is negative and lies between zero and -1 ($-1 < r < 0$), and r^2 is <1.
- Figure A1.9e shows a typical chart where there is no relationship between the two variables. Both r and r^2 are close to zero.
- In Figure A1.9f and Figure A1.9g there are strong relationships between the variables, but they are not linear and since r and r^2 measure the strength of a linear relationship they are both close to zero.
- Figure A1.9h shows how situations can arise where high values of r and r^2 could lead to a false conclusion that there is a relationship between two variables when there is not. In this type of situation where there is one outlying value, it almost does not matter where that point is, it will always create a high value for r and r^2.

Figure A1.9f–h highlight the risk that quoting r or r^2 figures without charting the data may lead to incorrect and misleading conclusions. Scatter diagrams should always be drawn if r or r^2 figures are quoted.

Figure A1.9c–e raise the issue of how to decide whether a value for r, or equivalently r^2, is statistically significant, or whether it is likely that the value of r could have arisen by chance. The value of r at which we can conclude that there is a statistically significant relationship depends on the number of points, n, plotted. Fortunately there are tables of r that give the value of r required for a specific sample size, n, to conclude with certain confidence levels (typically 90%, 95% and 98%) that there is a relationship between the variables.

r^2 is also a useful statistic. Most data sets exhibit variation. A key purpose for drawing scatter diagrams is to identify how much of the variation can be explained by another variable. For example, in Figure A1.9a, all the variation in Y can be accounted for by X, whereas in Figure A1.9e, none of the variation in Y can be explained by X. The r^2 value tells us what proportion of the variability is accounted for. For example in Figure A1.9a,b all the variation is accounted for and so $r^2 = 1$, and in Figure A1.9e $r^2 = 0$, i.e. none of the variability is explained.

It would be possible to develop tables for r^2 in the same way that they have been developed for r, but this is unnecessary.

It is usual and helpful to display the regression line on scatter diagrams where there is a significant relationship between the variables. The regression line is fitted to maximize the value of r and the equation for the regression line, $Y = aX + b$, where a and b are constants, can be used to predict a value of Y for any desired value of X. Further information on regression analysis, including the associated analysis of variance can be found in basic statistical texts.

It is important to realize that just because two variables are related this does not imply that one 'causes' the other. For example, there may be a strong relationship showing an increase in journeys taken on public transport and increases in the price charged for the journey, but this does not mean that increasing the cost of a journey causes an increase in the number of journeys, or vice versa. However, sometimes both may be caused by a third factor.

In this discussion we have considered only straight-line relationships. The ideas and methods can be extended to non-linear and more complex situations, and the reader is referred to a suitable statistical text.

A1.4.2 Further analysis of plan versus actual

Returning to the variance analysis, Figure A1.10 is the same as Figure A1.8 with the regression line and r-value included.

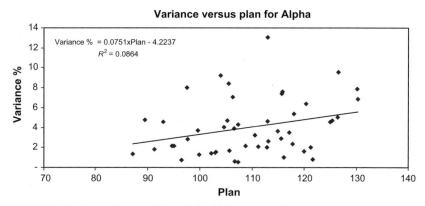

FIGURE A1.10   Scatter diagram of the relationship between variance and planned value for Alpha with regression line.

For this set of data $r^2 = 0.0864$, i.e. 8.6% of the variability in the variance % is explained by the variance % increasing with the Plan. This gives $r = 0.294$. Using tables of r with $n = 50$ data values, we only need a value of $r = 0.273$ to conclude that there is a greater than 95% chance that there is a relationship between the percentage variance and the planned value. Another way of explaining this is to say that there is less than a 5% chance of getting such a high value of r if the variables were not related.

The average Plan value for Beta is 167. We could estimate the average variance % that Alpha would have for an average plan of 167 by using the formula of the line we plotted:

$$\text{Variance } \% = 0.0751 \times 167 - 4.2237 = 8.38.$$

The variance % for Beta is 6.6%, and so we would conclude (subject to appropriate statistical tests) that Beta has a lower % variance than Alpha, having taken into account the size of the project.

We have made an important assumption in estimating the variance % for Alpha for a plan of 167. The Alpha data only extends to planned values of around 130. Therefore, the correlation analysis is strictly only valid for planned values up to 130. We have assumed that the increase of % variance would continue above 130 in the same manner as below 130.

We can also display more than one data set on a scatter diagram. Figure A1.11 shows the planned figure versus the % variance for both Alpha and Beta.

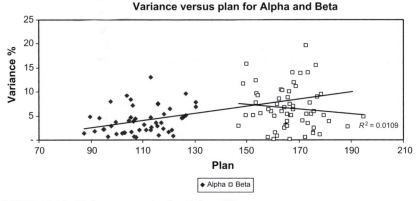

FIGURE A1.11   Variance versus plan for Alpha and Beta.

Figure A1.11 shows Alpha's data on the left with the regression line extended though to the Beta data. The regression line for Beta with the r^2 value is also shown on the right of the chart. $r^2 = 0.0109$, and $r = -0.1044$ which is not statistically significant and so we conclude that the variance % does not change with planned value for Beta. The fact that the extended regression line for Alpha lies above most of Beta's data suggests that if the growth in variance % were to continue for Alpha up to planned values of around 170, the % variance would be higher than for Beta, as identified above.

A1.4.3 Summary of variance analysis

This analysis of the variance data has demonstrated:

- The difficulty of trying to interpret tables of data.
- The use of histograms for analysing variability of data, and their limitation in their inability to identify, for example, trends or cycles.
- The use of run charts to identify trends, averages, outliers and cycles in data
- The use of scatter diagrams to analyse the relationship between variables.
- The use of r, r^2, and regression in analysing the relationship between variables, and the importance of drawing the scatter diagram and not just quoting the statistic.

A1.4.4 Bubble diagrams

Bubble diagrams can be used as an extension of several different types of chart including scatter diagrams, spider charts, and bar charts. The purpose of the bubble is to allow an extra variable to be displayed. Figure A1.12 is an example of a bubble diagram applied to a scatter diagram.

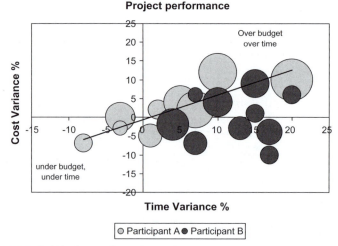

FIGURE A1.12   Bubble diagram for project performance.

The diagram shows project cost and time variance against plan for two participants of a benchmarking study. The vertical axis shows percentage cost variance calculated as

$$\text{Cost variance } \% = 100(\text{actual cost } - \text{budget})/\text{budget}.$$

The horizontal axis shows the percentage time variance calculated as:

Time variance $\% = 100$(actual time $-$ planned time)/planned time.

The size of the circles, or bubbles, reflects the budget of the project. Larger projects are represented by larger bubbles and smaller projects by smaller bubbles. The bottom left quadrant reflects under-budget and under-time projects, while the top right quadrant reflects over-budget over-time projects.

Participant A is shown as grey bubbles. The solid line passing through the points is the regression line and indicates that the cost and time variances increase together. The size of the bubbles suggests that small projects tend to be under-budget early delivery, while larger projects tend to be over-budget and over time. In contrast, Participant B, who reports more projects than participant A, has no under-time projects, no obvious relationship between time and cost variance and the project size is not related to either time or cost variance.

A1.5 CONTROL CHARTS

The control chart is a natural extension of the run chart as it is simply a run chart with control limits. The control limits are statistically determined limits showing the boundaries within which the data will lie. The limits are called the upper (UAL) and lower action limits (LAL) or sometimes 'control' rather than 'action' limits (there is a minor difference between them which is beyond the scope of this text). Figure A1.13 is the control chart of Beta's planned values. As with the run chart, we can see each individual point, joined up, with a line representing the average.

We can also see that there is no trend, outlier or other unusual aspect to the chart. Well-behaved processes such as these are said to be 'in control' or more correctly, if less frequently, 'in a state of statistical control'.

The top chart in Figure A1.14 shows a typical process in a state of control. All the values are:

- Randomly scattered, i.e.
  - there are no trends
  - no seasonality or other cycling of data
  - no obvious patterns
- Distributed around the mean, i.e.
  - approximately half the values are above the mean, and half below.
  - with more values closer to the mean than further away
- Within the (control/action) limits,

In contrast the remaining charts in Figure A1.14 exhibit out-of-control conditions.

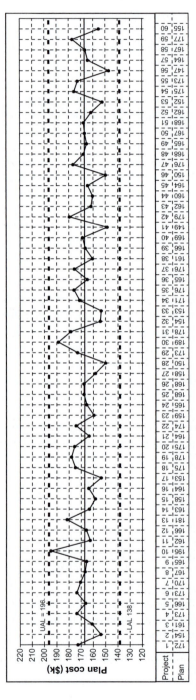

FIGURE A1.13 Control chart of Beta's planned costs.

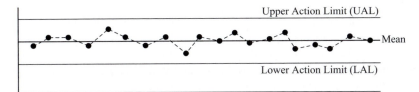

In a process that is in control, points are:
Randomly scattered, clustered around the mean, within limits, with no obvious

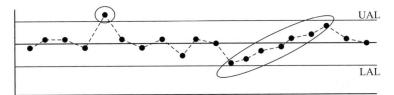

In a process that is not in control, points will:
be either outside the control limits, or show a trend or…

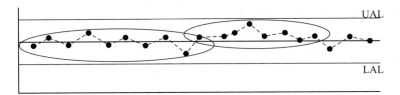

… cycle, or have a run above/below the average or…

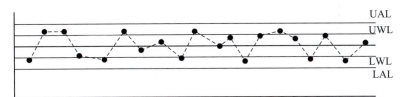

… not be distributed around the mean

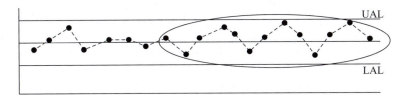

… or show some other pattern

FIGURE A1.14   Control chart patterns.

- The limits are determined so that only one in every 1000 values should lie outside the limits by chance. Therefore, if a point falls outside the limits we suspect that there has been a process upset.
- Runs of seven or more consecutive points increasing (or decreasing) in values are unlikely to occur and are therefore likely to signal a process upset.
- A run of 7 or more alternating high/low values are unlikely to occur naturally. This pattern is typical where the value being measured has two separate sources such as work shifts, and values are being sampled alternately from both processes.
- A run of 7 or more consecutive values above (or below) the average. This often indicates that the process average has changed.
- Few points near the mean, with points concentrated further away. In Figure A1.14, one chart has two extra lines drawn in addition to the action limits. These are called the upper/lower warning limits (UWL, LWL) and, when used, are set so that 95% of all values lie between them. In situations where the data do not lie close to the mean, it may be because the data from two processes are being mixed. If the data were drawn as a histogram we would see a double peak.
- More difficult to identify, and fortunately less common, other more or less regular patterns may occur indicating that the process is not in a state of control.

Being in control does not imply that the process is in some sense 'good'. Project budgets may always be 50% overspent – i.e. in (statistical) control, but consistently bad. A process being in control is important because if a process is in a state of control:

✓ We can predict future performance.
✓ We can begin process improvement activities.
✓ From the point of view of benchmarking, we can make comparisons between participants. If a process is not in control it is meaningless to compare the results with another process.

If a process is not in control we can consider it chaotic, and specifically, we are unable to predict future performance. Therefore it is not appropriate to compare the results of the process with anything else. The tacet assumption in benchmarking is that the data being compared are in a state of control. Where there are planned out of control conditions, such as manning level changes, expansions, or a re-organization, participants will normally mention that these are occurring. The difficulty arises if unknown changes are occurring. If the organization is not using control charts for monitoring, such changes may well go unnoticed.

Figure A1.15 is a control chart of the percentage variance for Beta, this time calculated as:

$$\text{Variance } \% = 100(\text{actual} - \text{plan})/\text{plan}.$$

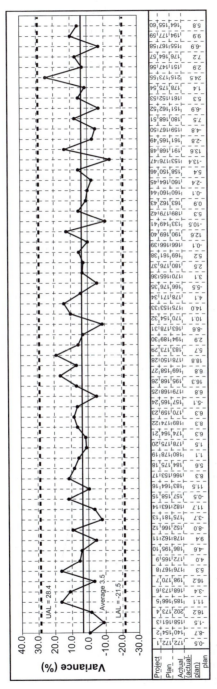

FIGURE A1.15   Control chart of Beta's variances (percentage).

Like Figure A1.13, it is in a state of control and we are able to say that if the process does not change:

- On average actual values will continue to be 3.5% above the plan.
- Any one actual will be between 28.4% above and -21.5% below the plan.

A1.6 BAR CHARTS

A bar chart displays the relationship between two variables one of which is numeric, the other of which is a category.

Figure A1.16a is a bar chart depicting the reasons for emergency hospital admissions. The reasons for admission are given along the bottom of the chart, and the number of admissions for each reason is given on the vertical axis. The chart shows that the most common reason for admission, for the period over which the data were collected, was ischaemic heart disease.

It is also possible to show the numbers of emergency admissions at each of several hospitals using a grouped bar chart, as in Figure A1.16b.

Each group of bars represents the data for one particular reason from each of the five hospitals. This facilitates the comparison of admission reasons between different hospitals. For example:

- General symptoms is the least common reason for emergency admissions at all hospitals, and number wise there is not a great difference between the hospitals.
- Ischaemic heart disease is the most common reason overall, but there is a large variability between hospitals.
- Hospital C has the highest number of admissions overall, but has fewest admissions due to 'other acute lower respiratory infection'.

With further inspection, other comments could be added.

Another development of the basic bar chart is the stacked bar chart. Using the same data as in Figure A1.16b, Figure A1.16c compares the total admissions for each hospital.

Each bar in the chart represents a hospital while the stacks of each bar indicate different reasons for admission. This style of chart allows easy comparison of total admissions, and it is easy to see that hospital C has the highest admissions of around 170, with the other four hospitals having between about 85 and 105. With so many items in the stack, analysis of the stacks is difficult, particularly when produced in black and white. If the number of categories is kept to less than about four, it is possible to compare each element of the stack. However, as an example, a cursory glance will show that 'Pregnancy with abortive outcome' is higher for hospital A than other hospitals, including hospital C which has a higher overall admission rate.

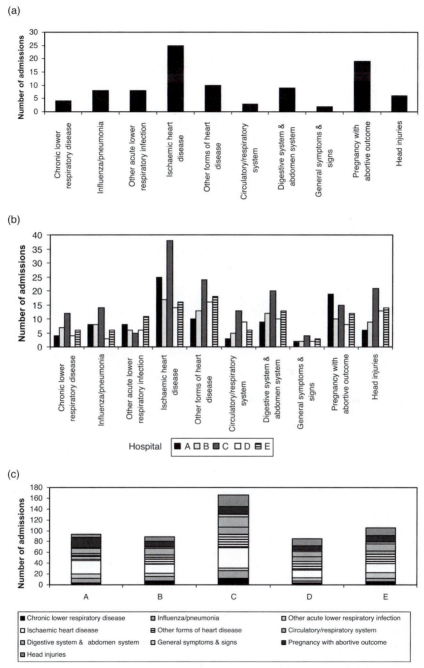

FIGURE A1.16   (a) Bar chart showing the reason for emergency admission. (b) Grouped bar chart showing the reason for emergency admission for five hospitals. (c) Stacked bar chart of the five hospitals showing the reason for emergency admission. (d) Stacked bar chart showing the reason for emergency admission for five hospitals. (e) Stacked bar chart of the five hospitals showing by percentage the reason for emergency admission.

(d)

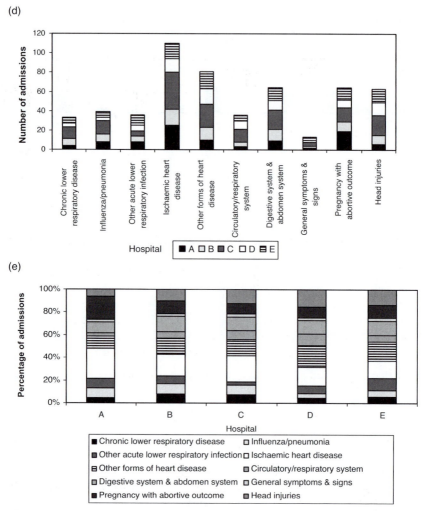

(e)

FIGURE A1.16 (*Continued*).

It is also be possible to plot the bars by 'reason for admission' as shown in Figure A1.16d, allowing comparison of the major causes of admission over the five hospitals.

Returning to the chart in Figure A1.16c, if we want to compare the percentage of different reasons for admissions rather than the number, we could re-scale the element of each stack not as the number, but as the percentage of admissions for each reason, as in Figure A1.16e.

Using Figure A1.16e it is much easier to compare the reasons for admissions in each hospital. For example:

- 'Pregnancy with abortive outcome' forms a higher percentage of admissions for hospital A than for other hospitals.
- 'Head injuries' is lower for hospital A than for other hospitals.

Again we note that with so many items in the stack it becomes difficult to differentiate between them.

To summarize, with this set of data there are eight charts that we could draw:

1. Number of admissions by reason for admission (one chart for each hospital) (Figure A1.16a).
2. Number of admissions by hospital (one chart for each reason for admission) (not drawn).
3. Number of admissions by hospital grouped by reason for admission (Figure A1.16b).
4. Number of admissions by reason for admission grouped by hospital (not drawn).
5. Number of admissions stacked by reason by hospital (Figure A1.16c).
6. Number of admissions stacked by hospital by reason (Figure A1.16d).
7. As 5 but reported by percentage (proportion) (Figure A1.16e).
8. As 6 but reported by percentage (proportion) (not drawn).

With so many charts even for a simple situation, it is easy to get swamped with charts and lost in the analysis.

Bar charts have a variety of purposes including:

- ✓ To provide different views of the data and enable important features to be identified.
- ✓ To aid in communicating to others the conclusions reached during the analysis. It is important to select the format of the bar chart that focuses on the conclusion you want to draw. (i.e. grouped or stacked, reported by number or proportion). The question to ask is: 'what is the best way to illustrate this conclusion with a chart?'
- ✓ To show the 'bridge' between a participant's performance and a target performance (⇨ Chapter 14).

A1.7 RANKED BAR CHARTS AND PARETO CHARTS

While a bar chart is better at conveying the information in data than a table, it can frequently be improved further by simply ordering the bars in decreasing (or increasing) frequency. This is illustrated by Figure A1.17a, which is a ranked bar chart of Figure A1.16a. We see immediately the major and minor reasons for admission. In addition, it is easy to see that the top two reasons are much higher than the remaining eight.

When used for process improvement purposes (e.g. if our aim it to focus attention on improving the hospital's care of emergency admissions for specific admission groups) we would focus on the top two causes. These are called the 'vital few', as they are likely to give us a higher return on investment than focusing on the other 'useful many' causes.

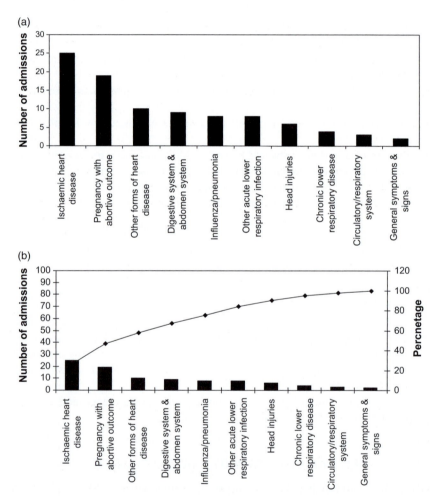

FIGURE A1.17   (a) Ranked bar chart of the reason for emergency admission. (b) Pareto chart of the reason for emergency admission.

The ranked bar chart is the first step towards creating a Pareto chart, which is shown in Figure A1.7b. The Pareto chart is a chart specifically designed to separate the 'vital few' contributors from the 'useful many'. In addition to the ranked bar chart, we calculate and plot the cumulative percentage of the categories. For example, in Figure A1.17b there are a total of 94 emergency admissions. The most frequent is due to ischaemic heart disease with has 25 admissions. This equates to $100(25/94) = 26.7\%$. The next most frequent is pregnancy with abortive outcome with 19 admissions. The top two reasons account for $25 + 19 = 44$ admissions, which equates to $100(44/94) = 46.8\%$. After all the causes are added, the total of 100% is reached, as indicated on the right hand side axis. Often, though not so clearly in this case, the line representing the cumulative totals is very steep to begin with and then flattens. It is the items to the left of where the line flattens that are deemed to be the vital few.

The concept behind the Pareto chart is important. Without the concept we may be tempted to try to reduce all the causes of breakdown, accidents, failures, rejects etc. What the Pareto chart and the principle behind it encourage us to do is to recognize that there is unlikely to be one single problem or one single solution. It encourages us to break down the problem and look for and work on the major causes of the problem. It is far more likely that we will be able to solve these vital few causes one at a time rather than trying to find one solution to solve all problems.

From the viewpoint of explaining the relative causes of an effect, if here is no clear breakpoint delineating the vital few it may suggest that the perceived causes are not the root causes of the effect. For example, if it were supposed that rejects at good inward-inspection were due to specific suppliers, a Pareto chart showing the percentage of each supplier's good that are rejected could be drawn. If the chart were almost flat, it would suggest that the fault did not lie with specific suppliers. If, however, the chart did distinguish the 'vital few' from the 'useful many' then it may indeed suggest that a few suppliers were responsible for a disproportionate number of rejects. Analysis could begin to identify how to improve their deliveries.

A1.8 RADAR (SPIDER) CHARTS

A radar (also known as a spider) diagram can be considered simply as a different way of displaying the same information as a bar chart. Figure A1.18 is a spider diagram of the same data as displayed in Figure A1.16a which is reproduced for ease of comparison.

The axis for the radar chart runs from the centre of the chart, zero, to the circumference at 25. Each point represents the frequency of the reason for admission.

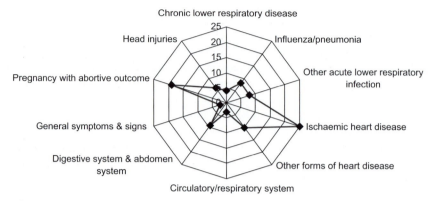

FIGURE A1.18   Spider diagram of admissions.

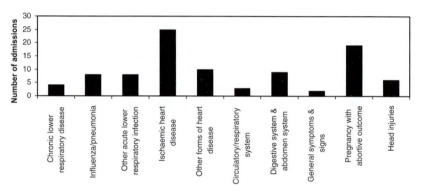

FIGURE A1.16a   Bar chart showing the reason for emergency admission (Reproduced).

In benchmarking radar charts are frequently used to summarize the relative position of a participant within the study. As an example, consider the summary data in Figure A1.19 from a study in which there are 12 participants and the key comparison areas are on time delivery, complaints, production cost per unit, overhead cost per unit, and safety incidents per million man hours. For each of these areas the table provides the following information:

- Alpha's performance level.
- The average value for all participants.
- Alpha's ranked position. '1' indicates the best performer, '2' the second best and so on.
- Percentage transposes the position into a percentage.
- The best performance value.
- The percentage by which Alpha is adrift from the best performance.

	Alpha	Average	Position	Percentage	Best	% of best
Late deliveries %	2.1	3.5	1	8.3	2.1	0
Complaints per thousand customers	13.15	7.2	10	83.3	5	163
Production cost per unit	150	187	3	25.0	136	10
Overhead cost per unit	124	153	2	16.7	118	5
Safety incidents per million man hours	2.8	2.3	7	58.3	1.36	106

FIGURE A1.19   Selection of summary data from a benchmarking study.

In each of these areas the ranked position for participant Alpha is calculated as 1st, 10th, 3rd 2nd and 7th. The corresponding radar chart is given in Figure A1.20.

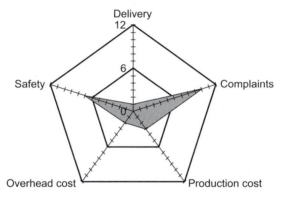

FIGURE A1.20   Radar chart for Alpha's ranked performance.

For example, Alpha achieves 2.1% late deliveries, compared to an average in the study of 3.5% and has the lowest late delivery rate resulting in position 1. Conversely, Alpha receives 13.15 complaints per thousand customers, which is above the average of 7.2 and ranks 10th of the 12 participants.

Since there are 12 participants, the axis has runs from 0 to 12. The best performance in an area has a rank of 1 and therefore a score of zero is unobtainable. The points closer to the centre are the better performances while those further away (complaints and safety) indicate areas of concern. As in Figure A1.20, the area enclosed by the points on the chart can be shaded to facilitate quicker appreciation of the shape. As with all charts, it is also worthwhile experimenting with other presentation possibilities. In particular, participants often like to know where the average or certain quartile performance levels lie. Figure 8.3 shows the median as a thickened line with other gridlines marked as crosses on the axes.

If desired, the ranking could start from zero as the best performance up to 11 for the worst performance so that for a top performance the value would be plotted at the centre of the chart. There are other possibilities for drawing this type of chart, the rank could be calculated as a percentage position rather than as an absolute. For example, position 10 could be plotted as an absolute position of 10 on the scale 0–12 or as $100(10/12) = 83$ on a scale from zero to 100.

The disadvantage of plotting the rank or percentage of rank (and using ranks in general) is that it gives no indication as to how much higher or lower the performance is compared to, for example, the best or average performance. To overcome this problem, we could plot the performance relative to a standard. The standard could be the best or average performance, a target, or some other level.

The order in which the performances are plotted can also be chosen to help guide interpretation. In Figure A1.20 the two cost elements have been drawn together. By grouping metrics like this it is possible to gain a general impression that as all the cost elements are close to the centre costs are low. While with only five elements shown in this chart the gains over grouping may seem small, if for example, there were three or four elements for each of cost, HSE (health, safety and environment), and customer satisfaction, it would be easier to appraise performance in the three areas if the elements were grouped.

An alternative method of grouping is possible where the metrics have two opposite poles. For example, higher maintenance should lead to fewer break-downs. In these cases maintenance levels and breakdowns could be plotted opposite each other on the chart. Figure A1.21 is an example of expenditure on maintenance, safety and personal development (HRD) with the opposite axes displaying the impact in terms of reliability, safety incidents and productivity. The results for two participants, Alpha and Beta are shown on the chart.

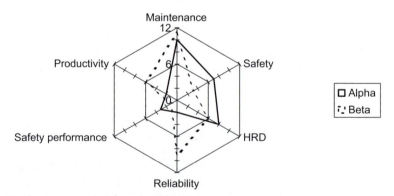

FIGURE A1.21   Radar chart for Alpha and Beta showing their rank out of 12 participants.

The rankings can be organized in different ways, for example, with the better performances being charted towards the centre and worse performances towards the outside. Using this method, high costs should relate to good performance, as shown by Alpha whose high maintenance expenditure ranking is 10 and whose good reliability is ranked 2nd. Similarly, Alpha's other expenditures are high resulting in good performances. As a result, Alpha's chart is 'fat' on the expense side and 'thin' on the performance side of the chart. Beta's results show that while they incur high maintenance expenditure, rank 11, their reliability is only ranked 9th, i.e. it appears that they do not reap the benefits of their high expenditure. Similarly, while their safety expenditure is ranked 3rd, their performance is ranked 1, i.e. they do not spend a lot, but incur few incidents. Both their HRD expenditure and performance is average, ranked 6th.

Just as it is possible to draw stacked bar charts, so it is possible to draw stacked radar charts. Similarly, it is possible to add a further variable by replacing the dots on the chart by bubbles.

A1.9 PIE CHARTS

Pie charts are an alternative to bar charts. Figure A1.22 below shows the same information as the bar chart in Figure A1.16a.

Where the pie is split into many slices, as in this case, it is usually easier to read the equivalent bar chart. There are few practical advantages to the pie chart, the chief one being that they can usually be arranged to take up less space than a bar chart. As with most charts is it possible to highlight items of interest, in this case by using a different colour for the largest slice of the pie.

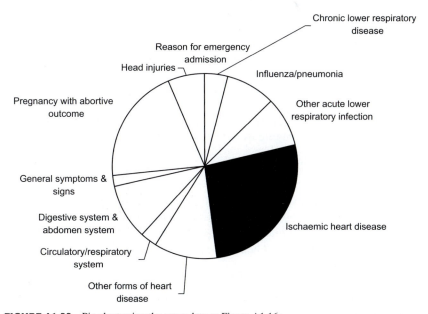

FIGURE A1.22   Pie chart using the same data as Figure A1.16a.

A1.10 FORCE FIELD ANALYSIS

Force Field Analysis is a useful tool for displaying forces for and against a proposal, idea or implementation. The concept behind Force Field Analysis is very simple and is illustrated in Figure A1.23 for getting agreement to begin using benchmarking as an improvement tool within an organization.

Issue: Gaining agreement to trial benchmarking		
Driving Forces	**Restraining Forces**	**Action**
Independent study highlights our poor performance	Internal resistance to the findings of the report	Identify and address specific resistance
Customer Liaison Manager		Find out who the Liaison Manager can influence that is against benchmarking.
Customers encouraging us to benchmark		Can we obtain specific quotes from customers? Ask a close customer to give a presentation to management on benchmarking.
Opportunity to improve productivity		etc.
etc.	Cost of benchmarking activities	
	Previous study seen as disaster	
	Finance manager	
	etc.	

FIGURE A1.23   Force Field Analysis example.

The first step is to draw up a table listing the issue at the top, and in the left hand column all the driving forces (i.e. reasons why) we want our preferred action to be implemented. In this example, some of the driving forces include:

- There has been a recent independent study highlighting our poor performance.
- The customer liaison department manager is keen to try benchmarking.
- Customers are encouraging us to benchmark.

Restraining forces include:

- Some people reject the finding of the report.
- It will cost money to benchmark (training, data collection, etc.).
- A study 3 years ago resulted in disaster when senior managers visited several other organizations but nothing changed internally.
- Finance director is against benchmarking because of the cost.

The next step is to consider how we can either:

- Convert all the restraining forces into driving forces.
- Minimize the impact of restraining forces.

And log any plans in the Action column. For example:

- To address the resistance to the findings of the report the action might be to identify what specific findings are disputed and why, and then decide what action to take next.
- The action regarding the Liaison Manager might be to find out why he supports benchmarking and who he might be able to influence amongst those who do not support benchmarking.
- Try to get quotes from specific customers. If there are customers that are particularly important, or with whom the organization works a lot, it may be appropriate for them to give a presentation to management on the benefits of benchmarking.

Appendix A2

Querying the Quartile

INTRODUCTION

One of the difficulties faced by many benchmarking data analysts is how to determine an appropriate target performance. Should it be based on the best performer? Or a technical limit perhaps? Perhaps we ignore the best performer, using the argument that their data might be incorrect, or that they may be in a special situation not appropriate to others. Should we then select the second best performer as our target?

Some analysts have attempted to overcome these and other difficulties by setting a clear performance target: top quartile. By selecting a target based on a group of the best performers the analyst protects themselves from the accusation that the target performance has been set at an unrealistic level. It can be safely assumed that a top quartile target should be achievable by all participants.

2.1 UNDERSTANDING QUARTILES

Before discussing the goal to be top quartile, we need to explain what quartiles are. The concept is quite simple:

- The best 25% of performers are deemed to be top quartile,
- The second 25% of performers in the 2nd quartile,
- The third 25% of performers is in the 3rd quartile and
- The remaining 25% of performers are in the 4th quartile.

Example 1

Consider the following set of failure rates submitted to a benchmarking study by 20 different companies:

7, 8, 12, 13, 16, 17, 18, 18, 22, 24, 25, 25, 26, 27, 30, 30, 31, 33, 34, 35.

Because there are 20 companies, there will be 5 companies in each quartile (Figure A2.1a). Therefore

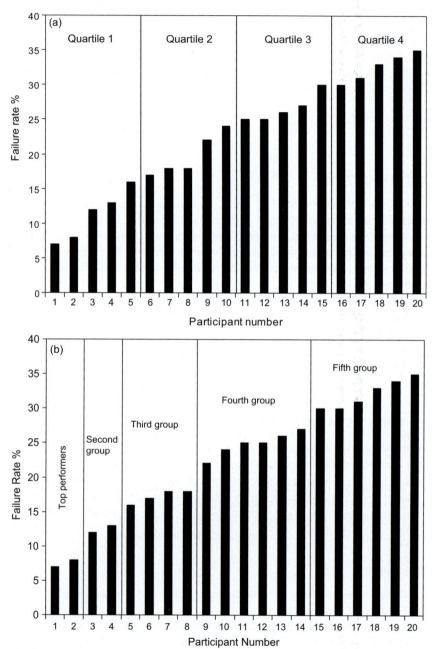

FIGURE A2.1 (a) Bar chart of the performance of 20 participants split by quartile. (b) Bar chart of the performance of 20 participants split by performance levels.

- The companies with failure rates of 7, 8, 12, 13, 16 will be in the top quartile.
- The companies with failure rates of 17, 18, 18, 22, 24 will be in the second quartile.
- The companies with failure rates of 25, 25, 26, 27, 30 will be in the third quartile.
- The companies with failure rates of 30, 31, 33, 34, 35 will be in the fourth quartile.

There are a number of practical difficulties with quartile calculations, for example:

✗ Notice that some of the failure rates are the same (e.g. there are two values of 18 and 25). As Figure A2.1a shows one of the participants with failure rate 30 is in quartile 3 and the other in quartile 4. Is seems inappropriate to place participants with very similar performances in different quartiles.
✗ When the number of participants points is divisible by 4, as in the illustration, we can always put the same number into each quartile. Where the number of data points is not divisible by 4, it is not possible.

There are also some dubious conclusions that are likely to be drawn by simply referring to quartile position, and we may also miss some opportunities for learning. For example:

- The top quartile figures are 7, 8, 12, 13, 16. If all that is required is that we be in the top quartile, achieving a 16 is as good as a 7 as both are in the top quartile. The participant reporting 16 can conclude that as they are top quartile, there is little opportunity for improvement at this time. Not only is it misleading to imply that these two performances are in some sense equally good, but we also miss the opportunity to learn. The top two participants with values 7 and 8 are likely to be significantly better performers (or at least in some way different) than the next participants at 12 and 13, and these in turn may well be significantly better performers than the other quartile 1 participant, reporting 16. If the main purpose for comparison is learning and improvement, then a better approach would be to group companies according to performance levels, for example as shown in Figure A2.1b, and aim to look for reasons for lower/ higher failure rates. In Figure A2.1b, the participants with the lowest two failure rates, 7 and 8 are grouped together. The next band also consists of two companies, 12 and 13. Then follows a group of four participants with values between 16 and 18 and so on.
- There is another question we may need to ask. This example is about failure rates. However, if it related to maintenance effort, it could be that the participants reporting 7 and 8, the lowest amount of maintenance effort,

are not doing enough maintenance and risking the integrity of the equipment: it is often dangerous to look at one figure by itself.

- It is not unusual to see one performance much better than all the others. When this happens, it is more likely that this is due to mis-reporting rather than genuine high level of performance and should always be investigated.

For these reasons, it is more appropriate to attempt to group performances together, as illustrated in Figure A2.1b, rather than arbitrarily putting the line between participants that may really have little or no performance difference. Perhaps better still, is to identify a good performance level, somewhere around 7 or 8 in this example, and consider setting that as the target.

There is another concern with the goal of being top quartile. It is important because many people do not recognize the implication. With ANY set of data 25% of the data will lie in the top quartile, 25% in the second quartile etc. BY DEFINITION. This DOES NOT mean that the top quartile performances are significantly different from the second or other quartile performances. The following simple example explains:

Example 2

Suppose we initiate a dice throwing competition in which we want to throw high values. We have one 6 sided dice and ask each of 40 people to throw the dice 5 times. The highest possible total score is $5 \times 6 = 30$ and the lowest is $5 \times 1 = 5$. The top quartile will be those 10 people with the highest total. We would not conclude that they are 'better' at throwing dice than the remaining 30 people. If the experiment is repeated with the same people, we would expect two or three of the 10 people in the top quartile in the first experiment to be in the top quartile in the second experiment. If the experiment were carried out a third time, it would not be surprising if one person was in the top quartile for all three experiments. It is only if the same person is in the top quartile for perhaps a fifth and sixth time that we would suspect that they are either cheating or have a method of throwing a dice that is likely to result in a high number.

We can generalize this argument and state that if there is no underlying difference in performance and that differences in results are purely random, then the probability of being in the top (or any other) quartile consecutively is given in the following table.

No. of times in top quartile	1	2	3	4	5
Probability	1 in 4	1 in 16	1 in 64	1 in 256	1 in 1024

This is a very obvious example if only because there is no skill involved. However, it does illustrate that being in a certain quartile of itself may have little meaning and we need to look beyond that.

Example 3

In recent years the performance of unit trusts, investment and similar funds has been reported in financial magazines. Performances are frequently presented as tables giving performances of various funds, grouped by fund type. One such set of tables includes the ranked performance over the last three months. For example, fund A may be ranked 15 out of 90 funds in the same group, and would therefore fall into quartile 1. Twelve consecutive quartile results for one such randomly selected fund reporting every 3 months are: 1, 4, 1, 1, 1, 3, 3, 4, 1, 4, 1, 1. The seeming randomness of these results adds weight to the theoretical argument above.

SUMMARY

✓ While setting the target performance to be the top quartile is quick and easy, measuring against and understanding their severe limitations should be understood. It is more important to identify the gaps between the best performer(s) and the other performers and then investigate the reasons for these gaps. Only once we understand the reason for the performance gaps can we evaluate whether a particular performance was 'good' or not. Understanding the reason for the gap is also fundamental in beginning the improvement process.

ⓘ For a more detailed discussion of variation see Mastering Statistical Process Control.

Glossary

Absolute value	A numerical value ignoring the sign. E.g. -3 and $+3$ have the same absolute value, i.e. 3.
Activity benchmarking	See Benchmarking types.
Aim	See Aspirational statements.
Aspirational statements	Summary statements of intent of an organization. Those discussed include:
Aim	The aim of an organization summarizes what an organization is trying to achieve.
Mission	A brief description of a company's fundamental purpose. A mission statement answers the question, "Why do we exist?" The mission statement articulates the company's purpose both for those in the organization and for the public.
Values	Describes what qualities the organization value.
Vision	Describes where an organization wants to be at some time in the future.
Availability	Time when a resource is available for use, also called uptime.
	Availability = total time $-$ downtime.
	See also Downtime.
Balanced scorecard	A group of metrics which together monitor all key aspects of an organization's performance with respect to its mission.
Baldridge	Name of the American who developed the Business Excellence Award which is named after him.
Baldridge award	See Excellence Award.
BEM	Business Excellence Model.
Benchmarker	The person or group carrying out a benchmarking study.

Benchmarking	A method of measuring and improving organizational performance by comparing with other organizations.
Benchmarking agreement	See Participant Agreement.
Benchmarking classes	Describe the participants of a benchmarking study and include:
Competitive	Benchmarking carried out between competitors who by definition will be in the same industry.
Cross industry	Benchmarking carried out between organizations in different industries and therefore between non-competitors.
External	Benchmarking carried out between organizations.
Industry	Benchmarking carried out between organizations in the same industry who may or may not be competitors.
Internal	Benchmarking carried out within an organization.
Non-competitive	Benchmarking carried out within the industry between non-competitors.
Benchmarking club	A benchmarking club is a group of organizations that benchmark together on a regular basis. Whilst the membership of the club is likely to change over time, most participants remain in the club for many years, reflecting the stance that benchmarking should be a regular not a one-off activity.
Benchmarking methods	There are many benchmarking methods. Those discussed in this book are:
Business excellence models	Carried out by an assessor scoring aspects of an organization against set criteria as defined in a Business Excellence Model. Results from participants can be compared to ascertain which has superior performances in each aspect being benchmarked.
Database	Where an organization compares its performance with performances held on a database. Usually the database is held by a consultant.

One-to-one	Benchmarking between two participants. Perhaps the most common type of benchmarking. Often one organization will visit another with the aim of learning how they achieve specific results.
Public domain	Carried out in the public domain, for example by consumer magazines and the media.
Review	Carried out by a person or team visiting each participant in turn. The team may include members from some or all participants and/or consultants.
Survey	Most commonly carried out by an independent organization engaged to survey customer opinions of the client and its competitors.
Trial	Carried out by testing or trialling products and/or services of participants. Participants may or may not know that they are being benchmarked.
Benchmarking partner	See Participant.
Benchmarking types	Loosely defines the scope of a benchmarking study. Those considered include:
Activity	Benchmarking of a specific activities such as inspection or planning rather than, for example, outcomes such as product or service quality and customer satisfaction.
Facility	Benchmarking of complete site or facility such as a factory, airport or refinery.
Functional	Benchmarking of a function such as the purchasing department or warehouse. The function may extend over more than one facility and more than one process.
Generic	Benchmarking between different functions or processes and focuses on achieving specific results, e.g. fast turnaround.
Process	Benchmarking of a process such as purchasing, warehousing or maintenance.
Product/service	Benchmarking of a product or service or combination (i.e. the output of a process), usually as experienced by the customer. Examples include domestic goods, cars, travel (e.g. by boat, plane or train).

Project	Benchmarking of activities carried out for a limited time to create a specific product or service such as a construction projects and software development.
Best in class	Refers to the best performer for the metric being considered.
Best practice forum	An event, typically of one or two days, at which the better performers (as identified in the data analysis) in different benchmarked topics will give presentations explaining how they achieve their performance levels.
Bridge bar chart	A bar chart designed to show the causes of differences between two performance levels, e.g. a participant's performance and a target performance.
Business excellence award	See Excellence Award.
Business excellence model	A set of interrelated criteria that aims to capture all key aspects of any successful organization. The model is designed such that the extent to which an organization adheres to these criteria reflects its success.
Business excellence model benchmarking	See Benchmarking Methods.
Categorization	See Normalization.
Charter	See Project Charter.
Class of participant	See Benchmarking Classes.
Code of conduct	Sets out the general principles of behaviour that guide the decisions of and interactions between those seeking to benchmark in such a manner as to respect the rights and expectations of all participants.
Competitive benchmarking	See Benchmarking Classes.
Confidentiality agreement	An agreement between participants of a benchmarking study that sets out how the data and information shared during the study may be used and disseminated by participants.
Control chart	A run chart with statistically determined control/action limits.
COPQ	Cost of Poor Quality.

Cost of poor quality study	A study in which the costs of poor quality are identified and measured. The results can be used to identify potential benchmarking studies.
Critical path	The group of activities within a project that must be completed sequentially that result in the longest total duration. This is therefore the shortest time in which the project can be completed.
Critical success factors	The relatively few factors in which it is critical that an organization succeed in order to realize its aspirations.
Cross industry benchmarking	See Benchmarking Classes.
CSF	See Critical Success Factors.
Data	Includes numeric data, textual responses to questions, plans, layouts and other information.
Database benchmarking	See Benchmarking Methods.
Data collection pack	Everything issued by the facilitators of the study to the participants in order to elicit the agreed data and information from the participant.
Downstream metric	See Metric.
Downtime	A metric that quantifies unavailability. Downtime can be:
Planned	Time when a resource has been planned to be unavailable for use.
Unplanned	Time when a resource is unavailable for use and the downtime has not been planned.
Effectiveness	See Metric.
Efficiency	See Metric.
EFQM	European Foundation for Quality Management.
Excellence award	Award associated with the corresponding Business Excellence Model. The award is not related to benchmarking.
Excellence model	An Excellence Model is a set of criteria common to all organizations that allow objective assessment of the degree of excellence of that organization. The results can be used to identify comparative strengths and weaknesses of organizations. There are many awards schemes in existence the best known being the Baldridge Award in America, the EFQM scheme in Europe and the Dubai Quality Award in the Middle East.
External benchmarking	See Benchmarking Classes.

External metric	See Metric.
Facilitator	A person or group that facilitates a benchmarking study. One of their tasks is likely to include facilitating meetings. Facilitators may be drawn from the participants or may be external consultants.
Facility benchmarking	See Benchmarking Types.
FAQ	Frequently Asked Questions.
Functional benchmarking	See Benchmarking Types.
Gap bar chart	A bar chart designed to show the differences between a participants performance levels. If often includes a bar representing a target (e.g. best in class) performance.
Generic benchmarking	See Benchmarking Types.
Global metric	See Metrics.
Hard metric	See Metrics.
Help desk	A support facility primarily aimed at answering participants queries on data requirements.
Idle time	Time that a resource is available for use but not in use.
Information exchanges	The activity of participants exchanging information on how they achieve their performance levels. In some situations information flow may only be from one participant to another, but in others information exchange will be reciprocal.
Initiator	The person or organization that initiates a benchmarking study.
Initiating organization	See initiator.
Internal benchmarking	See Benchmarking Classes.
Internal metric	See Metrics.
JV	Joint Venture.
Key performance indicators	The key metrics that indicate how well an organization is performing against its Critical Success Factors.
Kick off meeting	The inaugural meeting of a benchmarking study and also the meeting held before the commencement of regular benchmarking studies.
KPI	Key Performance Indicator.
Lag metric	See Metrics.
Lead metric	See Metrics.
Long term metric	See Metrics.

Loss	Financial or other loss that occurs due to unavailability of a resource.
Metric model	A diagram depicting how metrics relate to one another. Typical models include organization chart style costs and hours breakdowns and flow charts.
Metric structure	See Metric Model.
Metrics	Metrics are very similar to performance indicators, but whereas performance indicators are used specifically to indicate how well an organization is performing, a metric is any measure regardless of whether it is used to measure performance. There are many different types including:
Downstream	A metric that quantifies past performance (example: late deliveries).
Effectiveness	A metric measuring how well tasks are carried out. E.g. number of errors, failures, breakdowns.
Efficiency	A metric measuring how quickly/cheaply tasks are completed.
External	A metric by which people outside the organization or process measure its performance.
Global	A metric that measures corporate issues e.g. for a supermarket, sales per square metre.
Hard	A metric that is (theoretically) easy to measure such as financial measures, production rates and failures. (See also Soft metric).
Internal	A metric by which people in an organization or process measures its performance (see also External metric).
Lag	See metrics downstream.
Lead	See metrics upstream.
Local	A metric that measures local issues e.g. for a supermarket, customer view of the local branch.
Long term	A metric that measures log term issues e.g. investment.
Short term	A metric that measures short term issues e.g. daily sales
Soft	A metric that is (theoretically) difficult to measure such as employee satisfaction.

Upstream	A metric that helps predict future performance (example: customer orders help predict future profit).
Mission statement	See Aspirational Statements.
Mitigating circumstances	A known cause of relative poor (or superior) performance usually outside the control of the participant.
Modelling	See Normalization.
Modulus	See Absolute value
Non-competitive benchmarking	See Benchmarking Classes.
Non-productive time	The time that a resource is not being used, for whatever reason.
Normalization	A method that allows us to compare or combine data that we would not otherwise be able to compare. Methods include:
Categorization	Where data are grouped into categories. e.g. number of accidents for women aged 18-25.
Modelling	Where performance of individual data items are compared to a mathematical model developed from data.
Per unit (unitization)	Often thought of as rates or frequencies. E.g. cost per unit of production, incidents per hundred (or thousand etc.), percentage of failures.
Scoring	Where performance of different metrics are weighted and summed e.g. Business Excellence Models; final marks for academic courses may be weighted and summed across exams, course work and projects.
Selection	Where we select specific items, typically projects that we deem to be similar e.g. benchmarking two similar sized road construction projects.
Weighting Factors	Where packages of work are allocated relative weightings according to the size of package e.g. when comparing the cost to maintain different chemical plants we can assign a weighting factor to the equipment in each plant and compare cost of maintenance per unit weighting.
NPT	Non-Productive Time.
One-to-one benchmarking	See Benchmarking Methods.

OSHA	Occupational Health and Safety Administration
Pace setter	The Pace setter it that participant that is deemed to have the best overall performance, including consistency over time.
Pareto principal	The observation made by Wilfredo Pareto that 80% of the wealth in Italy was held by 20% of the people, later generalized by J Juran who observed that 80% of an effect was due to 20% of causes.
Participant	An organization or group taking part in a benchmarking study.
Initiator	The organization initiating the benchmarking study.
Potential	A group or organization with which we could benchmark.
Target	The organization or group with which the initiator of the benchmarking study wants to benchmark.
Participant agreement	A Benchmarking Agreement sets out the responsibilities and expectations of members of a specific benchmarking study. It may refer to related documents such as a confidentiality agreement or code of conduct and is intended to ensure the outcome of the study meets participant expectations.
Participant class	See Benchmarking Classes.
Partner	See Participant.
Per unit	See Normalization.
Performance indicators	Metrics that indicate how well an organization is performing. See also Key Performance Indicators.
Performance metric	See Metrics.
Phases of benchmarking	The three phases of benchmarking are: 1. Internal preparation. 2. Comparison of performance levels between participants. 3. Improvement activities.
PI	Performance Indicator.
Planned downtime	See Downtime.
Potential participant	See Participant.
ppm	parts per million.
Process benchmarking	See Benchmarking Types.

Product benchmarking	See Benchmarking Types.
Project benchmarking	See Benchmarking Types.
Project charter	The Project Charter (similar to Terms of Reference) is a formal short (often one page) document that encapsulates the scope, objectives, resource requirements and authorization of a project.
Project proposal	The detailed plan developed immediately prior to inviting participants to join the study.
Protocol	See Code of Conduct.
Public domain benchmarking	See Benchmarking methods.
Quartile	The performance above which a certain percentage of participants perform. The quartile most use is the top quartile which is the performance above which the top 25% (quarter) of participants perform. The second quartile is the performance value at which 50% of participants perform.
r	Correlation coefficient.
Reliability	Measures unplanned downtime.
Review benchmarking	See Benchmarking methods.
RIDDOR	Reporting of Injuries, Diseases and Dangerous Occurrences Regulations
Run chart	A plot of a process characteristic, usually over time.
Scope	Defines what is to be benchmarked i.e. the activities, products, services, groups, facilities to be included in a benchmarking study.
Scoring	See Normalization.
Selection	See Normalization.
Service benchmarking	See Benchmarking types.
Service level agreement	An agreement between a group and its customer stating what level of services will be provided.
Short term metric	See Metrics.
SLA	Service Level Agreement.
Soft metric	See Metrics.
Survey benchmarking	See Benchmarking Methods.
SWOT	Strengths, Weaknesses, Opportunities & Threats analysis.
Target participant	See Participant.
Target performance	A performance level believed to be achievable by a typical participant.
Team charter	See Project Charter.

Technical limit	A (usually) theoretically determined performance value which is seldom if ever attained. It could reflect, for example, the performance level that would be achieved if there were no failures, delays, interruptions etc. The advantage of using a technical limit as a target is that it is objective and not dependent on the participants.
Top quartile	See Quartile.
TQM	Total Quality Management.
Trial benchmarking	See Benchmarking methods.
TWF	Total Weighting Factor.
Unavailability	Time when a resource is not available for use = downtime.
Unplanned downtime	See Downtime.
Upstream metric	See Metrics.
Utilization	Time during which a resource is being used. Usually calculated as a percentage of total time.
Values statement	See Aspirational statements.
Vision statement	See Aspirational statements.
Weighting factor	See Normalization.
WF	Weighting Factor.

Index